T0207779

Springer Undergraduate Mathematics Series

Advisory Editors

Mark A. J. Chaplain, St. Andrews, UK

Angus Macintyre, Edinburgh, UK

Simon Scott, London, UK

Nicole Snashall, Leicester, UK

Endre Süli, Oxford, UK

Michael R. Tehranchi, Cambridge, UK

John F. Toland, Bath, UK

The Springer Undergraduate Mathematics Series (SUMS) is a series designed for undergraduates in mathematics and the sciences worldwide. From core foundational material to final year topics, SUMS books take a fresh and modern approach. Textual explanations are supported by a wealth of examples, problems and fully-worked solutions, with particular attention paid to universal areas of difficulty. These practical and concise texts are designed for a one- or two-semester course but the self-study approach makes them ideal for independent use.

More information about this series at http://www.springer.com/series/3423

Jeremy Gray

Change and Variations

A History of Differential Equations to 1900

 Springer

Jeremy Gray
School of Mathematics and Statistics
Open University
Milton Keynes, UK

ISSN 1615-2085   ISSN 2197-4144   (electronic)
Springer Undergraduate Mathematics Series
ISBN 978-3-030-70574-9   ISBN 978-3-030-70575-6   (eBook)
https://doi.org/10.1007/978-3-030-70575-6

Mathematics Subject Classification: 01A50, 01A55, 01A60, 34-03, 35-03

© The Editor(s) (if applicable) and The Author(s), under exclusive license to Springer Nature
Switzerland AG 2021
This work is subject to copyright. All rights are solely and exclusively licensed by the Publisher, whether
the whole or part of the material is concerned, specifically the rights of translation, reprinting, reuse
of illustrations, recitation, broadcasting, reproduction on microfilms or in any other physical way, and
transmission or information storage and retrieval, electronic adaptation, computer software, or by similar
or dissimilar methodology now known or hereafter developed.
The use of general descriptive names, registered names, trademarks, service marks, etc. in this publication
does not imply, even in the absence of a specific statement, that such names are exempt from the relevant
protective laws and regulations and therefore free for general use.
The publisher, the authors and the editors are safe to assume that the advice and information in this book
are believed to be true and accurate at the date of publication. Neither the publisher nor the authors or
the editors give a warranty, expressed or implied, with respect to the material contained herein or for any
errors or omissions that may have been made. The publisher remains neutral with regard to jurisdictional
claims in published maps and institutional affiliations.

This Springer imprint is published by the registered company Springer Nature Switzerland AG
The registered company address is: Gewerbestrasse 11, 6330 Cham, Switzerland

Preface

The Shape of the Book

This is a book on the history of mathematics; its basic dynamic is historical and therefore, up to a point, chronological. It follows the progress of a number of ideas that grew, sometimes came together, and often developed rich and fascinating branches and applications. At its core is an account of how the calculus of Newton and Leibniz—the calculus of functions of a single variable—led to attempts to develop a calculus of functions of several variable and how these new mathematical methods contributed to the study, first of ordinary, and then of partial differential equations. In each case, the rationale for that work was chiefly to develop general methods that could tackle problems in geometry and mechanics (the motions of solids and liquids under the action of forces).

The physical world being a complicated place, most of the applications involved partial differential equations, and here the story soon also became complicated. The first-order partial differential equation in two independent variables was initially difficult to solve, and this posed problems for the study of more than two independent variables and for equations of higher order. Important work on the first-order case was done by Lagrange and Monge before Cauchy was finally able to show that such equations almost always have a solution. But the second-order case almost immediately confined itself to three special cases, somewhat as Euler had suggested, and all of them, as we would say, linear. The first, and simplest, is the wave equation (the prototype hyperbolic equation), successfully tackled by d'Alembert. Euler regarded the one later known as the elliptic case (the key example being the Laplace equation) as being beyond current methods. Finally, the case we call parabolic fell through a gap in his approach, and strangely little was said about it before Fourier dealt with the canonical example: the heat equation. At this point, a significant departure from the theory of ordinary differential equations opened up: the need to pay attention to initial or boundary conditions. However, this issue was to remain obscure for several decades.

Euler quickly showed that linear ordinary differential equations with constant coefficients can be solved systematically. Other types of ordinary differential equations were studied in the eighteenth century, but the story is piecemeal, and instead, I chose to give just one example of the history of ordinary differential equations: the hypergeometric equation from Gauss to Riemann, Schwarz, and Poincaré. This is one of the glories of the subject, bringing together early ideas about group theory, complex function theory, and the then-novel hyperbolic or non-Euclidean geometry.

So how is all this material organised in this book? Chapter 1 connects the calculus to problems in ordinary differential equations and is mirrored by Chaps. 3 and 5 in which the calculus of several variables is developed and the first *partial* differential equations are studied. Then it is a fairly straight run through topics in partial differential equation theory in Chaps. 6, 8, 10, 13, 17–20. This allows us to see how the work of Euler, d'Alembert, and a few others rewrote Newton's *Principia Mathematica* for the eighteenth–twentieth centuries. The story of the hypergeometric equation occupies Chaps. 11 and 14–16 because it must start in 1812 with Gauss and because once it gets going it seems ridiculous to break it up. What intervenes here is Chap. 12 on Cauchy's demonstration of the existence of solutions to ordinary differential equations and Chap. 13 on Riemann's geometric version of complex function theory, which is needed for the subsequent three chapters.

What of the chapters not yet referred to? Chapter 2 describes the start of the calculus of variations, and Chap. 7 takes that subject further into the eighteenth century. Chapter 4 documents other successes of the partial differential calculus in studying natural phenomena other than the wave equation. (There is also a surprising link to the hypergeometric equation.) Chapters 9 and 21 are opportunities for revision; when I gave the course I used these lectures to discuss the assessment on the course so far.

The remaining chapters move into what may be less familiar material. Riemann's study of shock waves; Riemann and Weierstrass on minimal surfaces; the work of Thomson and Stokes on the telegraphist's equation and the laying of the trans-Atlantic cable; a look at the first ninieteenth-century attempts to rigorise the calculus of variations; the eventual introduction of the fundamental trichotomy (elliptic, parabolic, hyperbolic) for second-order linear partial differential equations and the first general existence theorems in the elliptic and hyperbolic cases including Hadamard's insistence of the distinction between initial and boundary value problems. Two chapters look at how Jacobi used Hamilton's ideas to create Hamilton-Jacobi theory and subsequent attempts to geometrise mechanics, and the connection to the solution of first-order partial differential equations.

All this material has a certain coherence that is worth spelling out. Ordinary differential equations grew out of, or alongside, problems in evaluating integrals, which is why we still talk, confusingly, of integrating a differential equation and its solutions as its integrals. It was soon recognised that the solution to an ordinary differential equation was a family of functions and an individual solution could be specified by means of some initial conditions. So, it was natural when differential equations with several independent variables were investigated that the earliest researchers (Jean le Rond d'Alembert, Leonhard Euler, Pierre Simon Laplace, and Joseph-Louis

Lagrange) thought of these partial differential equations in the same way, and looked for techniques that would produce a formula for the general solution (however, they seldom also discussed an auxiliary process of fitting the general solution to some initial conditions). Part of the story here is the gradual recognition that this is not the right way to think of partial differential equations. Rather, it is a dialogue between the general methods and the initial or boundary conditions that is central, and which underpins the crucial distinction between the elliptic and hyperbolic types to which formal methods are blind. As we shall see, this explains the problematic way in which complex variables were first used.

It is also interesting to see how questions of rigour enter the story in a way that is immediately important and does not appear as the whim of an analytic pedant. The ad hoc methods that can be used to solve a partial differential equation (such as the separation of variables) naturally raise the question of the uniqueness of the solutions that is important in applications. The need for power series to converge forces a heavy reliance on (real or complex) analytic methods that has, ultimately, to be outflanked.

Advice to Students

There are several important things being described in this course, and it may help to remember what they are before immersing yourself in the technical details. Newton's work is a case in point. Even though we shall only skim its surface, it is clear that this is a remarkable achievement, one that it took more than a century to confirm, and some of the best work of the twentieth century to surpass. Newton's account of the motion of the Moon and the planets does not rest on the calculus, still less differential equations, but everyone after him turned to the calculus, and Euler gave everyone the means to write celestial mechanics that way ever after. Mathematical accounts of fundamental physical processes—gravity, the production and spread of sound, the propagation of heat and of electric signals—are among the successes of the theory of partial differential equations.

But for the calculus to do this work, mathematicians have to have the confidence that it does work. This is partly a matter of rigour, and indeed it is satisfying to see that so many of the questions that are dealt with in courses on pure analysis arose in contexts where a practical, or at least a physical, answer depended on the quality of the reasoning. Less obviously, but perhaps more interestingly, it is worth seeing what even the best mathematicians did with difficult problems. Arguments have to be rigorous—ultimately. Before then they have to be some or all of convincing, intelligible, general, plausible, and applicable. Likewise, solutions have to be a number of things apart, ideally, from being right: among the criteria are, on one occasion or another, computable, accurate enough, intelligible, complete, and unique. Just as an argument might get to the heart of a problem or somehow merely work, a solution can be truly informative or merely a formula. All of this is on display here. And, of course, sometimes an equation with no answer in sight can still seem to be a valuable advance.

In this sense, the most momentous event on display here in the first half of the course is that the calculus, in the form of differential equations, both ordinary and partial, can deliver so much insight. The comparable change in the second half of the course, as I hinted above, is the transformation in what a solution is taken to be. For partial differential equations this is the rise to equal importance with the equation of the boundary or initial conditions, coupled as it is with a profound classification of these equations into types. The need for rigour played its part in these developments when appeals to the so-called generality of analysis and its supposed algebraic or formal basis began to fail.

This is not a set of lectures in which epsilons and deltas, ns and Ns dance ever more intricately, but this should not suggest that when mathematics is applied—whatever that might mean—standards drop. Mathematicians were doing their best at all times to get it right, although we can observe different ways in which they honoured that commandment. The difficult mathematics here comes from the difficulty of the problems: a partial differential equation is a difficult thing to understand, harder than an ordinary differential equation, and harder than many an early investigator realised. Qualitative arguments are often harder than quantitative ones, if less technical.

The challenge you face is to get a sense of that struggle, of the difficulty, and how it was tackled.

Being a historian of mathematics means attending to mathematics on its own terms as well as ours, and seeing it in the context of its time. What was known, what was thought to be true? When a mathematician tackles a problem, you ask: How had other problems like this one been tackled, how were they tackled after this one? Is the analysis of the problem convincing, is the solution informative? What, in the end, were these people trying to do?

Historiographical Remarks

There are several existing accounts of the history of calculus, and a number of specialist books and articles on particular aspects of that history. The contributions of Newton and Leibniz, Euler, Lagrange, Fourier, Cauchy, Riemann, Weierstrass, Poincaré, and Hadamard have been studied in some depth; various topics, such as the wave equation, the heat equation, Laplace's equation, and the Dirichlet problem have been looked at in some detail, although not always after the original breakthroughs were made. But there is no general history of differential equations, ordinary or partial. Histories of the calculus dwell on the story of the rigorisation of the calculus and the creation of modern (or, rather, nineteenth century) mathematical analysis, but tend to marginalise the story of what made the calculus valuable: the capacity it gives mathematicians and scientists to formulate and solve problems across the fields of physics and geometry. Historians have tended to forget that what made the calculus worth all the efforts to understand it was not ideas about infinitesimals, differentials, limits, and the like that were introduced to explain and justify it, but its many successes in providing an understanding of the natural world, from the motion of the planets to the transmission of electric signals, and in extending the powers of geometry.

There have been a few notable departures from this scholarly regime in recent years. Craig Fraser and Jesper Lützen have steadily enriched our understanding, and other historians (Tom Archibald, June Barrow-Green, Umberto Bottazzini, Christian Gilain, and Tom Hawkins, among them) have dealt with various aspects of the development of the theory of differential equations as it entered into the larger pictures they were exploring.

What This Book is Not

The largest omission is the work of Maxwell and the equations named after him, but it seemed to me that the modern theory, and the physical experiments that it explains, are largely unknown to mathematics students and would have entailed too great a detour to bring to life. In addition, Maxwell's ideas about the physics involved are not the modern ones—and famously, no Continental physicist claimed to understand them—so it would have been impossible to do them justice in the space available. Disappointed readers should consult Buchwald's *From Maxwell to Microphysics*. For much the same reasons, I was unable to deal with hydrodynamics and the Navier–Stokes equations, but readers may always turn to Darrigol's *Worlds of Flow*.

Another topic that is wholly missing is the use of perturbative methods. Most differential equations, and systems of such equations, that arose in practice could only be tackled by the method of undetermined coefficients. The idea was to start from a simplified version of the problem at hand that, however, admitted an exact solution, and to seek the solution to the solution to the actual problem by adding more terms to cope with the increased complexity. These might take the form of power series, or later trigonometric series, which were fitted to what data there was, especially in the important subject of astronomy, and their coefficients adjusted to refine predictions and explain other effects.

Another important topic that it would be good to have included is Sturm-Liouville theory, but there is already an excellent historical account in Lützen's *Joseph Liouville (1809–1882): Master of Pure and Applied Mathematics*, and I thought it better to add to the stock of historical information about the development of differential equations.

I would very much have liked to have concluded the course with Poincaré's ideas about flows on surfaces, and his brilliant extension of these techniques in his famous memoir on the three-body problem, which would have made an attractive connection back to Newton's *Principia Mathematica*, but there simply was no room. However, there are existing accounts of this subject.[1]

And, I admit, the wish to try to say something about the history of partial differential equations, surely the largest omission in the history of modern mathematics, also played a part in my decisions about what to include.

---

[1] See most informatively (Barrow-Green 1997), and also (Gray 2013), and (Verhulst 2012).

So, dear reader, if your favourite topic is not here, and especially if there is not a good modern history of it, then the opportunity is there for you to write it and in that way fill a gap in the literature. There are short overviews of the historical development of differential equation (for example in (Kline 1972)) and there are detailed treatments of selected topics, as I have tried to acknowledge and benefit from. There is ongoing work by a number of historians of mathematics, but the fact remains that the history of mathematics is tied closely to the history of pure mathematics through a shared interest in foundations, and the history of classical applied and applicable mathematics lags behind. This would be merely unfortunate, were it not for the fact that it is through differential equations that the calculus largely justified its existence—geometry being another, but smaller, vital field.

Without histories of differential equations, we lack a significant part of the history of mathematics. We cannot properly explain to our students where we are coming from and how we got here, we cannot explain the significance of mathematics to historians of science, and we are hindered in our attempts to rescue philosophical accounts of mathematics from the grip of foundationalists, who see only set theory and logic.

There are considerable losses. We are likely to leap from Newton, Leibniz, and the invention of the calculus straight to Cauchy, with perhaps a glance at Lagrange's unsuccessful earlier attempt at rigorising the calculus. In this way, the entire eighteenth century is largely forgotten and is only dealt with in fragments. The study of partial differential equations is reduced to what I jokingly refer to as solving the only four partial differential equations that exist: the general first-order partial differential equation, Laplace's equation, the heat equation, and the wave equation.

This book is an attempt to fill in some of the gaps.

Sources and Their Uses

There is inevitably an absence of material in English on this material. Newton, of course, has been generously put into English when he did not write it himself, J. M. Child translated some of Leibniz's considerable and mostly unpublished writings of relevance in (Child 1920), and for the eighteenth century, there is the remarkable and growing resource of the Euler Archive, where almost all of the original work of Euler can be found along with many substantial translations. Among the nineteenth-century mathematicians, almost all of Riemann's work is now in English, as are Hilbert's remarks in his Paris address on Mathematical Problems; and Hamilton and Green naturally wrote in English. The rest remains in Latin, French, and German, and a richer study would embrace Italian and Russian.

Source books have done something to ease the students' paths: Struik's on the period 1200–1800 and Birkhoff's rather freer translations of nineteenth-century work are very helpful, and more can be found in the book by Fauvel and Gray (referred to here as F&G). Historians' translations of shorter extracts can also be found in their papers.

I have therefore added to the collection of works translated into English some items from Cauchy (on the existence of solutions to first-order partial differential equations) Darboux (on the telegraphist's equation), Schwarz (on analytic maps from a half-plane or disc to a polygon, his alternating method, and part of his paper on the hypergeometric equation), and a passage from the introduction Picard wrote to one of his papers.

As for illustrations, I had originally planned to include pictures of most of the important mathematicians whose work is discussed in this book, but copyright issues posed obstacles in a number of cases. However, these days a great many pictures, very often accurately identified, are available on the internet.

Advice to Instructors

This book is the fourth and last of my books based on courses I taught, each over a period of four years, at the University of Warwick. Together, they cover the emergence of a fair amount of mathematics in the standard syllabus at many universities today. They have been published at a time when the prospects for courses in the history of modern mathematics in Britain have become poor, I believe for two reasons. First, in Britain as in many places, there seem to be few prospects for anyone wanting to shape a career as a historian of mathematics; students know this, and they seldom entertain the idea at the graduate level.[2] Second, there are problems for anyone wanting to run a course in the subject: problems of language, problems of sources, problems with assessment. These four books are offered partly as a way around the second problem, and that accounts for their content, specifically the three chapters on assessment. I wanted to show that there are ways to assess student's grasp of the history of mathematics that are not simply exercises in old mathematics, and the result is the adaptation of what my Open University colleagues and I did to more advanced topics.

There are many reasons why a course in the history of mathematics at university can benefit students. It humanises the subject, demonstrates the intent behind many discoveries, and helps to explain why we have the mathematics we do. It always seemed to me that any history of mathematics course best belonged in the students' final year, when they already know enough mathematics for the history to get a proper look in. In a world in which few students go on to do research in straight mathematics, but many go on to be mathematicians in a huge variety of environments, I believe that a historical overview of part of the subject offers at least as much value as any other specialism.

Of course, I do not claim that any of the four volumes is *the* course to adopt. It might well make sense to use any of the books selectively. This one might yield a course on partial differential equations, or a short course on the hypergeometric

---

[2]Nor is their much room for history of mathematics in the tightly determined school mathematical syllabus.

equation, for example. It will depend on the audience. And I would cheerfully admit that almost every chapter here is too long to be a lecture; indeed I never taught it that way. Each chapter is a resource, and in the absence of other material for the student to read, I thought it best to provide enough for readers to engage with. There is more than enough for three lectures a week here, but not too much for a week's study.

Acknowledgements

I thank June Barrow-Green and Robin Wilson, my former Open University colleagues, for sharing with me their knowledge and expertise over many years and in particular for their many discussions of the early history of the calculus; I thank June also for her guidance in everything to do with celestial mechanics. I thank Mario Micallef for patiently educating me in the ways of partial differential equations; he is a famously good teacher and I hope that shows here. I thank my students at Warwick University for their appreciation of the courses on which this book is based.

I also thank the two anonymous referees from the history of mathematics community for their helpful comments and corrections. I thank another anonymous referee for finding numerous small errors and typos and for making a number of more substantial suggestions for improvements that I have tried to incorporate in this book. And I thank J. D. Verhulst for his very thorough reading of the book, and his numerous suggestions large and small, even those that, with reluctance, I could not adopt. The book is a selection of topics from a large field; it could be nothing else. Finally, I thank Remi Lodh, my editor at Springer, for his constant drive to improve the book and his many improvements to the diagrams and figures in particular.

Milton Keynes, UK   Jeremy Gray

Topics Discussed in This Book

1. Newton's laws of motion (in the *Principia* and in Euler's form, Chap. 4 and Appendix A)
2. Debeaune's differential equation (Chap. 1)
3. The brachistochrone (the curve of quickest descent) (Chap. 2)
4. Euler's solution method for linear ordinary differential equations with constant coefficients (Chap. 1)
5. The partial differential equation for the vibrating string and its solutions (Chap. 3)
6. Euler's equations for the propagation of sound and for fluid mechanics (Chap. 4)
7. Characteristics for first-order partial differential equations (Chap. 5 and Appendix B)
8. Lagrange's account of first-order partial differential equations (Chap. 6)
9. The Euler–Lagrange equations and Lagrange's methods in the calculus of variations (Chap. 7)
10. Lagrange's generalised coordinates (Chap. 7)
11. Monge's solutions to first- and second-order partial differential equations (Chap. 8)
12. The heat equation and its solutions in Fourier series and as a Fourier integral (Chap. 10)
13. The hypergeometric equation as studied by Gauss, Riemann, Schwarz, and Poincaré (Chaps. 11, 14–16)
14. Cauchy's existence theorems for first-order ordinary differential equations (Chaps. 12 and 31)
15. Cauchy's existence theorem for first-order partial differential equations and the Cauchy–Kovalevskaya theorem (Chaps. 17 and 31, and Appendix D)
16. Potential theory, Green's functions, and adjoint equations (Chap. 18 and Appendix D)
17. The Dirichlet problem and Schwarz's alternating method (Chaps. 19 and 31)
18. The telegraphist's equation (Chap. 20)
19. Riemann on shock waves (Chaps. 22 and 31)

Contents

List of Figures

Chapter 1
The First Ordinary Differential Equations

1.1 Introduction

The chapter is in two parts. In the first, we look briefly at the discovery of these methods for finding tangents, which is, of course, part of the seventeenth-century discovery of the methods of calculus. The first inverse tangent problem—the precursor of differential equations—was asked early on, in 1638, and it is interesting to see that neither Descartes nor Leibniz could properly solve it.

In the second part, we see how the calculus, in Euler's hands, led to the development of methods for solving various kinds of ordinary differential equation, including Debeaune's. The work of Euler and Bernoulli on vibrating rod and hanging chain led to a breakthrough in the study of linear differential equations and the introduction of the idea of a basis of solutions. Even more importantly, Euler was able to adapt the methods of the calculus to the study of mechanics, and so was able to express Newton's laws of motion for the first time as differential equations.

1.2 Origins: Inverse Tangent Problems

In the 1620s and 1630s, various mathematicians—Pierre Fermat, René Descartes, and Gilles Personne de Roberval among them—began to develop methods for finding tangents to curves, either at a given point on the curve or from an arbitrary point and to the curve. With the success of these methods, it became possible to think of raising and answering the opposite question, that of finding a curve given some properties of its tangents.

The person with the honour of having formulated the first inverse tangent problem is Florimond Debeaune. Debeaune was a wealthy member of the nobility in his hometown of Blois, where he was born in 1601 and where he became a counselor at the Court of Justice. He also had a reputation as a high-quality lens grinder, and

© The Author(s), under exclusive license to Springer Nature Switzerland AG 2021
J. Gray, *Change and Variations*, Springer Undergraduate Mathematics Series,
https://doi.org/10.1007/978-3-030-70575-6_1

Fig. 1.1 Debeaune's problem

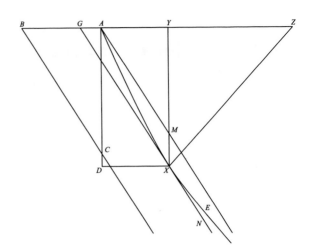

in 1639, Descartes wrote to him to ask him to design a machine that would make hyperbolic lenses. The project failed, but they remained in touch, and Debeaune went on to write the *Notes brièves* that were published in 1649 in the first Latin edition of Descartes's *La Geometrie*. In this work, he showed that the equations $y^2 = xy + bx$, $y^2 = -dy + bx$, and $y^2 = bx - x^2$ represent a hyperbola, a parabola, and an ellipse, respectively.

Debeaune was led to propose it in 1638 as a result of his study of Descartes's *La géométrie*—so soon did Descartes's ideas begin to transform geometry.[1] He raised it out of his interest in explaining mathematically why a plucked string vibrates as it does, specifically, in explaining why the frequency with which the string vibrates is independent of the force with which it is struck.[2] It was one of four problems that he presented to the mathematical community, and it has come down to us in the form of a letter to Roberval.

1.2.1 Debeaune's Problem

Debeaune stated the problem this way.[3]

> Let there be a curve AXE whose vertex is A, axis AYZ, and the property of this curve is that, having taken any point on it you wish, say X, from which the line XY is drawn as a perpendicular ordinate to the axis, and having taken the tangent GXN through the same

---

[1] Descartes' *La Geometrie* was published in 1637 as one of a number of appendices in his *Discours de la Méthode*.

[2] Debeaune to Marin Mersenne, March 1639, Mersenne *Correspondance* VIII, 348, in F&G 11. B1(b). Marin Mersenne was a Minimite friar who operated an informal postal service for the communication of letters across Europe about science and mathematics.

[3] Debeaune, letter to Roberval, sent to Marin Mersenne, Mersenne *Correspondance* VIII, 142–143, in F&G 11.B1(a).

point X, and extended the perpendicular XZ to it at X until it meets the axis, there will be the same ratio of ZY to YX as a given line, like AB, has to the line $YX - AY$ (Fig. 1.1).

Draw the axis, AYZ, a curve, AXE, a tangent, GXN, at some point, X, on the curve, and erect the perpendicular, XY, as shown. Locate the line segments ZY, YX, and AY, and the line segment AB that provides a unit of length. Debeaune's problem asks to find the curve, AXE, with the property of its tangents that

$$\frac{ZY}{YX} = \frac{AB}{YX - AY} \ .$$

Debeaune probably expected an answer in the form of a recipe for constructing points on the curve geometrically, rather than as an equation in some system of coordinates, but in any case, he was to be unlucky. Roberval showed that line through B drawn at 45° to the axis is an asymptote to the curve, and in October 1638, Descartes gave a mechanical description of how the curve might be drawn approximately, which was sufficient to confirm Roberval's result, but no one could answer the challenge before Debeaune died in 1652.

It has proved the characteristic of inverse tangent problems—differential equations as they became called—that they can be easy to state but very difficult to solve. One merit of the calculus was to be that it not only provided a way of stating inverse tangent problems but it also provided a set of rules for manipulating the problem symbolically until it could (quite often) be solved, at least in the sense that the solution curve could be described via equations or formulas.

1.2.2 Other Inverse Tangent Problems

Inverse tangent problems often arose naturally in the contemporary study of physical and astronomical problems.

For example, in the 1670s, Claude Perrault, who is best remembered as the architect of the east wing of the Louvre Palace in Paris, asked for the curve traced by a heavy weight drawn behind someone walking along a straight line. The solution is a curve called the *tractrix* (see Fig. 1.2) and had previously been considered by both Newton and Leibniz, although they did not identify it as a solution to this problem, and later by Huygens. More formally, it is the curve with the property that the length of the tangent from a point on the curve to a fixed line is a constant.

In his *Principia Mathematica* [206], Newton investigated the paths of particles that moved subject to forces directed at a central point. Later, the paths of particles moving under gravity and encountering various forms of air resistance were studied by Newton and others; these too arose as the answers to inverse tangent problems. In these cases, it is the instantaneous direction of acceleration that is known, not the instantaneous velocity, so the problem is not strictly an inverse tangent problem but a generalisation.

Fig. 1.2 The tractrix

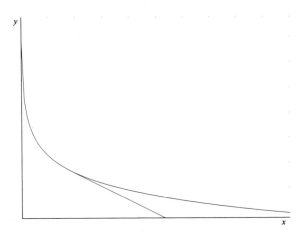

In 1696, Johann Bernoulli challenged the mathematical community to find the curve along which a sliding bead would descend most quickly between two given points—as we shall see in more detail in Chap. 2 the answer is a cycloid with a vertical tangent at the starting point (a cycloid is the curve traced by a point on the rim of a wheel rolling along a straight line). Five mathematicians responded, among them Newton, who stayed up all night and answered the question the day he received it; Bernoulli said he recognised the solution as Newton's as one recognises the lion by his paw (it had taken Bernoulli 2 weeks).

As these dates indicate, these problems were very difficult, and the calculus enabled some progress to be made on a broad front. Unsurprisingly, it was not always quite as simple as that.

1.3   From Inverse Tangent Problems to Differential Equations

In the 1690s and early 1700s, Johann Bernoulli became the leading exponent of the calculus, which he and his older brother had learned by corresponding with Leibniz and reading his published papers. He went to Paris in 1691 and got himself hired to teach the Marquis de l'Hôpital the new calculus, and as a result, the first book on the differential calculus appears with de l'Hôpital as the author. [4]

From the book, we can see that Bernoulli's definition of integration is interesting, for he defined it as Newton had done as the inverse of differentiation and not, as Leibniz did, as an infinite sum, and he gave several methods for finding areas. Then Bernoulli turned to inverse tangent problems and solved a variety of examples. The concluding, and arguably most important, part of the book was an exposition of how

---

[4] See L'Hôpital [186].

problems in geometry or mechanics can be translated into the language of calculus. Calculus was still very new, and showing how to express problems using it could be the hardest part of a mathematician's work.

In summary, translation procedure would go as follows.

1. Set up a system of x and y coordinates with respect to which the solution to the problem can be expressed as a curve, and then formulate the problem in terms of equations involving these coordinate variables.
2. Interpret the problem as a statement about the relationship between neighbouring points on the curve; this will take the form of an equation involving differentials.
3. Pass back from the differentials to the finite quantities (x and y) and so determine the precise form of the equation that describes the solution curve.

This method was not invented by Bernoulli. Leibniz had already tackled Debeaune's problem this way, however clumsily—Leibniz was prone to careless errors. Bernoulli raised his approach to the level of a systematic *method* for tackling many such problems, and this moment of transition is marked by a change of name. Henceforth, inverse tangent problems became called *differential equations*, because in step (2) they were literally expressed as equations involving differentials. That the new name stuck shows how closely the new methods of the calculus became associated with problems involving instantaneous change or changes from point to point along a curve.[5]

None of this would be of any use if the resulting differential equation could not be solved. Now, to Bernoulli and his contemporaries, a solution ideally meant a geometrical description of the required curve. This is a global description of a curve, such as is usually given for a circle, a conic section, or a few other curves such as the cycloid. The calculus, however, did not always lend itself to providing such a thing. When stage (3) has been carried out successfully, the solution is expressed as an equation in coordinates x and y that defines a curve (depending on some initial conditions). But to mathematicians of the late seventeenth century, a further step was required in which the curve was characterised by some property by which it could be recognised, much as we today routinely gloss a curve defined by a quadratic equation as a particular sort of conic section. This amounts to reversing stage (1) and is much harder than traversing it, and since that was often hard enough, going backwards often proved to be too difficult. Indeed, why should the solution curve be any kind of known curve? However, if this step is not taken, the curve can at best be drawn pointwise, and important properties of it might remain undetected.[6] Gradually, mathematicians began to accept equations as the solution and not to look beyond them; and the more they did so the more mathematics became more formal and algebraic, and less geometrical in nature.

The catenary problem is a good example. This asks for the shape of a heavy-weighted chain, and so it is of obvious interest to bridge builders. Galileo had sug-

---

[5]For the same reason, British mathematicians spoke more and more of fluxional equations because Newton had expressed his ideas in terms of fluxional quantities, such as the rates of change of quantities.

[6]This was an issue that Descartes recognised when he put forward his ideas about geometry in 1637.

gested in 1638 it would "assume the form of a parabola", but in 1646, Huygens
showed that this is incorrect.[7] The problem was first solved by Leibniz and Johann
Bernoulli independently in 1691, and then raised again by Jakob Bernoulli in the *Acta
Eruditorum* in 1701 as a challenge to the mathematical community.[8] This invites the
question of what Jakob Bernoulli thought he was doing asking a question that had
been dealt with successfully a decade before, and Paolo Fregulia, whose account is
our guide here, speculates that his aim was "to put this solution in a more theoretical
general context", namely, isoperimetrical problems.[9]

Johann Bernoulli's solution, written in 1701 but published only in 1706, proceeded
by first expressing the force on an infinitesimal piece, AB, of the chain in terms
of the weight of the string hanging below A, then by calculating the force at the
neighbouring point B, and then by arguing that since AB does not move, the effect
of the forces at A and B must cancel out. So the raw ingredients are the length, s, of
the chain from the far end E to the point A—which is proportional to its weight—and
the differentials dx and dy which come in because the string is curved and so the
forces at A and B do not point in quite the same direction. The differential equation
Bernoulli obtained is
$$\frac{dy}{dx} = \frac{s}{a},$$
where a is a constant. Stage (ii) was completed when the variable s, which depends
on the values of x and y, was eliminated in favour of an explicit expression involving
x and y. We omit the details of how this was done, and pass straight to the resulting
equation:
$$dy = \frac{a\,dx}{\sqrt{x^2 + 2ax}}.$$

This completes stage (ii). To carry out stage (iii), Bernoulli noticed, as Newton had
done much earlier, that the simplest kind of differential equation one could hope to
get was of the form

$$(\text{something in } x)\, dx = (\text{something in } y)\, dy$$

because you could then hope to integrate both sides. The method of trying to arrange
for this to happen he called the *method of separation of variables*. When the variables
do not separate, Bernoulli enriched the method by suggesting that one looks for
new variables with respect to which the differential equation does separate. This
meant setting aside all qualms about the nature of differentials and manipulating
them formally just as one does finite quantities in elementary algebra. Bernoulli's
techniques are simple algebraic devices—his insight was in seeing that such methods

[7]See Galileo *Two New Sciences* [113], 149, and Huygens [149]. For a historical account, see
Bukowski [26].
[8]See Bernoulli [8]. The solution is the curve called a catenary—the name derives from the Latina,
catena, for a chain—with equation $y = \cosh x$.
[9]See Fregulia and Giaquinta [109]; the quoted remark is from a private communication.

Fig. 1.3 Bernoulli's
formulation of Debeaune's
problem

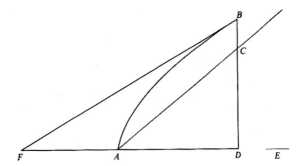

are applied in their new setting. In so doing, he was following the lead of Newton
and Leibniz.

These ideas are well illustrated by Bernoulli's discussion of Debeaune's problem,
which he gave in one of a series of lessons to the Marquis de l'Hôpital in 1691.[10]

> Another such example is the problem set to M. Descartes by M. Debeaune, the solution to
> which is not in his works but can be found in his *Letters* (vol. III, No. 71). The solution of
> it does not appear to be very easy according to our method, indeed at first sight the problem
> appears impossible by this method. But we shall see that by a change of variables it becomes
> easy to separate them, and that this problem can be solved completely once the quadrature
> of the hyperbola is given, for the curve is mechanical (Fig. 1.3).
>
> The problem goes like this: a line AC makes an angle of half a right angle with the axis
> AD, and E is a given constant line segment; what is the nature of the curve AB in which
> the ordinates BD are to the subtangents FD as the given E is to BC?
>
> *Solution.* Let $AD = x$, $DB = y$, $E = a$, suppose by hypothesis that $dy : dx = a : (y - x)$,
> then $adx = ydy - xdy$. From this equation the nature of the curve is to be found, either
> by integration or by rewriting y with dy on one side and x with dx on the other, for then
> two areas can be found and by comparing them the nature of the curve can be found. But
> the equation just found cannot be integrated, nor can x and dx be separated from y and
> dy; however, it can be changed into another by substituting the value of another variable.
> Therefore let $y - x = z$, $y = x + z$ and $dy = dz + dx$. The equation just found transforms
> into this: $adx = zdz + zdx$ or $adx - zdx = zdz$ and $dx = zdz : (a - z)$. Therefore these
> two variables separate, and we are led to the curve on multiplying by a, $adx = azdz :
> (a - z)$.
>
> [...]
>
> *Corollary* I. The curve AB has its asymptote parallel to AC.
>
> *Corollary* II. The space [i.e. area] $ADB = xy + ax - \frac{1}{2}yy$.

We see that Bernoulli first stated the problem, then he introduced coordinates
(stage (i)) and then differentials (stage (ii)). Aware in advance that the method of
separation of variables does not apply to the differential equation $adx = ydy -
xdy$, he introduced the change of variable $y - x = z$, which implies $dy - dx = dz$,
thereby extending a method used for finite quantities to differentials. Now he had an
equation in the variables x and z in which the variables *do* separate,

---

[10]See Bernoulli, J. *Opera Omnia* 3, 1742, 423–424, and F&G 13.B1.

$$dx = \frac{z}{a - z} dz.$$

As was thought necessary at the time, he did not accept the solution to the differential equation that results from integrating both sides,

$$x = -z - a \ln(a - z),$$

but went on to give a geometric interpretation of the result. Helpful though that can be, the tradition was to lapse in the face of too many cases where it could not profitably be done.

However, if we look at some ideas that Bernoulli wrote down only a few years later, in 1702, we can see the growth of formalism.[11] By now, he was quite clear that expressions such as $\frac{dx}{x + f}$, where f is a constant, are differentials of logarithms, and therefore that

$$\int \frac{dx}{x + f}$$

is a logarithm. He now introduced a variety of changes of variables to deduce that certain previously encountered integrals can be expressed in terms of either logarithms or circular arcs. His list includes the integral

$$\int \frac{dx}{\sqrt{x^2 + 2ax}},$$

which arose in the catenary problem, and the logarithm lurking in Debeaune's problem. Bernoulli regarded his changes of variable as enabling him to pass from circular arcs to arcs of hyperbolas and back, which made his geometric interpretations of analytic formulas more flexible. Interestingly, they involved him in introducing complex numbers, which was to occasion some confusion later on.

1.4   Differential Equations

We turn now to study how Euler rewrote the calculus.

The Leibnizian form of the calculus, which was the form adopted by the mathematicians of continental Europe, was initially seen as a set of algorithms for handling problems about curves. These algorithms work because they apply to formal expressions involving variables, and the two basic operations, differentiation and integration, d and $\int$, obey rules such as

---

[11] See Bernoulli [11] and F&G 13.B2.

$$d(uv) = udv + vdu \text{ and } d\int u = u.$$

The connection between these formal operations and geometry arises from their geometrical interpretations—for example, d has to do with finding tangents, and $\int$ with areas.

Euler rewrote the calculus by regarding it being about *formal expressions* and replacing the concept of a curve with that of a *function*. Calculus is about expressions that can be differentiated and integrated. It is only about curves in so far as they can be described by formal expressions, which allow one to use differentiation to find tangents and so forth, and the solution to a differential equation is not to be expressed as a curve but as an explicit or implicit function of the coordinates.

As is well known, Euler's analysis of sine, cosine, and the exponential functions is unified by the function idea. He expressed them as power series and treated them formally or algebraically, there is very little geometry. In particular, his controversial solution to the problem of defining the logarithm of a negative number proceeded by defining log as the inverse function to exp.[12]

Euler's emphasis on the calculus, and indeed much of mathematics, as a science of formal expressions widely restructured mathematical theory. His treatment of Debeaune's problem provides another example. Euler went from the differential equation

$$dx = \frac{zdz}{(a - z)}$$

to an answer in this form:

$$x + z + a \log(a - z) = constant.$$

He saw this equation as the answer, and saw no need for a geometrical interpretation. Indeed, in his definitive account of the integral calculus (published between 1768 and 1770) he did not even mention Debeaune's equation by name when he gave a complete account of how to solve all differential equations of the form

$$(\alpha + \beta x + \gamma z)dx = (\delta + \varepsilon x + \mu z)dz$$

(this equation reduces to Debeaune's on setting $\alpha = a, \beta = 0 = \delta = \varepsilon, \gamma = -1, \mu = 1$).

We shall see that Euler's mathematics is full of investigations of objects defined by differential equations or integrals. Problems are expressed as equations (finite or differential) and solved by finding power series expansions or other algebraic reformulations. If one sees mathematics as having three aspects—problems, methods, and results—then one might say that Euler very often saw problems algebraically and

---

[12] See Euler [78].

solved them algebraically, expressing his results either in finite terms or as infinite series.[13]

1.5   Linear Ordinary Differential Equations

It had been known from the time of Newton and Leibniz that the calculus provides good answers and not-so-good answers. A three-way correspondence between Euler, Johann Bernoulli, and Johann's son Daniel provides a good illustration of how this difficulty was addressed in the context of finding the shapes of a vibrating, clamped rod (setting aside the question of how the shape varies in time).

Daniel raised this problem in a letter to Euler of 18 December 1734 (setting aside the question of how the shape varies in time). He wrote to Euler again on 4 May 1735 to say that he had found a differential equation that describes its shape, but that the only solutions he could find to the equation, which involved sines and exponentials, seemed inappropriate. Euler replied with a solution in the form of a power series, which he wrote up and presented as a paper [71], E40 to the St. Petersburg Academy of Sciences. This is not a good answer. As so often, the power series is unilluminating, and it concealed from Euler the fact that the rod can vibrate in several distinct ways.

Then, in 1739, Euler spotted a much better approach and found a much better answer, which he described in a letter that he wrote to Johann Bernoulli on 15 September,[14] and more fully in a paper (E62) published in 1743.

In this letter, he proposed a simple, general method for all differential equations of a form he described, which we could call linear ordinary differential equations of arbitrary order and with constant coefficients. His method reduces these problems to the solution of a polynomial equation and establishes that the answer to the differential equation is always given as a sum of exponentials, sines and cosines. He wrote:

I have recently found a remarkable way of integrating differential equations of higher degrees in one step, as soon as a finite [algebraic] equation has been obtained. Moreover this method extends to all equations which, on setting dx constant, are contained in this general form:

$$y + \frac{ady}{dx} + \frac{bddy}{dx^2} + \frac{cd^3y}{dx^3} + \frac{dd^4y}{dx^4} + \frac{ed^5y}{dx^5} + \text{etc.} = 0.$$

To find the integral of this equation I consider this equation or algebraic expression:

$$1 - ap + bp^2 - cp^3 + dp^4 - ep^5 + \text{etc.} = 0.$$

If possible this expression is resolved into simple real factors of the form $1 - \alpha p$: if, however, this cannot be done resolve it into factors of two dimensions of this form $1 - \alpha p + \beta pp$, which resolution can always be done in reals, for whatever form the equation may have

---

[13]One person's method can be somebody else's problem, and so forth, but the trichotomy is no less useful even so.

[14]See Eneström [66], 33–38, Cannon and Dostrovsky [28], F&G 14.A1(a), and Euler [99], OO213.

it can always be put in the form of a product of factors either simple, $1 - \alpha p$, or of two dimensions $1 - \alpha p + \beta pp$, all real. This resolution being done, I say that the value of y is a finite expression in x and constants, obtained from all the members which have been factors of the algebraic expressions, and singular members supply singular terms of the integral. Certainly the simple factor $1 - \alpha p$ gives as member of the integral $Ce^{x/a}$, and a composite factor $1 - \alpha p + \beta pp$ gives this member of the integral

$$e^{-1\alpha x/2\beta}\left(C \sin A.\frac{x\sqrt{4\beta - \alpha\alpha}}{2\beta} + D \cos A.\frac{x\sqrt{4\beta - \alpha\alpha}}{2\beta}\right)$$

where for me $\sin A.$ and $\cos A.$ denote the sine and the cosine of arcs in a circle of radius $= 1$: however it is to be noticed that if the expression $1 - \alpha p + \beta pp$ cannot be resolved into simple real factors, when $4\beta > \alpha\alpha$, still the integrals are real.

Let the following be taken as a suitable example

$$y dx^4 = K^4 d^4 y, \quad \text{or} \quad y - \frac{K^4 d^4 y}{dx^4} = 0;$$

this gives rise to the algebraic expression $1 - K^4 p^4$, whose real factors are these three $1 - Kp, 1 + Kp, 1 + K^2 p^2$; and from these spring the integrals of the equation

$$y = Ce^{-x/K} + De^{x/K} + E \sin A.\frac{x}{K} + F \cos A.\frac{x}{K};$$

in which expression, because a four-fold integration has been done in one operation, there are four new constants as the nature of the integration demands. If it would please you, most excellent sir, I shall write down the method of proof on another occasion.

It is not clear how Euler came upon his brilliant idea. It falls out, however, as soon as one tries to see if the differential equation is solved by functions of the form $y = e^{-px}$. Because

$$\frac{dy}{dx} = -py, \quad \frac{d^2 y}{dx^2} = p^2 y, \quad \text{and so on,}$$

when $y = e^{-px}$ is substituted into the equation, the resulting equation is

$$e^{-px}(1 - ap + bp^2 - cp^3 + dp^4 - ep^5 + \text{etc}). = 0.$$

The expression e^{-px} is never zero, so it can be divided out, and therefore, as Euler claimed, $y = e^{-px}$ is a solution of the differential equation if p is a solution of the polynomial equation.

To find the values of p, Euler claimed that the polynomial equation can always be factored into linear terms of the form $1 - \alpha p$ and quadratic terms of the form $1 - \alpha p + \beta pp$. This is an example of what came to be called the fundamental theorem of algebra, which was widely believed at the time, but not proved. Euler then solved these equations for p and found $p = 1/\alpha$ and $p = \frac{\alpha \pm \sqrt{\alpha^2 - 4\beta}}{2\beta}$, respectively. Notice that the second expression is equal to $\frac{\alpha \pm i\sqrt{4\beta - \alpha^2}}{2\beta}$.

The first case leads to the solution $y = e^{-x/\alpha}$. The second case leads to the solution

$$\exp\left(\frac{-\alpha}{2\beta} \mp \frac{i\sqrt{4\beta - \alpha^2}}{2\beta}\right) x = \exp\left(\frac{-\alpha x}{2\beta} \mp \frac{i x\sqrt{4\beta - \alpha^2}}{2\beta}\right).$$

Now he already knew that

$$e^{p+iq} = e^p(\cos q + i \sin q),$$

so he saw that

$$\exp\left(\frac{-\alpha x}{2\beta} \mp \frac{i x\sqrt{4\beta - \alpha^2}}{2\beta}\right) = \exp\left(\frac{-\alpha x}{2\beta}\right)\left(\cos \frac{x\sqrt{4\beta - \alpha^2}}{2\beta} + i \sin \frac{x\sqrt{4\beta - \alpha^2}}{2\beta}\right).$$

The term involving the sine function remains a solution when multiplied by any constant, and so the factor of i can be removed by multiplying by i, and Euler's solutions are finally obtained.

As we remarked, Daniel Bernoulli had already noticed that exponentials, sines, and cosines were among the solutions, but Euler was the first to see that every solution could be written in terms of them. As a result, his new solution to the differential equation for the vibrating rod is much better than his earlier power series solution, because it becomes possible to see what some of the solutions actually look like, and this is very instructive. For example, each of the next four functions is separately a solution: $y = e^{-x/K}$, $y = e^{x/K}$, $y = \sin(x/K)$, and $y = \cos(x/K)$, which can be called its *basic modes of vibration* of the rod. Moreover, the new approach brings to light that the shape of the rod at any instant is a certain sum of these basic modes with constant coefficients.

Furthermore, because the rod is fastened to the wall and protrudes, say, horizontally, and the mortar is secure and immovable, any solution is subject to the two initial conditions that when $x = 0$ necessarily $y = 0$ and $\frac{dy}{dx} = 0$. This eliminates some combinations of the basic modes, and the allowed solutions (at any moment of time) are all of the form

$$\alpha e^{Kx} + \beta e^{-Kx} - (\alpha + \beta)\cos Kx - (\alpha - \beta)\sin Kx.$$

This also explained what Daniel Bernoulli had noted experimentally: a thin rod clamped to a wall can be made to emit several different sounds as it is plucked, and indeed, several different sounds at once because it can be in several distinct shapes.[15]

Euler informed his 72-year-old former professor Johann Bernoulli of his claims about this class of differential equations but did not send him a proof. Not to be outdone, Bernoulli replied with a proof in early December 1739, but his approach is interestingly old-fashioned. He first showed how to reduce the differential equation to a polynomial equation, much as Euler had done. However, Bernoulli adhered to the geometric language that Euler's work would gradually drive out, and interpreted the solutions, which he wrote in the form $y = n^{x/p}$, as "logarithmic curves whose

---

[15]Bernoulli used a needle.

subtangent is to be found".[16] In addition, Bernoulli regarded the equation $p^4 - K^4 = 0$ as the same as $p = K$ and remarked that whereas he had one solution, Euler had exhibited several. He commented that for this to be the case "my logarithms will be impossible or imaginary, but it is also the same in your solution, allowed to be more general, for you must let K be impossible or non-real". This shows that complex numbers were puzzling when they occurred in problems involving real quantities, but were nonetheless accepted, perhaps as something that needed to be better understood.

Euler's attitude to what he would consider an answer to a differential equation makes one crucial advance over Johann Bernoulli's. The methods they used to solve differential equations were not so very different: changes of variable, cunning substitutions, and so on, although certainly Euler's insight that reduces these differential equations to polynomial equations was a breakthrough. But changing what was considered an acceptable answer was an essential ingredient in advancing the calculus— and Euler seems to have had some success in convincing Johann Bernoulli, too, of the value of thinking in this way.

The first two volumes of Euler's *Institutionum Calculi Integralis* (E342 and E 366) give a good indication of what he could do by the late 1760s (the book was presented to the St. Petersburg Academy in August 1766; volume 1 was published in 1768, volume 2 in 1769). Euler investigated a number of different kinds of ordinary differential equations, looking for simplifications and for general methods. Quite an amount of insight into complete solutions, particular integrals, and initial conditions is accumulated. In volume 2 Chap. 8 Euler considered the second-order ordinary differential equation, and began by solving the linear equation by the method of undetermined coefficients. (A particular type of this equation is the hypergeometric equation, which we shall investigate in Chap. 11).

Given the equation

$$\frac{d^2y}{dx^2} + M\frac{dy}{dx} + Ny = X,$$

in which M, N, and X are functions of x, Euler began, as usual, by taking particular cases. His first example was

$$x^2(a + bx^n)\frac{d^2y}{dx^2} + x(c + dx^n)\frac{dy}{dx} + (f + gx^n)y = 0.$$

He looked for a solution of the form

$$x^\lambda(A + Bx^n + Cx^{2n} + \cdots),$$

and by looking at the lowest power of x was led to the equation

---

[16]The subtangent to a curve at a point P is the distance from the point where the tangent meets the x-axis to the point on the x-axis vertically above or below P. So, if P has coordinates (x, y) and $\frac{dy}{dx} = p$ at P, then the subtangent is $\frac{y}{p}$.

$$\lambda(\lambda - 1)a + \lambda c + f = 0.$$

Once this is solved, the constants $B, C, \ldots$ can all be found recursively in terms of A.

Euler also looked for solutions in which the powers of x decrease, at the cases of both real and imaginary values of λ, and the more difficult case when the values of λ are either the same or differ by an integer and the method provides only one solution, not two. In this case, Euler found another solution with a logarithmic term.

It seems that linear differential equations of higher order were beyond Euler's reach not because of the method of series but because of problems with the correspondingly higher order equation for λ. There were, of course, no explicit methods for solving the quintic equation, and nor was there a secure proof of the so-called fundamental theorem of algebra. So although Volume 2, Sect. 2, Chap. 2 of Euler's book covers the third-order linear equation with constant coefficients, the solutions of which are of the form $e^{\lambda x}$ for the values of λ that satisfy the corresponding cubic equation, and Euler dealt with both the case of distinct roots and the case of repeated roots, the extension of the analysis to higher order equations became lost in the details.

It had not been a hundred years since Leibniz had struggled to master one inverse tangent problem, but by the 1760s, Euler had a theory of many different kinds of differential equations. It embraced differentials of various degrees; homogeneous equations; solution methods that introduced multipliers or employed the method of infinite series or relied on a method of successive approximation. The formal side of the calculus was evidently being deployed, so much so that examples appear for the first time only on page 355—surely a good sign that we are in the presence of a theory rich enough to keep mere examples at bay.

1.5.1 A Note on the Adjoint Equation

I mention here that Euler's work was extended by the young Joseph-Louis Lagrange in a 200-page memoir [171] that he published in *Miscellanea Taurensis*. Lagrange established that a linear equation of order n of the form

$$Ly + M\frac{dy}{dt} + N\frac{d^2y}{dt^2} + \cdots = T,$$

where $L, M, N, \ldots, T$ are functions of t, will have n solutions (independence is implied but not stated) as the method of undetermined coefficients would suggest, and proceeded to investigate interesting cases and methods for reducing the order of the equation. This led him to discover that one can associate to a given ordinary differential equation another that, if solved, enables one to reduce the order of the original ordinary differential equation by one. Repeating this trick on the new equation returns almost to the original one; it actually comes back as

$$Ly + M\frac{dy}{dt} + N\frac{d^2y}{dt^2} + \cdots = 0.$$

In the nineteenth century, the second equation came to be called the adjoint equation of the first equation, and so one could say that Lagrange had proved that the adjoint of an equation is (the homogeneous form of) the original equation.

It will be enough to demonstrate the trick on the second-order equation

$$Ly + M\frac{dy}{dt} + N\frac{d^2y}{dt^2} = T.$$

Lagrange multiplied both sides by an unknown function $z = z(t)$ and integrated, to get

$$\int Lyz\,dt + \int M\frac{dy}{dt}z\,dt + \int N\frac{d^2y}{dt^2}z\,dt = \int Tz\,dt.$$

He integrated the terms involving M and N by parts, so the equation becomes

$$\int Lyz\,dt + Myz - \int \frac{dMz}{dt}y\,dt + Ny'z - \int \frac{dNz}{dt}y'\,dt = \int Tz\,dt,$$

and, integrating by parts again,

$$\int Lyz\,dt + Myz - \int \frac{dMz}{dt}y\,dt + Ny'z - y\frac{dNz}{dt} + \int \frac{d^2Nz}{dt^2}y\,dt = \int Tz\,dt,$$

which he rearranged with respect to y, when it becomes

$$y\left(Mz - \frac{dNz}{dt}\right) + \frac{dy}{dt}Nz + \int \left(Lz - \frac{dMz}{dt} + \frac{d^2Nz}{dt^2}\right)y\,dt = \int Tz\,dt.$$

So, if z is chosen to be a solution of the equation

$$Lz - \frac{dMz}{dt} + \frac{d^2Nz}{dt^2} = 0,$$

then the original differential equation reduces to

$$y\left(Mz - \frac{dNz}{dt}\right) + \frac{dy}{dt}Nz = \int Tz\,dt,$$

which is of degree one less than before. As Lagrange pointed out, the adjoint equation is simpler than the original one because it is homogeneous, and if it can be solved the original equation is also simpler because has been reduced to one of lower degree.

Lagrange then turned to other questions: the motion of fluids, the vibrating string, the motion of the planets. All in all, it is a formidable paper, familiar with the work of d'Alembert on fluids and Euler on the vibrating string, although Lagrange sided more with d'Alembert than Euler over the generality of the solutions of the wave equation.

1.6   Exercises

1. Solve Debeaune's problem and show that the solution with the stated initial conditions has the line $x + y = 0$ as an asymptote.
2. Solve Euler's equation

$$x^2(a + bx^n)\frac{d^2 y}{dx^2} + x(c + dx^n)\frac{dy}{dx} + (f + gx^n)y = 0.$$

What qualitative features of the solution are apparent to you (if any)?

Questions

1. What qualitative features of a function are apparent to you from its power series representation? Consider, for example, the series for exp, cos, and sin. Is it at all obvious that the series for cos and sin define periodic functions?
2. How would you attempt to graph the equation for the tractrix if you did not know of its origins as a problem in physics? In the light of your experience, how well do you think a mathematician of the late seventeenth century could claim to understand a curve knowing only its equation?

Chapter 2
Variational Problems and the Calculus

2.1 Introduction

Inspired by calculus, which made problems look simple that not long before no one had dared to raise, mathematicians began to ask a variety of questions about curves. We met some in the previous chapter that led to inverse tangent problems, but others were to lead to a new branch of the calculus and ultimately to new principles for the study of mechanics. As such, if they were solved at all they were solved by ingenuity rather than a systematic method, but—as we shall see—the insights that were produced on the way were often deep and lasting.[1]

Very likely the oldest, most attractive, and most famous problem of the kind we are about to discuss is known as Dido's problem, which asks for the shape of the largest area bounded by a straight line and a curve of given length, or, in some versions, the greatest planar area enclosed by a curve of given length. According to various mythological sources, Dido fled the city of Tyre, perhaps around 825 BCE, and came to a place on the North African coast where she was granted permission to have much land as a strip made from oxhide could enclose. She cut the hide long and thin, and enclosed an area upon which the city of Carthage was founded—but which problem she solved, and what her solution was, mythology does not make precise.[2]

The problem was brought to people's attention in one of Lord Kelvin's popular lectures in 1893, and the mathematician Adolf Kneser solved it his textbook [160] in this form: Find the curve of given length joining two points A and B that, together with the chord AB, encloses the greatest area. The solution is a circular arc with

---

[1] This chapter follows Fregulia and Giaquinta [109] to which readers are referred for much fascinating information.

[2] In Virgil's *Aeneid*, Dido then falls in love with Aeneas, who was seeking a new home after the destruction of his native Troy, but when he abandons her she commits suicide, calling down endless hate upon him. This was the origin of the Punic war centuries later between her city, Carthage, and Rome, the city founded by descendants of Aeneas. Virgil's poem is most likely incompatible with what we know of the Trojan wars.

© The Author(s), under exclusive license to Springer Nature Switzerland AG 2021
J. Gray, *Change and Variations*, Springer Undergraduate Mathematics Series,
https://doi.org/10.1007/978-3-030-70575-6_2

the line as a chord, as we shall see in Sect. 26.3. But he returned to it in a paper [161], where he pointed out that the problem is far from being the simplest in the calculus of variations. It is entirely possible that Dido could have taken the problem to mean: Find the curve of given length joining two points on the coast that encloses the greatest area, and this allows for the end points to be variable. In this case, too, the solution curve must be a circular arc, but it is not clear that the maximum area is necessarily attained.[3]

2.2 Bernoulli's Problems

In June 1696, Johann Bernoulli added a challenge to the mathematical community to a paper of his in the journal the *Acta Eruditorum*:

> Given points A and B in a vertical plane to find the path AMB down which a movable point M must, by virtue of its weight, proceed from A to B in the shortest possible time.

This problem, the problem of quickest descent or, to give it the name for the curve that Bernoulli gave it from the Greek, the brachistochrone, is emblematic of the topic. Bernoulli added that the solution was not the straight line going A and B but was in fact a curve well known to mathematicians, and promised to publish the solution if no one could find it. He also sent the problem in a letter to Leibniz on 9 June 1696, who replied on 16 June with a solution and the suggestion that Bernoulli delay publishing the solution because the journal travelled only slowly across Europe.

In May 1697, Bernoulli published his solution, and one by his brother Jakob, as well as discussions of the problem by Tschirnhaus and de l'Hôpital. Leibniz withdrew his solution, saying that it was too close to the ones proposed by the Bernoulli brothers.

Before we look at Johann Bernoulli's solution, we should note that the problem was not original with him. It had been proposed before, by Galileo, in his *Dialogues Concerning Two New Sciences*, where he gave a fallacious argument that purported to show that the solution was an arc of a circle.[4]

This mistake comes at the end of the long dialogue on the third day of the *Two New Sciences*, where Galileo recorded his lasting contribution to the study of motion: his laws of falling bodies.[5] He had studied the times taken by balls to roll down inclined planes of various slopes, thus slowing their rate of descent to lengths of time that could be measured by accurate, regular counting. He was led to proclaim that a uniformly accelerated body will fall as far in an interval of time as one moving with a constant velocity that is the average of the initial and final speeds of the first body. Furthermore, the distance covered by the accelerating body in equal intervals of time increase with the square of the time.

---

[3] We look at a solution to this problem on Sect. 26.3.1 below.

[4] See Galileo [113] Theorem XXII, Proposition XXXVI, Scholium.

[5] For an extract, see F&G 10.B4.

Fig. 2.1 Galileo's law of falling bodies

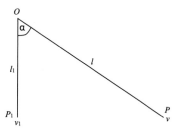

Then he noted (see Fig. 2.1) that "If a body falls freely along smooth planes inclined at any angle whatsoever, but of the same height, the speeds with which it reaches the bottom are the same". Crucially for present purposes, he immediately remarked (Theorem III, Proposition III)

> If one and the same body, starting from rest, falls along an inclined plane and also along a vertical, each having the same height, the times of descent will be to each other as the lengths of the inclined plane to the vertical.

We would see this by resolving the velocity along the slope into its horizontal and vertical components. One ball falls from O to P_1, a distance of l_1, which it reaches with velocity v_1. Another ball rolls from O to P, a distance of l, which it reaches with velocity v. We have, by conservation of energy,

$$v_1 = v, \quad l_1 = l \cos \alpha,$$

so

$$\frac{v_1}{l_1} = \frac{v}{l \cos \alpha}.$$

We also know that, because the acceleration is uniform, the time to fall from rest through a distance h is h times half the velocity at h, so

$$t_1 = l_1 \frac{2}{v_1} \quad \text{and } t = l\frac{2}{v},$$

so

$$\frac{t_1}{t} = \cos \alpha = \frac{l_1}{l},$$

as Galileo claimed.

Galileo's argument was not that different: he measured the magnitude of the force acting on the ball on the slope by the weight of a ball hanging vertically down from the top of the slope that was attached to the first ball by a cord and such that the two balls did not move.

The quantification of velocity and acceleration by Galileo was to be exactly what was needed to study motion with the advent of the calculus.

Fig. 2.2 Snell's law of
refraction

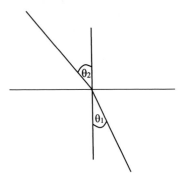

Equal significance is attached to Fermat's argument that the path of light in a
varying medium is the one that takes the least time. In the late 1630s, he had already
been in an argument with Descartes about the refraction of light and the explanation
of Willebrod Snell's law for refraction. Snell's law states that when a ray of light
leaves one medium and enters another there is a constant ratio r, determined by the
two media, such that

$$\sin \theta_1 / \sin \theta_2 = r,$$

where θ_1 and θ_2 are the angles the light makes with the normal at the point of crossing
in the two media (see Fig. 2.2).

In 1657, Fermat learned of the Greek mathematician Heron's ideas about why
light travels in straight lines (in a constant medium) and saw that he could adapt
them to explain refraction. He supposed that light travels at one speed in air, say v_1,
and another, slower, speed in water, say v_2, and then wrote down the time of travel
between a point in the water to a point in the air on the assumption that it travelled
along straight lines in each medium. His ad hoc techniques for finding the minima
of certain quantities were up the task, and he deduced that in the present set-up,

$$\frac{\sin \theta_1}{\sin \theta_2} = \frac{v_1}{v_2}.$$

We can argue the same conclusion slightly more rigorously. We choose a point
A_1 that is a_1 units below the surface of the water, and a point A_2 that is a_2 units
above it, and we suppose they are d units apart horizontally. We suppose that the
light travels along a straight line from A_1 to a point P on the water surface and then
along a straight line to A_2. With angles with the normal at P as given, we have for
the horizontal distance (Fig. 2.3)

$$a_1 \tan \theta_1 + a_2 \tan \theta_2 = d.$$

The distance travelled in the water is $a_1 \sec \theta_1$, and in the air is $a_2 \sec \theta_2$, and so the
total time taken is

Fig. 2.3 Fermat's deduction

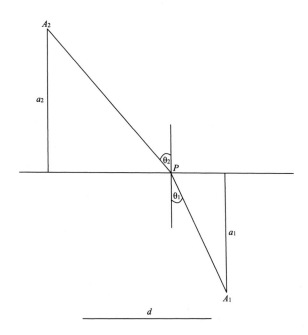

$$T = \frac{a_1 \sec \theta_1}{v_1} + \frac{a_2 \sec \theta_2}{v_2}.$$

From the first equation, we find by differentiating that

$$a_1 \sec^2 \theta_1 + a_2 \sec^2 \theta_2 \frac{d\theta_2}{d\theta_1} = 0.$$

From the second equation, we deduce that

$$\frac{dT}{d\theta_1} = \frac{a_1 \sin \theta_1 \sec^2 \theta_1}{v_1} + \frac{a_2 \sin \theta_2 \sec^2 \theta_2}{v_2} \frac{d\theta_2}{d\theta_1}.$$

For the shortest time, we require that

$$\frac{dT}{d\theta_1} = 0.$$

Eliminating $\dfrac{d\theta_2}{d\theta_1}$ from these equations, we find that

$$0 = \frac{a_1 \sin \theta_1 \sec^2 \theta_1}{v_1} + \frac{a_2 \sin \theta_2 \sec^2 \theta_2}{v_2} \left(\frac{-a_1 \sec^2 \theta_1}{a_2 \sec^2 \theta_2} \right),$$

which simplifies to

$$\frac{\sin \theta_1}{\sin \theta_2} = \frac{v_1}{v_2}.$$

The constant in Snell's law is revealed to be the ratio of the velocities of light in the two media.

So Fermat's principle that light takes the least time to travel between two points led him to a theoretical derivation of Snell's law—but on what basis? Can this principle really be fundamental, or is not the case that light behaves as it does for some (possibly unknown) reason and that this reason implies the principle? Can a principle such as least time act as a cause? These questions were not to be answered for a long time, and Bernoulli himself is our source for the information that[6]

> Leibniz in the *Acta Eruditorum*, 1682, pp. 285 et seq., and soon after the famous Huygens, in his *Treatise on Light*, p. 40, have demonstrated this more comprehensively and by most valid arguments, have established the physical, or better the metaphysical, principle which Fermat seems to have abandoned.

2.3 The Bernoullis' Brachistochrones

Johann Bernoulli tackled the brachistochrone problem by what was to become a standard method in mechanical questions: replace the problem by a number of discrete problems and let these problems crowd together and their number increase indefinitely until they tend to the original problem of interest.[7]

In this case, Bernoulli considered the path of light through a sequence of horizontal layers of translucent material, each layer having a different density. As he put it, they are made of "a diaphanous matter of a certain density decreasing or increasing according to a certain law". At each boundary, Snell's law applies and so the path of light through these media can be determined.

How he had this idea is not known, but it is clear enough that by adjusting the density of the diaphanous layers a wide variety of paths can be obtained, just as a varying law of acceleration can. So "In this way we can solve the problem for an arbitrary law of acceleration, since it is reduced to the determination of the path of a light ray through a medium of arbitrarily varying density".

Then by looking at an infinitesimal moment, Bernoulli deduced that if the moving particle goes from a point (x, y) to a point $(x + dx, y + dy)$ and its velocity increases from v to $v + dv$ then, by Snell's law,

$$\frac{dx}{dz} = \frac{1}{a}v,$$

---

[6]See Fregulia and Giaquinta [109], 40.

[7]See Bernoulli [9]. There is an English translation of this paper and Jakob's in Struik *Source Book* 391–399.

for some constant a, where dz is the amount of motion along the tangent, so $dx^2 + dy^2 = dz^2$. This gave him that

$$\frac{dx}{dy} = \frac{v}{\sqrt{a^2 - v^2}}.$$

He then took from Galileo's law of falling bodies that $v = \sqrt{ay}$, and so

$$dx = \sqrt{\frac{y}{a - y}} dy,$$

and so, he said, the brachistochrone is a cycloid.

Neither for him nor for us is this deduction entirely easy to make. We can write

$$x = \int_0 \sqrt{\frac{y}{a - y}} dy.$$

Set $y = a \sin^2 t$, so $dy = 2a \cos t \sin t\, dt$, and the integral becomes

$$x = 2a \int_0^{y=z} \sin^2 t\, dt = -\frac{a}{2} \sin 2t + at = \frac{a}{2} (2t - \sin 2t).$$

We can write

$$\sin^2 t = \frac{1}{2}(1 - \cos 2t)$$

and so express the solution curve parametrically in the form

$$x = \frac{a}{2} (2t - \sin 2t), \quad y = \frac{a}{2}(1 - \cos 2t).$$

This is the equation of a cycloid that starts at the origin. Moreover, this cycloid has a vertical initial tangent, because when $t = 0$ $\frac{dx}{dt} = 0 = \frac{dy}{dt}$ but $\frac{d^2x}{dt^2} = 0$ and $\frac{d^2y}{dt^2} = 2a \neq 0$.

The constant a can be determined from the separation of the end points, and once that is given the cycloid is unique.

Jakob Bernoulli's solution was different and seems to have influenced Euler a generation later. He argued that if a path between points A and B is the path of quickest descent then it must be the path of quickest descent between any two of its points.[8] For, if it was not the path of quickest descent between two intermediate points C and D, say, then that path could be replaced with a quicker one and this

---

[8]If the points A and B do not lie in a vertical line then the particle must start with some non-zero velocity.

would shorten the time of descent from A to B as well, which was assumed to be a minimum.

He then formulated this insight as an infinitesimal statement about a piece of the quickest path as compared to some nearby path, drawing on Galileo's laws of motion. By an argument here suppressed he arrived at the same integral, and therefore the same solution, as his brother.

2.4   Geodesics on Surfaces

A geodesic on a surface is a curve of the shortest length that joins two points on the surface and lies entirely in the surface. In the plane, a geodesic is a straight line; on the sphere, it is an arc of a great circle (the circle cut out by the plane that passes through the given points and the centre of the sphere).

In 1697, Johann Bernoulli challenged the mathematical community to investigate geodesics on curved surfaces. His insight into the problem had to do with the best approximating plane to a curve. He considered a geodesic on a surface and passing through a point P, and he looked at two points P' and P'' on the geodesic that tend to P. He argued that the plane through these three points on the geodesic then tends, as these points tend together to the point P, to the plane containing the tangent to the geodesic at P that is perpendicular to the surface at P.

How might we come to believe that? We might argue that if the geodesic is traversed at a constant speed then its normal is perpendicular to the curve, the normal and the tangent between them define the plane in question, and for a geodesic, the normal is also perpendicular to the surface. Or, we might argue that the claim is true for a sphere, so it is true for the best approximating sphere to the surface at P and because these surfaces are arbitrarily close in the limit what is true for the sphere is true for the surface. Of course, we need another argument for surfaces that are saddle-shaped near P.

Bernoulli's argument was something like the first of ours, but in reverse. From the original geometric insight, he deduced that the curvature vector is proportional to the normal vector at every point on a geodesic. Thus, he could interpret the requirement that the curvature vector of a curve in a surface be normal to the surface as an equation for a geodesic. Even so, this did not lead to a solution to the problem except in special cases, and the problem lay fallow for 30 years until Bernoulli proposed it to the young Leonhard Euler in 1728.

It is worth noting that it follows from Johann Bernoulli's characterisation of a geodesic that force-free motion along a surface is along a geodesic if force-free is taken to mean no forces acting on the surface (or, if you prefer, there are no forces acting that have a non-zero component in the tangent plane).

Euler published a short paper on geodesics in 1732, in answer to a question from Johann Bernoulli. He considered a geodesic GMH on a surface, where the points are infinitely close together and M is the midpoint, and the plane through M parallel to the (y, z)-plane cuts the surface in the curve IMK.

Euler took Cartesian coordinates in space—one of the earliest times this had been done—and wrote down the distances GM and MH on the assumption that they were well approximated by infinitesimal line segments. He said that the coordinates of the points were

$$G = (a, b, c), \quad M = (a + \alpha, y, z), \quad H = (a + 2\alpha, f, g),$$

so, by the three-dimensional Pythagorean theorem, distances between the points are

$$GM = \sqrt{\alpha^2 + (y - b)^2 + (z - c)^2}, \quad MH = \sqrt{\alpha^2 + (f - y)^2 + (g - z)^2}.$$

For this to be a minimum as M varies the differential of $GM + MH$ must vanish, and this implies that

$$\frac{(y - b)dy + (z - c)dz}{\sqrt{\alpha^2 + (y - b)^2 + (z - c)^2}} = \frac{(f - y)dy + (g - z)dz}{\sqrt{\alpha^2 + (f - y)^2 + (g - z)^2}}. \tag{2.1}$$

It remained for Euler to eliminate the arbitrary infinitesimals α, y, z, f, and g. After some work, here omitted, he obtained a second-order ordinary differential equation that he was able to solve in simple cases, when the surface is a cylinder, a conoid (a cone on a plane curve), or a surface of rotation.

2.5 Exercises

1. Obtain a formula for the radius of curvature of a curve and show that for a curve traversed at unit speed the radius is the reciprocal of the magnitude of the acceleration.
2. Use Bernoulli's insight in Sect. 2.4 to find the geodesics on a sphere, a cylinder, and a cone.

Questions

1. Information about tangents to a curve and its radii of curvature translate into information about velocity and acceleration along a curve. Why do you think this often struck mathematicians in the eighteenth and early nineteenth centuries as enough?

Chapter 3
The Vibrating String and the Partial Differential Calculus

3.1 Introduction

The study of problems involving more than one independent variable and the extension of the calculus to deal with these problems were significant advances of the first half of the eighteenth century. The first significant success was d'Alembert's mathematically correct description of the vibrating string, which has become famous as being the first partial differential equation to be solved, and although that title that can be disputed on a technicality, this should not be allowed to mask the real breakthrough his analysis achieved. Here we examine what he did to formulate and solve a problem in two independent variables and show how it enabled many basic phenomena of musical sounds to be explained.[1]

In the case of two independent variables, d'Alembert and Euler found it natural to introduce formal complex variables, but this raised questions they were unable to answer about the implications of using them.[2]

3.2 Early Investigations into the Partial Differential Calculus

The first person to extend the calculus systematically to two independent variables was Nicolaus I Bernoulli in unpublished work in 1719. He had been led to it through

---

[1]This account overlaps with the accounts in Barrow-Green, Gray and Wilson [4] and Gray and Micallef forthcoming.

[2]I offer a blanket warning that the dates of publications in the eighteenth century can be confusing. It was usual for a member of an Academy to present a paper, which would then be published in a journal of the Academy, but the process was slow, and it is common for a paper published two or three years after it was presented. Generally, I have referred to papers by their publication dates.

© The Author(s), under exclusive license to Springer Nature Switzerland AG 2021
J. Gray, *Change and Variations*, Springer Undergraduate Mathematics Series,
https://doi.org/10.1007/978-3-030-70575-6_3

his work on a problem of contemporary interest, the determination of a family of orthogonal trajectories to a given family of curves. Each curve in the orthogonal trajectory is specified by a parameter α; the coordinates of a point on a given curve are functions of a parameter t.

We know from unpublished manuscripts that Bernoulli deduced the equality of mixed partial derivatives from the following observation[3]: moving from an initial point t on a curve with parameter α to the point $t + dt$ and then along the orthogonal trajectory to a point on the curve with parameter $\alpha + d\alpha$ is the same as first moving to the curve with the parameter $\alpha + d\alpha$ and then changing t to $t + dt$. Euler argued much the same way independently in his "De differentiatione" of 1730, which he also left unpublished for several years. When he eventually did publish this result in another paper [72] Nicolaus wrote to Euler in 1743 to say that he had not published it himself because he regarded it as an axiom "which I thought to be obvious to anybody from the mere notion of differentials".[4]

The impetus to extend the calculus to functions of two variables had a second source in the study of ordinary differential equations, specifically when mathematicians were led to consider inexact differentials. These are expressions of the form $a(x, y)dx + b(x, y)dy$ that cannot be written as $d(g(x, y))$. Alexis Claude Clairaut, in his paper [43] began by noting that if

$$a(x, y)dx + b(x, y)dy = d(g(x, y))$$

then necessarily, by the equality of mixed partial derivatives, $\frac{\partial a}{\partial y} = \frac{\partial b}{\partial x}$, and conversely, or so he claimed, if this condition is met then the differential is exact. This is only true if the functions $a(x, y)$ and $b(x, y)$ are defined everywhere in a simply connected region, as the counter-example $a(x, y) = x/(x^2 + y^2)$, $b(x, y) = y/(x^2 + y^2)$ shows; this differential is not exact. But Clairaut deduced the theory from a consideration of monomials of the form $x^m y^n$ because he believed that every function of two variables is expressible as an infinite sum of such monomials.[5]

When the differential is not exact he proposed to look for a factor $\mu(x, y)$ such that

$$\mu(x, y)a(x, y)dx + \mu(x, y)b(x, y)dy$$

is exact, and this led him to the partial differential equation

$$\frac{\partial(\mu a)}{\partial y} = \frac{\partial(\mu b)}{\partial x},$$

or, equivalently,

---

[3]Only in the nineteenth century did mathematicians rephrase this as a necessary and sufficient condition.

[4]See Engelsman [68] for the details. The quote from Nicolaus I Bernoulli occurs on p. 106.

[5]See Clairaut [44], 45. Clairaut developed his ideas in competition with Alexis Fontaine. For a discussion of their rivalry, and the full context, which includes investigations into the shape of the Earth see Greenberg [129].

$$\mu \frac{\partial a}{\partial y} + a \frac{\partial \mu}{\partial y} - \mu \frac{\partial b}{\partial x} - b \frac{\partial \mu}{\partial x} = 0.$$

Clairaut was able to give a number of ways of finding the integrating factor μ in particular cases.

Clairaut had heard about Euler's work on the subject from Daniel Bernoulli and wrote to Euler about it on 17 September 1740, enclosing copies of some of his papers. Euler wrote back on 19 October to say that he was very pleased with them, and this started a productive association between the two, a highlight of which is Clairaut's analysis of the motion of the Moon that was one of the decisive papers in the Continental acceptance of Newtonian gravity in 1749.[6]

All this work led to the emergence of partial differential equations as a topic of investigation. The first important problem to be solved in the theory of partial differential equations was the problem of the vibrating string, and historians place the emphasis here, rather than on the question of integrating factors, because it marks a real if intangible shift towards the full acceptance of two independent variables. Clairaut had seen the question of finding an integrating factor as a question about differentials and not as a question in the subject of partial differential equations; such a subject did not exist and he was not inspired to create one.

But even the story of the wave equation tells us that a new field of enquiry was only gradually being born. However, natural it might seem for someone trying to create a general theory of second-order partial differential equations to take the wave equation as a major example, d'Alembert never wrote down the wave equation when he studied the motion of the vibrating string in his [52]. Even at this stage, the problem of the vibrating string was a problem in the partial differential calculus rather than in the theory of partial differential equations, which was still to be created.

3.3 D'Alembert: The Vibrating String and the Wave Equation

The problem of the vibrating string had by then attracted the attention of mathematicians for over a century because of its close connections to music. Every musician knows that a violin string makes a predictable sound and that tightening the string raises its pitch, as does shortening it. In 1638, Marin Mersenne had stated this law for determining the frequency of vibration of a string:

$$\nu = \frac{\sigma}{\ell} \sqrt{T} \, ,$$

where ν denotes the frequency, ℓ the length, and T the tension in the string and σ is a constant (determined by the nature of the string). However, neither he nor anyone he

---

[6]See the discussion in Sect. A.2.

consulted could explain *why* this rule should be true. Nor could Christiaan Huygens, some years later. The first person to get anywhere with it was Brook Taylor in 1713.

Taylor came from a musical family—he played the harpsichord—and his cryptic paper (Taylor [251]), earned him the reputation of being the first person to derive Mersenne's law mathematically. After writing up this theoretical account, he devised ingenious experiments designed to measure the rate at which harpsichord strings vibrate (they vibrate too fast for anyone to count).

Taylor began his paper by making two simplifying assumptions.

- The amplitude of oscillation of a string is independent of its frequency (volume is independent of pitch.)
- The string vibrates in such a way that all of the string crosses the axis simultaneously.

The second, and far from plausible, assumption enabled Taylor to argue that each point of the string behaved like a simple pendulum, and that each point on it moves up and down with the *same* period. He then argued that the force at each point is determined by the curvature of the string at that point, and that it is equal to the force that would cause a simple pendulum to oscillate with the same period as the string. As a result, Taylor was able to determine the shape of the string and its frequency of vibration, and to derive Mersenne's law.

In fact, Taylor's second assumption is wrong for all but the simplest oscillations, nor does it follow that each point of the string behaves like a simple pendulum. Even so, his analysis was the accepted one until it was replaced by d'Alembert's account.

3.3.1 D'Alembert's Breakthrough

In a paper written in 1747 and published in 1749, d'Alembert (see Fig. 3.1) assumed that the string was of uniform thickness. He then wrote[7]:

> Let t be the time elapsed from the moment when the string started to vibrate: it is certain that the ordinate PM can only be expressed by a function of the time t and of the abscissa or the corresponding arc s or AP. Let, therefore, $PM = \varphi(t, s)$, that is, let it be equal to an unknown function of t and s.[8]

So in d'Alembert's account, the height of the string above the x-axis at time t is given by a function $\varphi(t, s)$, where the variable s denotes arc length along the string. He then explicitly assumed that the vibrations of the string are so small that the length of the string from one point to another is "reasonably equal" to the difference in the x coordinates of the points. This made the mathematics tractable, at the price of considerably restricting the analysis.

D'Alembert set $d\varphi = p\,dt + q\,ds$, where $p = \frac{\partial \varphi}{\partial t}$ and $q = \frac{\partial \varphi}{\partial s}$. He referred to Euler [72] for the equality of mixed partial derivatives, and stated that

---

[7] See Alembert [52].
[8] Quoted in Struik *Source Book*, 353.

Fig. 3.1 Jean le Rond
d'Alembert (1717–1783),
Artist unknown, after
Maurice Quentin de la Tour

$$dp = \alpha dt + v ds, \quad dq = v dt + \beta ds, \tag{3.1}$$

where

$$\alpha = \frac{\partial^2 \varphi}{\partial t^2}, \quad v = \frac{\partial^2 \varphi}{\partial t \partial s}, \beta = \frac{\partial^2 \varphi}{\partial s^2}.$$

He then followed Taylor, and argued on physical grounds that the acceleration of a point on the string is proportional to $\pm \frac{\partial^2 \varphi}{\partial s^2}$, where the sign is positive if the curve is concave towards the x-axis and negative if it is convex.

Next, he argued that the acceleration at a point depends only on position, so $\frac{\partial^2 \varphi}{\partial s^2} = \beta$ (this uses the identification of string length with the x coordinate, so the oscillations must be very small). Then, by looking at the position of the string at two moments of time an amount dt apart, he argued that a point will have moved an amount αdt. He brought these two observations together by referring to Newton's *Principia*, where Newton had discussed motion under gravity, to deduce that (with respect to a suitable choice of units)[9]:

$$\alpha = \beta.$$

Had he transcribed his remark into the notation of second partial derivatives he would have written the wave equation,

$$\frac{\partial^2 \varphi}{\partial t^2} = c^2 \frac{\partial^2 \varphi}{\partial s^2},$$

[9]Struik notes (*Source Book* 354, n. 5) that the reference is to *Principia* Book I, Sec, X, Prop. LII, where Newton reworked Huygens's discussion of the pendulum, and that Taylor had done the same.

where $\dfrac{\partial^2 \varphi}{\partial t^2}$ is the second derivative of φ with respect to t with s regarded as a constant, and $\dfrac{\partial^2 \varphi}{\partial s^2}$ is the second derivative of φ with respect to s with t regarded as a constant, but evidently that is not how d'Alembert thought of it.

To solve the equations,

$$dp + dq = (\alpha + v)(dt + ds) \text{ and } dp - dq = (\alpha - v)(dt - ds),$$

we note that they can be written in the form

$$dp + dq = (\alpha + v)d(t + s) \text{ and } dp - dq = (\alpha - v)d(t - s)$$

when it becomes clear that $p + q$ is a function of $t + s$ and $p - q$ is a function of $t - s$. d'Alembert wrote

$$p = \Phi(t + s) + \Delta(t - s), \text{ and } q = \Phi(t + s) - \Delta(t - s),$$

and deduced that the general shape of the string was given by

$$\varphi = \psi(t + s) + \Gamma(t - s),$$

where ψ and Γ are arbitrary functions; $\varphi(t + s) = \psi_t(t + s) + \Gamma_t(t - s)$ and $\Delta(t + s) = \psi_s(t + s) - \Gamma_s(t - s)$. Here, the suffices denote differentiation with respect to t and s.

This was a dramatic moment: the first time that the most powerful branch of the calculus, that of *differential equations*, was shown to extend to problems with more than one independent variable.[10] A path now seemed to be open to tackle the many problems in several variables in which the natural world would surely abound.

But the success soon brought with it a profound disquiet. D'Alembert's solutions were anything of the form

$$\varphi(t, s) = f(ct + s) + g(ct - s),$$

where f and g are *arbitrary* functions. Upon reflection, the functions f and g should be capable of being differentiated twice (and of course for each value of t the graph of φ depicts a string fastened down at each end, as the original problem requires). This solution is very general, which is as it should be, because the string can be released from any initial shape and with any initial velocity. As d'Alembert noted "this equation includes an infinity of curves".[11]

---

[10]Earlier, d'Alembert had reformulated Daniel Bernoulli's study of the small oscillations of a hanging chain as a partial differential equation, but he had not been able to solve it. See D'Alembert [50], 171.

[11]Quoted in Struik, *Source Book*, p. 355.

But just how general could such a solution curve be was a question that, once raised, was to become one of the most famous mathematical controversies of the century. For a further discussion of it, see any good history of mathematics in the eighteenth century.

In his [53], d'Alembert looked for solutions of the form

$$\varphi(t, s) = F(t) \times G(s) .$$

This reduced his differential equation in two independent variables to two differential equations each in a single variable, as follows.

If we substitute $\varphi(t, s) = F(t) \times G(s)$ into the equation

$$\frac{\partial^2 \varphi}{\partial t^2} = c^2 \frac{\partial^2 \varphi}{\partial s^2},$$

we obtain

$$\frac{\partial^2 F}{\partial t^2} G(s) = c^2 \frac{\partial^2 G}{\partial s^2} F(t),$$

which implies that

$$\frac{1}{F(t)} \frac{\partial^2 F}{\partial t^2} = c^2 \frac{1}{G(s)} \frac{\partial^2 G}{\partial s^2}.$$

But a function of t can only be equal to a function of s if they are both constant, which we shall call $-k^2 c^2$. Then we obtain the ordinary differential equations

$$\frac{d^2 F}{dt^2} = -k^2 c^2 F(t) \text{ and } \frac{d^2 G}{ds^2} = -k^2 G(s).$$

The solutions to these equations are of the form:

$$F(t) = \cos kct \text{ or } \sin kct, \quad G(s) = \cos ks \text{ or } \sin ks.$$

Here, c is a constant determined by the material in the string, the tension in the string, and its shape, and k^2 is some constant, as yet undetermined, related to the frequency of the string's vibration. (We can now see that only choosing a negative quantity for the above constant makes physical sense: if you follow through the above argument with $+k^2 c^2$ you see that hyperbolic functions are obtained that cannot match the presumed boundary conditions and are liable to grow impossibly large.)

For the first time, recognisable solutions had appeared: functions of the form $\cos kct \cos ks$ (or $\cos kct \sin ks$, and so on) are solutions of the wave equation, although by no means the most general.

We shall shortly look ahead by three years to see how Euler also found that the wave equation has very general solutions. His solution is also important because it is one of the first occasions where the equality of mixed partial derivatives was used and understood. But first, we attend to the physics of musical sounds.

3.3.2 Mersenne's Law and Modes

First, it is clear that d'Alembert's new ideas led to the first satisfactory deduction of Mersenne's law. We take the solution

$$\varphi(t, s) = \cos kct \times \sin ks$$

and look at a particular point on the string—that is, at the situation for a fixed value of s—then as time varies this point moves according to the equation

$$\varphi = \cos kct \times \text{constant}.$$

This means that it oscillates with a frequency, kc, that is the same whatever point of the string is taken. So, a given string vibrates with a specific frequency.

To see that the frequency k is inversely proportional to the length of the string, observe that both ends of the string must be fixed, so $\varphi = 0$ when $s = \ell$. Therefore, $\sin k\ell = 0$ so $k\ell$ must be a multiple of π, say $k\ell = N\pi$. So $k = N\pi/\ell$ and the frequency is inversely proportional to the length of the string, as Mersenne had claimed.

Musicians knew that halving a string results in a note an octave above the basic note (it doubles the frequency). This phenomenon of modes was explained in the manner of d'Alembert by Euler in his [80], Sect. 41. The solutions

$$\varphi = \cos \pi ct/\ell \times \sin \pi s/\ell$$

correspond to the choice $k = \pi/\ell$ (with $N = 1$) and

$$\varphi = \cos 2\pi ct/\ell \times \sin 2\pi s/\ell$$

correspond to the choice $k = 2\pi/\ell$ ($N = 2$). In the second case the string, which has twice the frequency, behaves as though it were two strings each of half the original length and joined at a fixed point in the middle. This explains how the same string can be made to play certain different notes without being tightened or changed in length, and indeed that it will naturally vibrate in a variety of ways—but not in any way: the tones it can emit—its *harmonics*—are all the notes whose frequencies are multiples of a basic frequency.

Another musical phenomenon, which had puzzled Mersenne, is that a string can emit several notes at once. But it is easy to show that if $\varphi(t, s)$ and $\psi(t, s)$ are solutions of the wave equation, then so is any sum of the form $a\varphi + b\psi$ where a and b are arbitrary constants. So a string may vibrate in two or more ways simultaneously, emitting two or more different notes as it does so. Euler and Daniel Bernoulli had noticed the same behaviour in their analysis of the vibrating clamped rod in the mid-1730s. It makes it clear, also, that Taylor's assumption that the whole string crosses the axis simultaneously must be wrong.

3.4   Euler Rewrites the Wave Equation

Although it will take us slightly ahead of the story, this is a possible place to look at how Euler treated the wave equation some years later, as he began to develop a theory of partial differential equations.

In his [80], E213, Euler first rederived the equation of the vibrating string in a way that he felt led more simply to the solution. He took the wave equation in the form $\left(\frac{\partial^2}{\partial t^2} - c^2 \frac{\partial^2}{\partial x^2} \right) y = 0$, factorised it as $\left(\frac{\partial}{\partial t} + c \frac{\partial}{\partial x} \right) \left(\frac{\partial}{\partial t} - c \frac{\partial}{\partial x} \right) y = 0$, and argued (in Sect. 25 of his paper) that the equation of motion of the string can be regarded as a system of two first-order differential equations.

More precisely, Euler considered (Sect. 25) a function y that satisfied this first-order equation

$$\frac{\partial y}{\partial t} = c \frac{\partial y}{\partial x}, \tag{3.2}$$

and argued that therefore

$$\frac{\partial^2 y}{\partial t^2} = \frac{\partial}{\partial t} \left(\frac{\partial y}{\partial t} \right) = c \frac{\partial}{\partial t} \left(\frac{\partial y}{\partial x} \right) = c \frac{\partial^2 y}{\partial t \partial x} =$$

$$c \frac{\partial^2 y}{\partial x \partial t} = c \frac{\partial}{\partial x} \left(\frac{\partial y}{\partial t} \right) = c^2 \frac{\partial}{\partial x} \left(\frac{\partial y}{\partial x} \right) = c^2 \frac{\partial^2 y}{\partial x^2}.$$

So a solution of Eq. (3.2) satisfies the wave equation

$$\frac{\partial^2 y}{\partial t^2} = c^2 \frac{\partial^2 y}{\partial x^2}.$$

Now, if $y = f(x + ct)$ let $z = x + ct$ and write

$$y = f(z), \quad z = g(x, t) = x + ct.$$

Note that

$$\frac{\partial g}{\partial t} = c \quad \text{and} \quad \frac{\partial g}{\partial x} = 1.$$

The chain rule for differentiation yields:

$$\frac{\partial y}{\partial t} = \frac{df}{dz}\frac{\partial g}{\partial t} = c\frac{df}{dz}$$

$$\frac{\partial y}{\partial x} = \frac{df}{dz}\frac{\partial g}{\partial x} = \frac{df}{dz}$$

so

$$\frac{\partial y}{\partial t} = c\frac{\partial y}{\partial x}.$$

Therefore, functions of the form $y = f(x + ct)$ are solutions of the wave equation. By an analogous argument, so are functions of the form $y = f(x - ct)$. Euler finished off by checking explicitly that functions of this form are solutions of the wave equation.

3.5   Formal Complex Methods

As we have seen, there was a close connection between certain partial differential equations and certain exact differentials that mathematicians such as d'Alembert, and Euler were learning to exploit. A major example of this method was published by d'Alembert in 1752 in another context. He began with two differentials similar to the ones above, imposed an exactness condition, and was led to a different constraint on a collection of functions. Although the only change in the differentials was one change of sign, the consequence was the arrival of what may be called formal complex methods in the theory of differential equations.

D'Alembert entered the manuscript of his *Essai d'une nouvelle théorie de la résistance des fluides* for a prize competition of the Berlin Academy, and when no prize was awarded he blamed Euler and their already poor relations worsened.[12] D'Alembert reworked the manuscript and published it as a book [54], which is where the differential equations of hydrodynamics were first written down.

In Chap. IV, Sect. 45 d'Alembert considered a two-dimensional flow in the (x, z)-plane. The general problem was very difficult, and so he turned (in Sects. 57–60) to address the simpler task of finding the conditions on functions M and N of x and z such that the differentials

$$M\,dx + N\,dz, \quad N\,dx - M\,dz \tag{3.3}$$

are exact. This resembles his account of the wave equation, but with a change of sign: $M = \alpha$, $N = v$ and with $\beta = -M$. As we shall see, this analogy was to be much appreciated by Euler.

---

[12]They only improved until 1759 when d'Alembert declined Frederick the Great's invitation to become President of the Berlin Academy and recommended Euler instead; neither was appointed and Euler soon left for St. Petersburg.

D'Alembert then argued that if the differentials are exact then there are functions p and q such that

$$Mdx + Ndz = dq, \quad Ndx - Mdz = dp.$$

Therefore, the differentials

$$(M + iN)(dx - idz) \quad \text{and} \quad (M - iN)(dx + idz)$$

are exact, and, on putting $du = dx - idz$ and $dt = dx + idz$, $M + iN = \sigma$ and $M - iN = \tau$, the expressions $\sigma\,du$ and $\tau\,dt$ become exact differentials. Therefore, $\sigma = M + iN$ must be a function of $u = x - iz$, and $\tau = M - iN$ must be a function of $t = x + iz$. This allowed him to "deduce the values of M and N", as he put it— they are

$$M = \frac{1}{2}(\sigma + \tau) \text{ and } N = \frac{1}{2i}(\sigma - \tau).$$

D'Alembert then gave what he called a simpler argument to the same effect. The definitions of p and q imply that

$$p_z = -q_x \text{ and } q_z = p_x,$$

and he deduced immediately that $qdx + pdz$ and $pdx - qdz$ are exact differentials and that therefore $q + ip$ is a function of $x - iz$ and $q - ip$ is a function of $x + iz$. (This is correct because $(q + ip)(dx - idz) = qdx + pdz + i(pdx - qdz)$ is exact.) So, he set $q + ip = F(x - iz)$ and $q - ip = G(x + iz)$ and separated out the corresponding expressions for p and q.

Now he had to show that this line of argument led to real-valued functions because his problem was connected to real-valued functions in the plane. So he said that for p and q to be real q must be a function of the form

$$\xi(x - iz) + i\zeta(x + iz) + \xi(x + iz) - i\zeta(x - iz),$$

where the functions ξ and ζ (regarded as power series, something d'Alembert always thought possible) have real coefficients (he omitted the similar result for p).

So d'Alembert obtained expressions for p and q in which, in his phrase, the imaginary quantities destroy themselves. Later eyes can see that d'Alembert had sketched a quick argument from the Cauchy–Riemann equations to the existence of formal complex functions, but d'Alembert did not; indeed, he made no further use of those equations in the rest of the memoir. Instead, he returned to his original problem and solved it by power series methods.

However, d'Alembert's work was very influential; there are many later references to "the method of d'Alembert" in the study of surfaces. Formal complex methods rely on a free transition from real to complex variables and functions, which are then handled by algebra and differentiation, but with no appreciation of what it is for a function to be *complex* differentiable. For example, $\xi(x - iz)$ and $\xi(x + iz)$ above are complex conjugates, as are $\zeta(x + iz)$ and $\zeta(x - iz)$, and so q is the real part of the complex-valued function $\xi(x - iz) + i\zeta(x + iz)$, and q is a harmonic function of x and z—but none of this was remarked upon by d'Alembert. The method provided a convenient notation, at the cost of requiring that any imaginary quantities could be made to vanish in the end, and leave only equations between purely real quantities.

3.6 Exercises

1. Find the equation of a plucked string released from an initial position formed by a straight line joining the point $(x, t) = (0, 0)$ to $(3\pi/4, 1)$ and a straight line joining $(3\pi/4, 1)$ to $(\pi, 0)$.
2. Either write a programme to show the motion of the string or find one on the web. How would you describe the motion of the string? How does it produce a sound?

Questions

1. In the absence of any clear distinction at the time between continuous, differentiable, and analytic, what could Euler have meant by a curve drawn by a free motion of the hand?
2. What sorts of solutions might Euler admit to the partial differential equation

$$\frac{\partial u}{\partial x} = 0 ?$$

3. It was quickly appreciated that if a single term of the form $\cos kct \sin ks$ is a solution of the wave equation then so is a sum of terms of this form, and even an infinite sum of terms of this form. This resulted in a number of 'Fourier series' being produced before Fourier, and speculation about whether every function can be written in this form. Daniel Bernoulli suggested that might be true, Euler disagreed, and Lagrange got close to providing a plausibility argument for the claim before retreating. Information about this debate can be found in Bottazzini [21] and in the commentaries on Euler's works. Given the range of functions known to them, what do you think an eighteenth-century mathematician might say was involved in deciding this claim?

Chapter 4
Rational Mechanics

4.1 Introduction

The study of many natural phenomena was opened up by the use of partial differential calculus. In particular, in the late 1750s, Euler was able to extend his methods to produce equations for the motion of an ideal fluid. He also studied the partial differential equations that describe the propagation of sound. This led him to a partial differential equation of lasting importance for the theory, and in the course of attempting to solve it, he also came up with an ordinary differential equation that was to be the most important and thoroughly analysed equation of its type in the nineteenth century (as we shall see in Chaps. 11 and 16). He also gave the first general formulation of rigid body mechanics, which rewrote and extended Newton's ideas and put them into something like their modern form.

4.2 Fluid Mechanics

The English word 'hydrodynamics' is derived from the title of a book Daniel Bernoulli published in 1738, his *Hydrodynamica*, which is in turn a word that he coined. His topic was an old one, the flow of water from a vessel through a pipe, and his advance was to determine how the pressure on the walls of a container of a volume of fluid in a container is affected by the velocity of the water. One presumably unfortunate result of his success was that his competitive father Johann Bernoulli then not only wrote and published his *Hydraulica* in 1742 but sought to pretend that it had been written in 1732, thus claiming priority over Daniel's work.

Much more important investigations were soon made: D'Alembert's *Traité de dynamique* [50], and in particular his *Réflexions sur la cause générale des vents* [51] which outlined a new theory of the tides, and of course his *Essai d'une nouvelle théorie de la résistance des fluides*, and Clairaut's memoir *Théorie de la figure de*

© The Author(s), under exclusive license to Springer Nature Switzerland AG 2021
J. Gray, *Change and Variations*, Springer Undergraduate Mathematics Series,
https://doi.org/10.1007/978-3-030-70575-6_4

Fig. 4.1 Leonhard Euler
(1707–1783) by Jakob
Emanuel Handmann, 1756

la terre (Theory of the shape of the Earth) [44] which discussed the shape of the Earth, regarded as a rotating fluid mass. Even so, it was Euler's work on fluids that has become the basis of all subsequent mathematical discussions of the motion of fluids.

Euler (Fig. 4.1) published three memoirs on the subject in 1757 (E225, 226, 227) and a further paper (E258) in 1761. They demonstrate very clearly both the power of the new methods and the difficulties that have to be overcome. He began E225 by noting that there was a consensus that the different behaviour of solids and liquids must be explained by expressing clearly the essential difference between the two. But, he said, this difference had never previously been understood. He proposed that it consisted in the fact that a solid can be held in equilibrium by two equal and opposite forces, whereas a fluid is only in equilibrium if it is held in place by an equal force at every point of its surface that acts perpendicular to the surface (he explicitly assumed that no forces are acting inside the fluid).

On this foundation, he derived the equations of motion for a perfect fluid, one that is incompressible and inviscid (without viscosity or 'stickiness'). Euler was particularly pleased to derive a theory of fluids based on the idea that a fluid is composed of infinitesimal solid bodies because this extended his version of Newton's mechanics to fluids.

In an incompressible fluid Euler's principle implies that the pressure in a body of fluid in equilibrium is known when it is known at a single point, and indeed that the pressure at a point depends only on the depth of the point.[1]

Euler analysed a fluid by introducing mutually perpendicular axes OA, OB, OC. This was his standard approach to all questions in dynamics. He resolved the force

---

[1] In later papers, he noted the changes that have to be made if the fluid is elastic or compressible.

of gravity acting at a point Z in the fluid along these three axes as follows: the component of the force in the directions OA, OB, and OC through Z he called P, Q, and R, respectively. He regarded P, Q, and R as functions of x, y, and z.

He next considered an infinitesimal volume of fluid in the form of a parallelepiped with one vertex at Z and of size $dx\,dy\,dz$. He took the density of the fluid to be q, so the forces acting on the parallelepiped were written down as $Pq\,dx\,y\,dz$ in the direction $ZL = OA$, $Qq\,dx\,dy\,dz$ in the direction $ZM = OB$, and $Rq\,dx\,dy\,dz$ in the direction $ZN = OC$. The pressure of the fluid above the parallelepiped, which Euler supposed to form a column of height p, is the other force involved.

Euler then wrote[2]

$$dp = L\,dx + M\,dy + N\,dz,$$

where, accordingly,

$$\frac{\partial L}{\partial y} = \frac{\partial M}{\partial x}, \quad \frac{\partial L}{\partial z} = \frac{\partial N}{\partial x}, \quad \frac{\partial M}{\partial z} = \frac{\partial N}{\partial y}.$$

He then considered the change in pressure between opposite sides of the parallelepiped and deduced that

$$L = Pq, \quad M = Qq, \quad N = Rq.$$

Therefore,

$$dp = q(P\,dx + Q\,dy + R\,dz).$$

Because the left-hand side can be integrated, Euler deduced from this that the right-hand side can also be integrated, and so (following some earlier remarks by Clairaut)

$$\frac{\partial Pq}{\partial y} = \frac{\partial Qq}{\partial x}, \quad \frac{\partial Pq}{\partial z} = \frac{\partial Rq}{\partial x}, \quad \frac{\partial Qq}{\partial z} = \frac{\partial Rq}{\partial y}.$$

Much now depends on whether the fluid has a constant density or if the density varies with the depth. The analysis became complicated and so Euler turned to the study of particular cases, such as the theory of the barometer. Here, he observed that a wind must arise whenever the heat at equal heights is different and that a study of the equilibrium figures of a fluid suggests that some of them might approximate the shape of a planet.

Euler returned to his general analysis in his next paper, E226. Here, he considered the motion of an infinitesimal cube in the fluid. It will be helpful to sketch his argument initially without the mathematical details.

---

[2] Euler wrote $\left(\dfrac{dL}{dx}\right)$ where we have written $\dfrac{\partial L}{\partial x}$; in this period, partial derivatives were often written as ordinary derivatives enclosed in round brackets.

In an infinitesimal moment of time, the infinitesimal cube is stretched or squashed in the directions of the axes, although its volume cannot change because the fluid is assumed to be incompressible. Euler considered that what drives the motion is the difference in pressure between each pair of faces of the infinitesimal cube. Consider, for example, what is involved in saying that an infinitesimal cube does not sink under gravity. The pressure on the bottom face differs from the pressure on the top face by an amount equal to the weight of the liquid in the cube, which is equal to the volume of the cube times the density of the liquid. If these quantities were not equal, the pressure difference would cause the cube to move. This pressure difference manifests itself as a force, and this will bring about a change in the velocities of each particle of the fluid.

If we put all this together, we expect to find equations, one in each of the x-, y-, and z-directions, that say that a pressure difference on a small cube of fluid is equal to an acceleration in that direction multiplied by the mass of the cube. We indeed expect to see that the acceleration in the fluid will be described as the acceleration at each point and some measure of the stretching and squashing of the cube. This is exactly what Euler found.

Euler supposed that at a point in the fluid with coordinates (x, y, z) the velocity was (u, v, w), where each of u, v, and w is a function of x, y, x and the time t. At a nearby point with coordinates $(x + dx, y + dy, z + dz)$ the velocity is given by

$$u + \frac{\partial u}{\partial x}dx + \frac{\partial u}{\partial y}dy + \frac{\partial u}{\partial z}dz$$

in the x-direction, and by similar expressions for the velocities in the y- and z-directions.

As for the pressure differences, if p is the pressure, the difference in pressure across the faces separated by a distance dx is $\frac{\partial p}{\partial x}dx$.

The differences in velocities are of two kinds. The point originally at (x, y, z) has moved to $(x + udt, y + vdt, z + wdt)$, and each of u, v, and w has changed. For example, by standard Taylor series arguments, the change in u is from u to

$$u + \frac{du}{dt}dt = u + \frac{\partial u}{\partial x}\frac{dx}{dt}dt + \frac{\partial u}{\partial y}\frac{dy}{dt}dt + \frac{\partial u}{\partial z}\frac{dz}{dt}dt,$$

with similar expressions for v and w. Note that $dx/dt = u$, etc., so the increase in u in a time dt is

$$\frac{\partial u}{\partial x}udt + \frac{\partial u}{\partial y}vdt + \frac{\partial u}{\partial z}wdt.$$

To obtain the equations of motion, Euler interpreted the rule that force equals mass times acceleration as saying that acceleration equals force divided by mass, and took the mass of an infinitesimal cuboid to be its volume times its density. He assumed that the density, ρ, was constant, which is a good approximation for water in many situations.

When all this is put together, the result (see E226, p. 286) is these three equations:

$$\frac{\partial u}{\partial t} + u\frac{\partial u}{\partial x} + v\frac{\partial u}{\partial y} + w\frac{\partial u}{\partial z} = -\frac{1}{q}\frac{\partial p}{\partial x} \tag{4.1}$$

$$\frac{\partial v}{\partial t} + u\frac{\partial v}{\partial x} + v\frac{\partial v}{\partial y} + w\frac{\partial v}{\partial z} = -\frac{1}{q}\frac{\partial p}{\partial y} \tag{4.2}$$

$$\frac{\partial w}{\partial t} + u\frac{\partial w}{\partial x} + v\frac{\partial w}{\partial y} + w\frac{\partial w}{\partial z} = -\frac{1}{q}\frac{\partial p}{\partial z} - g \tag{4.3}$$

These are the differential equations for the motion of the fluid: the minus signs arise because we (with Euler) are measuring gravity in the opposite direction to the corresponding axis and because pressure increases as depth increases.

There is also an equation stating that the volume of any part of the fluid does not change, which had been known a decade before to d'Alembert, in a paper Euler had read. Euler explained that better, and expressed the conclusion more elegantly, in his "Principia motus fluidorum" (Principles of the motion of fluids, E258), which he published in 1761. He now considered the motion of a particle of the fluid infinitesimally close to (x, y, z), and considered what would happen to an infinitesimally small pyramid of the fluid in an interval of time dt. It is moved to another infinitesimally small pyramid of the same volume, and by calculating these volumes and equating them Euler deduced that in the flow at any instant

$$\frac{\partial u}{\partial x} + \frac{\partial v}{\partial y} + \frac{\partial w}{\partial z} = 0.$$

This is known as the *continuity equation* for the flow of an incompressible liquid in space, and it expresses the idea that as the fluid flows it does not change its volume.

Euler returned to the subject in a paper (E258) he published in 1761 as part of a long investigation into the motion of fluids, and in this paper, the Laplace equation was written down for the first time.[3] Here, he again considered incompressible fluids of constant density in two or three dimensions. In two dimensions, he wrote the components of velocity of the fluid at a point (u, v), and by calculating the volumes of infinitesimal elements he deduced that for an incompressible fluid $u_x + v_y = 0$. He then claimed that $udx + vdy$ is exact (which it is only if the flow is irrotational, in later terminology) and defined S as its integral, writing

$$dS = udx + vdy + Udt.$$

From the continuity condition, he deduced that $vdx - udy$ is another complete differential.[4]

---

[3] How it acquired Laplace's name is a long story often told elsewhere.
[4] This is only true if the fluid is, in today's language, irrotational; Euler seems to have assumed this in this early work.

He then repeated this argument in three dimensions, found that $udx + vdy + wdz$ is exact, and introduced[5]

$$dS = udx + vdy + wdz + Udt.$$

In paragraph 67 he deduced that

$$S_{xx} + S_{yy} + S_{zz} = 0.$$

He then wrote "Since it is not obvious how in general this can be made to happen, I shall consider certain classes of possibilities", and found, as his first example, that $(ax + by + cz)^n$ will work for any n, provided $a^2 + b^2 + c^2 = 0$ if $n > 1$. Linear combinations of these will also work, and he wrote down all expressions of degree less than 6. Then he went back to his real business—the motion of fluids.

It does not seem that he went back to the two-dimensional case and deduced that powers of $x \pm iy$ are harmonic, or that he made use of this obvious deduction elsewhere in his work. Most likely it became part of the general education of mathematicians of the next generation without its significance being appreciated.

However, it is one thing to have some equations of motion, and another to solve them. A few minutes looking at water in a bath or at the wind makes it clear that even the simple fluids that Euler described can display very complicated motions, and in general Euler was able to deal only with special cases, although he did become the first person to describe motion in vortices in mathematical terms. Perhaps the best way that we can indicate the difficulties in this branch of mathematics is to point out that Euler's equations of motion for a perfect fluid are still far from being adequately understood, and the problems raised by the equations for a general fluid (the so-called Navier–Stokes equations) are among the "millennium problems" whose solutions could earn a mathematician a million-dollar prize from the Clay Mathematics Institute.[6]

The fact is that Euler was exceedingly prescient when he wrote, at the start of E226, that[7]:

> Having established in my previous Memoir the principles of fluid equilibrium in their most general form, regarding both the diverse nature of fluids and the forces that act upon them, I now propose to deal with the motion of fluids in the same way and to seek out the general principles on which the entire science of fluid motion is based. It will readily be understood that this is a much more difficult undertaking and involves studies of incomparably greater depth. Nevertheless, I hope to arrive at an equally successful conclusion, so that, if difficulties remain, they will pertain not to Mechanics but purely to Analysis, this science not yet having been brought to the degree of perfection necessary to develop analytical equations that embody the principles of fluid motion.

---

[5]The function S was later called the velocity potential by Helmholtz.

[6]See Carlson et al. [32].

[7]See Frisch's translation in the Euler Archive. He notes that Euler wrote "formules" where he has supplied "analytical equations".

4.2.1   Recent Discoveries About the Euler Equations

I cannot resist quoting from an astonishing passage in Cédric Villani's [261] *Birth of a Theorem* (2016, 91–93):

> Imagine you're walking through the woods on a peaceful summer's afternoon. You pause at the edge of a pond. Everything is perfectly calm, not the slightest breeze.
>
> Suddenly the surface of the pond becomes agitated, as though seized by convulsions; a few moments later, it is sucked down into a roaring whirlpool. And then, a few moments after that, everything is calm once more. Still not a breath of air, not even a ripple on the surface from a fish swimming beneath it. So what happened?
>
> The Scheffer–Shnirelman paradox, surely the most astonishing result in all of fluid mechanics, proved that such a monstrosity is possible, at least in the mathematical world.
>
> [...] It rests on the incompressible Euler equations, the oldest of all partial differential equations, used by mathematicians and physicists everywhere to describe a perfectly incompressible fluid without any internal friction. It has been more than two hundred fifty years since Euler derived his fundamental equations, and yet not all of their mysteries have been penetrated. Indeed, they are still considered to mark out one of the most treacherous regions of the mathematical world. When the Clay Mathematics Institute set seven 'millennium problems' in 2000, offering $1 million apiece for their solution, it did not hesitate to include the regularity of solutions to the Navier–Stokes equations. It was very careful, however, to avoid any mention of Euler's equations – a far greater and more terrifying beast.
>
> And yet at first glance Euler's equations seem so simple, so innocent, utterly devoid of guile or cunning. No need to model variations in density or to grapple with the enigmas of viscosity. One has only to write down the classical laws of conservation: conservation of mass, quantity of motion, and energy.
>
> But then . . . suddenly, in 1993, Scheffer showed that Euler's equations in the plane are consistent with the spontaneous creation of energy! Thanks to [several subsequent authors] we now realise that even less is known about Euler's equations than we thought.
>
> And what we thought we knew wasn't much to begin with.

4.3   Euler and the Propagation of Sound

In the mid-eighteenth century, the nature and the propagation of sound were poorly understood. Newton had written about it in the *Principia*, and Euler in a study of heat, but d'Alembert had dismissed both in his *Traité des fluides* Sect. 219, writing that

> The formula given without proof by Euler is very different from Newton's, and I do not know how he was led to it; as for Newton's formula, it is proved in the *Principia* but in perhaps the most obscure and difficult part of that work.

So it was ambitious of the young Lagrange to write as one of his first works a 112-page memoir on the subject ([169], from which the above quote was taken).[8]

---

[8]This work has an interesting account of the ideas of Euler, D'Alembert, and Daniel Bernoulli on the nature of solutions to the equation for the vibrating string.

His work in turn provoked Euler to return to the subject, and in his [92], he first observed that Newton's account was ingenious reasoning based on purely arbitrary hypotheses. Then he wrote that[9]

> All those who have dealt with this matter after Newton either have fallen into the same trap, or, wanting to delve into the true movement of the air, have rushed into intractable calculations, from which one could absolutely not draw any conclusions, and I must admit that I arrived at one or the other place whenever I undertook this research. I was therefore pleasantly surprised when I saw in this excellent book that I have just mentioned, that Mr. De La Grange has happily overcome all these difficulties, and that by calculations which could seem quite unintelligible. This is unquestionably one of the most important discoveries we have made for a long time in Mathematics, and one which may lead us to many others.
>
> In examining these prodigious calculations, I wondered at first if it would not be possible to achieve the same goal by an easier route, and after some effort I got there. I have therefore the honor to explain here the method that seems the most suitable for this study, but, as simple as it may appear, I must insist that it would not have occurred to me, if I had not seen the ingenious analysis of M. De La Grange.

None of which meant that the derivation was simple (and because it is difficult it is omitted here). The upshot was a system of three second-order partial differential equations not unlike the equation for the vibrating string that described infinitesimally small motions of the air that described the passage of a sound wave. This is a lateral wave—the particles of the air move in the direction of the sound (unlike the vibrating string, which oscillates transversely). Euler obtained these equations in his ([93], Sect. 43). If the air is homogeneous and the sound travels radially at the same speed in all directions, then with respect to coordinates centred at the source of the sound the displacement (x, y, z) of a particle at (X, Y, Z) is given by

$$x = Xs, \quad y = Ys, \quad z = Zs,$$

where s is a function of the time t and the radial distance $V = \sqrt{X^2 + Y^2 + Z^2}$. In this case, Euler's equations reduce to ([93], Sect. 45)

$$\frac{1}{2gh}\frac{\partial^2 s}{\partial t^2} = \frac{4}{V}\frac{\partial s}{\partial t} + \frac{\partial^2 s}{\partial V^2}, \tag{4.4}$$

where $2g$ is the acceleration due to gravity in Euler's units and h is a measure of the elasticity of the air.

Mathematically, the constant term $2gh$ can be absorbed by changing the time variable to $\sqrt{2gh}t$, at the cost of changing $\frac{4}{V}$ to $\frac{4\sqrt{2gh}}{V}$. This produces an equation of the form

$$\frac{\partial^2 s}{\partial t^2} - \frac{\partial^2 s}{\partial V^2} = \beta\frac{\partial s}{\partial t}.$$

A further change of variable will write this as

---

[9]Translation slightly modified from that of Ian Bruce in the Euler Archive.

$$\frac{\partial^2 s}{\partial t \partial V} + \beta \frac{\partial s}{\partial t} + \beta' \frac{\partial s}{\partial V} = 0,$$

which is propitious.

This is the form in which Euler studied it in his *Institutionum Calculi Integralis* vol. 3, Part 2, Sect. 4 (E385, 1770, Sect. 322).[10] Euler wrote it as

$$(x + y)^2 \frac{\partial^2 z}{\partial x \partial y} + m(x + y)\frac{\partial z}{\partial x} + m(x + y)\frac{\partial z}{\partial y} + nz = 0. \qquad (4.5)$$

He regarded $x + y$ as a new variable, and looked for a solution in the form $v = (x + y)^\mu F(x)$. On this assumption, the method of undetermined coefficients allowed Euler to calculate $F(x)$ as a power series, and he deduced a recurrence relation:

$$n + 2m\lambda + \lambda^2 - \lambda = 0$$

$$(n + 2m\lambda + 2m + \lambda^2 + \lambda)B + (m + \lambda)A = 0$$

$$(n + 2m\lambda + 4m + \lambda^2 + 3\lambda + 2)C + (m + \lambda + 1)B = 0$$

$$(n + 2m\lambda + 6m + \lambda^2 + 5\lambda + 6)D + (m + \lambda + 2)C = 0$$

from which he deduced $B, C, D \ldots$. If we write $A = b_0, B = b_1, C = b_2, \ldots$, Euler's conclusion was that

$$b_j = -\frac{m + \lambda + j - 1}{j(2m + 2\lambda + j - 1)} b_{j-1}.$$

But Euler also noticed that there are values where the series breaks off. These occur when j is an integer and $m + \lambda + j = 0$, which can occur when $\frac{1}{4} - m - n + m^2$ is a square.

The same can be done with y instead of x, and so Euler declared that the general solution was of the form

$$\sum_j a_j (x + y)^{\lambda+j} (f^{(j)}(x) + g^{(j)}(y)),$$

where, for example, $f^{(j)}$ denotes the jth derivative of f, which Euler was confident was a complete solution because it contained two arbitrary functions, $f(x)$ and $g(y)$.

Darboux's clear analysis of Euler's equation (4.4) and its solutions will be found at the end of this chapter.

---

[10]A similar equation for the propagation of sound was studied by Laplace, with less success, in a paper he presented to the Académie des Sciences in 1773 but which was published only in 1777.

4.4 Euler's Vision of Mechanics

In his two-volume *Mechanica* [70] Euler discussed the motion of point masses under forces and in resisting media (vol. 1), and their motion on surfaces (vol. 2). Significantly, the book is written throughout in the language of the (Leibnizian) calculus. He also sketched a plan for describing the motion of solid rigid masses, elastic bodies, fluids, and gases. Much of this was an unknown territory at the time and over the decades Euler's contributions to various parts of this programme greatly enlarged the reach of mathematics.

Let us observe in passing that in volume 2 of the *Mechanica* (p. 464, Sect. 832) Euler mentioned what may be one of the first partial differential equations[11]:

> Finally I have turned or rounded surfaces [of revolution], which are generated by the rotation of any curve about an axis ; if AX were such an axis, on putting x constant, the equation between y and z gives a circle with centre P. Whereby the equation for these has this form:
>
> $$dz = Pdx - \frac{y}{z}dy \quad \text{or} \quad zdz + ydy = zPdx,$$
>
> where Pz only depends on x; or
>
> $$Q = -\frac{y}{z} \quad \text{and} \quad P = \frac{X}{z}$$
>
> with X present as a function of x.

The partial differential equation appears here as a differential of the form $Pdx = Rdz + Qdy$ that is to be integrated.

Euler dealt with rigid bodies in his 'Découverte d'un nouveau principe de Mécanique' ([79], E177). In it, he gave a decisive reformulation of the theory of mechanics that brought it into line with the practice of the calculus as he understood it. Truesdell, in his *An idiot's fugitive essays in science* ([259], 317) called the paper "a great masterpiece", and correctly observed that "it has dominated the mechanics of extended bodies ever since". He went on

> This paper contains the first proposal of the so-called Newton's equations, $\mathbf{f} = m\mathbf{a}$ in rectangular Cartesian coordinates, as a "new principle of mechanics", the common origin of all the several other principles then in use.

Euler's plan for the paper began with a definition of a solid body as one whose parts do not move with respect to each other (unlike, say, a liquid). He then said that existing principles of mechanics would show that at any instant the motion of a solid body can be analysed in terms of the motion of its centre of gravity and the rotation of the solid around an axis through the centre of gravity. This would be done by showing how the forces on the body determine how the centre of gravity will move. However, new principles would be needed to understand the rotation, which is about a varying axis. A start could be made by analysing rotations about a fixed axis, but it

---

[11]Translation by Ian Bruce in the Euler Archive.

would be necessary to consider axes of rotation that do not pass through the centre of gravity of the body.

The new principle, upon which he proposed to base all of mechanics, should, he said, be derived

> from first principles, or rather axioms, on which all the theory of motion is based. The axioms relate to infinitely small bodies that can only have a progressive motion; and all other principles of motion must be deduced from these, those which serve to determine the motion of solids as well as of fluids; all other principles will be nothing but the application of these axioms in various ways.[12]

As he remarked, there were several such principles in use, and he proposed to derive them all from the new principle that he now put forward.

He began by considering an infinitely small body of mass M acted upon by some forces. Its motion can be described with respect to a fixed but arbitrary plane and considering the height, x of the point mass above this plane. The forces acting on the point mass in various directions can be expressed in terms of forces parallel to the plane and forces perpendicular to it; Euler let P be the force perpendicular to the plane. After a time dt, the point mass will be a distance $x + dx$ from the plane,

> and taking the element of time dt as constant, it will be the case that $2Mddx = \pm Pdt^2$, according as the force P tends to move the body away from or towards the plane. It is this single formula that contains all the principles of mechanics.[13]

Euler employed a system of units in which the quantity M is measured in units such that the point mass has a weight of M near the surface of the Earth, and accordingly the force P is then the weight of the body. If the body moves away from the plane with a speed dx/dt, and if this is the speed that it would acquire by falling through a height of h, then one has

$$\left(\frac{dx}{dt}\right)^2 = h, \quad \text{and so} \quad dt = \frac{dx}{\sqrt{h}}.$$

Euler next supposed that the motion of the point mass was measured with respect to three mutually perpendicular planes, and supposed that the forces acting perpendicularly to these planes were P, Q, and R, respectively. He then wrote down these equations of motion:

$$2Mddx = Pdt^2; \quad 2Mddy = Qdt^2; \quad 2Mddz = Rdt^2.$$

This is the first time Newton's equations of motion were expressed in the formalism of the calculus. Moreover, they have been expressed with respect to three mutually perpendicular but otherwise arbitrary axes—taking perpendicular axes as standard

---

[12] See Euler [79], 194.

[13] See Euler [79], 195. Note that Euler's conventions about units produce factors of 2 in formulas where our conventions do not.

came in with Euler, not, for example, Descartes. There is another important difference between Euler's formulation and Newton's: Newton spoke of bodies, Euler of infinitesimal elements out of which bodies are formed.

As Euler then noted, if no forces act then $P = 0$, $Q = 0$, $R = 0$ and so the above differential equations can be integrated and the point mass is shown to move in a straight line. This establishes that a body initially at rest remains at rest, and one initially in motion remains moving uniformly in the same direction unless it is acted upon by a force.[14]

Next, Euler analysed the motion of a body whose centre of gravity is fixed, and by a more complicated argument of the same kind as before he deduced that at any instant the body is rotating about an axis through the centre of gravity.

He then set about describing the motion in general. He supposed that there are three mutually perpendicular axes in the body (OA, OB, and OC) that meet at the centre of gravity O. (Unlike the earlier choice of coordinates, which were taken with respect to three fixed planes in space, these axes are fixed in the body and so are moving.) To deal with the fact that the axis about which the body is rotating itself changes with time, Euler showed that it is enough to know how the three axes OA, OB, and OC change with time, which depends on the shape of the body and the distribution of mass within it. However, although Euler was able to obtain the differential equations of motion, even he found them "too long", and he concluded this paper with a discussion of some special cases.

More than 10 years later, however, Euler returned to this question and showed in his book the *Theoria motus corporum solidorum seu rigidorum* (Theory of the motion of solid or rigid bodies) (E289) of 1765 that every rigid body has a set of axes with respect to which its behaviour is particularly simple.[15]

First, he gave a new account of how a body rotates about a fixed but arbitrary axis through its centre of gravity in terms of what are called today the 'Euler angles' of a rotation. Then he introduced the concept of the principal axes of rotation of the body. This is the crucial breakthrough that extended Newtonian mechanics from the study of point masses to arbitrary bodies—everything from the bones in our bodies to car chasses and orbiting satellites.

Euler began by describing how to describe and quantify the motion of a rotating body. This led him to define

> Sect. 422. The moment of inertia of a body with respect to some axis is the sum of all the products which arise, if the individual elements of the body are multiplied by the square of their distances from the axis.

The moment of inertia is an integral of terms dM and r^2 that are always positive, and so it is necessarily positive. Moreover, it can be calculated with respect to any axis, not merely an axis about which one might suppose that the body is 'really' rotating.

---

[14]Euler here seems to have shared a naive belief that rest and motion of any kind are somehow different, but the separation of rest from uniform motion here may have been a pedagogical position.

[15]Our account concentrates on Vol. 1, Chap. 5. There is an English translation of much of the book by Ian Bruce in the Euler Archive.

To calculate the motion of a body from the forces acting upon it, Euler proposed that one looks for the most appropriate set of axes to use for a given body. First, he said that one should take the axis IG through the centre of inertia that yields a minimum for the moment of inertia among all such axes. Then one should find orthogonal axes through the centre of inertia, I, and specifically for axes about which the moment of inertia is a minimum or a maximum. This is a calculus problem that leads to a cubic equation, which must have either one real root or three. However, Euler was unable to deduce from the equation itself that there are always three real roots, and only gave an obscure argument to support the claim that there are three real roots. Finally, Euler proclaimed that every rigid body has three axes mutually at right angles with respect to which the moments of inertia are either a maximum or a minimum. These he called the *principal axes*

446. The principal axes of any body are these three axes passing through the centre of inertia of this body, with respect to which the moments of inertia are either a maximum or a minimum.

Euler then showed how to analyse the motion of a rotating solid body in terms of its motion with respect to the principal axes, how the action of forces affects the motion, and how to solve many problems in the dynamics of rigid bodies.

4.5 Darboux's Account

Euler's study of Eq. (4.4), the equation for the passage of sound, was nicely illustrated by Gaston Darboux over a century later in his ([58], Vol. 2, Chap. 3). He wrote the equation in the form

$$\frac{\partial^2 z}{\partial x \partial y} - \frac{m}{x-y}\frac{\partial z}{\partial x} + \frac{n}{x-y}\frac{\partial z}{\partial y} - \frac{p}{(x-y)^2}z = 0,$$

and noted that the substitution $z = \theta(x, y)(x - y)^\alpha$ turns it into one of the same form for θ and in which

$$m' = m + \alpha, \quad n' = n + \alpha, \quad p' = p + \alpha^2 + \alpha(m + n - 1).$$

So it is possible to choose a value of α that makes $p' = 0$ and write the equation in the form

$$\frac{\partial^2 z}{\partial x \partial y} - \frac{\beta'}{x-y}\frac{\partial z}{\partial x} + \frac{\beta}{x-y}\frac{\partial z}{\partial y} = 0. \tag{4.6}$$

Routine calculations show that if $Z(\beta, \beta') = Z(\beta, \beta')(x, y)$ is a solution, then so is

$$Z(1 - \beta', 1 - \beta)(y - x)^{\beta + \beta - 1}.$$

If we set $t = y/x$, and $z = x^\lambda \varphi(t)$ then z is a solution of Eq. (4.6) if and only if $\varphi(t)$ satisfies the ordinary differential equation

$$t(1 - t)\varphi''(t) + (1 - \lambda - \beta - (1 - \lambda - \beta')t)\varphi'(t) + \lambda\beta'\varphi(t) = 0,$$

which is a differential equation that later became known as the hypergeometric equation and is arguably the most important ordinary differential equation in the history of mathematics.[16]

The canonical form of the hypergeometric equation is the equation

$$z(1 - z)\frac{d^2w}{dz^2} + (\gamma - (\alpha + \beta + 1)z)\frac{dw}{dz} - \alpha\beta z = 0. \tag{4.7}$$

One solution of this equation is given by the hypergeometric series

$$F(\alpha, \beta, \gamma, z) = 1 + \frac{\alpha\beta}{1.\gamma}z + \frac{\alpha(\alpha + 1)\beta(\beta + 1)}{1.2.\gamma(\gamma + 1)}z^2 + \cdots. \tag{4.8}$$

This series converges in the disc $|z| < 1$. Euler considered only the case in which the variable is real, and gave two accounts of the equation and the series respectively, one in four chapters of the *Institutionum Calculi Integralis* ([95], Vol. 2, Part I, Chaps. 8–11), and a later one presented to the St. Petersburg Academy of Science in 1778 and published posthumously as Euler [98].

It is easy, if unilluminating, to solve Euler's hypergeometric equation by the method of undetermined coefficients. Let us denote one solution of it by $F(-\lambda, \beta', 1 - \lambda - \beta, y/x)$ and another by

$$(y/x)^{\lambda+\beta} F(\beta, \beta' + \beta + \lambda, 1 + \beta + \lambda, y/x),$$

then the corresponding solutions of Eq. (4.6) are

$$z = z^\lambda F(-\lambda, \beta', 1 - \lambda - \beta, y/x),$$

$$z = x^{-\beta} y^{\beta+\lambda} F(\beta, \beta' + \beta + \lambda, 1 + \beta + \lambda, y/x).$$

As Euler had already noticed, special cases arise when λ is a positive integer.

There are also solutions obtained by the method of separation of variables, $z = X(x)Y(y)$. This leads to the equation

$$x + \frac{\beta X}{X'} = y + \frac{\beta' Y}{Y'},$$

so both sides must be a constant, α say, and so

---

[16]See Chaps. 11 and 16.

$$X(x) = (x - \alpha)^{-\beta}, \quad Y(y) = (y - \alpha)^{-\beta'},$$

and so

$$z = (x - \alpha)^{-\beta}(y - \alpha)^{-\beta'}.$$

Finally, as Darboux remarked, it was shown by Paul Appell in 1882 that if $\varphi(x, y)$ is an arbitrary solution of Eq. (4.6) then the most general solution is of the form

$$(cx + d)^{-\beta}(cy + d)^{-\beta'}\varphi\left(\frac{ax + b}{cx + d}, \frac{ay + b}{cy + d}\right),$$

where a, b, c, d are arbitrary constants and $ad - bc \neq 0$.

4.6 Exercises

1. The hypergeometric equation is the ordinary differential equation

$$z(1 - z)\frac{d^2w}{dx^2} + (\gamma - (\alpha + \beta + 1)x)\frac{dw}{dx} - \alpha\beta w = 0.$$

Show that the series

$$F(\alpha, \beta, \gamma, x) = 1 + \frac{\alpha\beta}{1.\gamma}x + \frac{\alpha(\alpha + 1)\beta(\beta + 1)}{1.2.\gamma(\gamma + 1)}x^2 + \cdots$$

is a solution of the equation, and find another, linearly independent solution.
2. The series reduces a polynomial if either $\alpha - 1$ or $\beta - 1$ is a negative integer, and is not defined at all if γ is a negative integer or zero (this case Euler excluded). Show that in all other cases the series is convergent for $x = a + bi$ provided that $a^2 + b^2 < 1$.

Questions

1. In what ways did Euler's approach to mechanics improve upon Newton's?
2. Find out what you can about the motion of the Moon. Would it be easy or difficult to use knowledge of the motion of the Moon to determine longitude at sea? A fascinating take on this is provided by the story of Harrison's chronometers; see what you can find out about them.

Chapter 5
The Early Theory of Partial Differential Equations

5.1 Introduction

As methods for dealing with two or more independent variables advanced it became possible to pose questions about partial differential equations. First, as with ordinary differential equations, mathematicians looked for formal methods that would lead to exact, general solutions, and this led to questions about the existence of integrating factors. Euler discussed the problem of finding integrating factors for expressions of the form $dx + a(x, y)dy$ in volume 1 of his *Institutionum Calculi Integralis* [94], and considered many types of cases without being able to show that one always existed.[1] However, as we shall see, a solution to this problem had been published a little earlier by d'Alembert. The method of characteristics was introduced for the first time by Euler and d'Alembert, and extended by Lagrange and Gaspard Monge, who applied it to first- and second-order partial differential equations.

5.2 Euler's General Theory of Partial Differential Equations

Euler began to think of creating a theory of partial differential equations in the early 1760s and outlined how he proposed to start in his [88, 89], where he looked at what later generations would call linear and quasi-linear first-order partial differential equations in two variables. For reasons of brevity, we shall pass to his later, more general account where he repeated most of this analysis and also discussed second-order linear partial differential equations. This is *Institutionum Calculi Integralis* Volume 3 of 1770.

---

[1] This volume was published in 1768. It and the next volume deal with ordinary differential equations, the third volume with partial differential equations. There is a useful English translation of all three books by Ian Bruce, available at the Euler Archive.

© The Author(s), under exclusive license to Springer Nature Switzerland AG 2021
J. Gray, *Change and Variations*, Springer Undergraduate Mathematics Series,
https://doi.org/10.1007/978-3-030-70575-6_5

He dealt with first-order equations of various kinds in Sect. 1 of Part 1 of the book by working through a long series of examples of steadily increasing complexity. He would present what we might call a theorem as a problem, show how to tackle it in general, comment on the solution, perhaps by obtaining it another way, and then work through some examples. As so often with Euler, it is not clear whether he was doing this solely on pedagogical grounds or if he was writing up a version of his own original route through the topic.[2]

In Chap. 2, Euler began with the simplest case of a first-order partial differential equation:

$$\frac{\partial z}{\partial x} = a,$$

where a is a constant, and the solution is $z = ax + f(y)$. He carefully pointed out that the arbitrary constant of integration that arises is an arbitrary function of y. Not only is it not necessarily given by an equation but also its graph may be a curve that is drawn by a free motion of the hand, or indeed by several such curves not connected in any way. This insistence on the extreme arbitrariness of the curve was original with Euler and not shared by all his contemporaries, and as we shall see, he specifically noted that the initial position of a vibrating string may be given by such a function.

In this chapter, Euler solved first-order partial differential equations of various forms. He wrote

$$p = \frac{\partial z}{\partial x} \text{ and } q = \frac{\partial z}{\partial y},$$

and set himself as Problem 21 the equation

$$px + qy = 0.$$

He wrote $dz = p\,dx + q\,dy$, eliminated q, and deduced that

$$dz = p(dx - (x/y)dy) = pyd\frac{x}{y}.$$

Both sides are therefore differentials of a function, and so he deduced that py must be a function of $\frac{x}{y}$, say

$$py = f'\left(\frac{x}{y}\right),$$

where f' is the derivative of an arbitrary function. Therefore,

$$dz = f'\left(\frac{x}{y}\right) d\left(\frac{x}{y}\right),$$

---

[2]The translation is taken, with slight modification, from Ian Bruce's version on the Euler Archive. I have changed Euler's notation to make it more like ours.

and by integrating both sides, he found that

$$z = f\left(\frac{x}{y}\right).$$

(It would seem that Euler's idea of arbitrariness has shrunk to implying differentiability.)

In this chapter, Problem 22, Euler got close to the general first-order linear partial differential equation (to use a more modern term). He wrote

a function $z(x, y)$ is sought for which, on writing $p = z_x$ and $q = z_y$ one has $q = pV$, where $V = V(x, y)$.

As before, he worked with differentials—these are differential equations, after all. He observed that the given equation and the identity $dz = pdx + qdy$ imply that

$$dz = p(dx + Vdy).$$

He then remarked

Now a multiplier M will be given, likewise a function of x and y, so that $M(dx + Vdy)$ is made integrable. Therefore there is put $M(dx + Vdy) = dS$, and also S will be given a function of the same x and y. Hence since there shall be $dz = \dfrac{pdS}{M}$, it is evident that the quantity $\frac{p}{M}$ must be equal to a function of S, whereby if we put $\dfrac{p}{M} = f'(S)$, there becomes $z = f(S)$ and thereupon will be

$$p = Mf'(S) \text{ and } q = MVf'(S).$$

We shall turn to the multiplier shortly, but first, let us unpack his solution. We can interpret the solution this way: *if* there is a function $M = M(x, y)$ such that $M(dx + Vdy)$ is integrable, and say the integral is $S(x, y)$ so

$$M(dx + Vdy) = dS,$$

then the solution to the partial differential equation is an arbitrary function of S. This is because

$$dz = p(dx + Vdy) = \frac{p}{M}dS = f'(S)dS,$$

so, integrating $f'(S)$,

$$z = \int \frac{p}{M}dS + G(x),$$

where $G(x)$ is an arbitrary function of x.

Euler's comments are interesting because they are circumspect. He went on

Corollary 5.1 *Therefore in this case the function sought z is found at once expressed in terms of x and y, because S is given by x and y. But it can come about the S gives*

rise to a transcending function, so that moreover by the methods so far known the multiplier M indeed cannot be found.

Euler's remark that "the multiplier M indeed cannot be found" may well mean that the multiplier might simply not exist. In his ([88], Sect. 20), he had said that this was too much to hope for, so his later opinion depends on what he made of d'Alembert's discussion, published in d'Alembert [56], which we shall look at shortly. Perhaps he meant, in agreement with d'Alembert, that the multiplier cannot be found explicitly.

With Problem 23 Euler reached what we would call the general first-order linear partial differential equation in two variables. He wrote the partial differential equations he was interested in the form

$$\frac{\partial z}{\partial y} = V \frac{\partial z}{\partial x} + U, \tag{5.1}$$

where U and V are functions of x and y (see Sect. 1, Chap. 5, Problem 23, Sect. 146). He then solved this equation by again recalling that $dz = p\,dx + q\,dy$, passing to the differential equation

$$dz = p(dx + V\,dy) + U\,dy,$$

and then looking for an integrating factor $M = M(x, y)$ such that $M(dx + V\,dy)$ is exact, say

$$M(dx + V\,dy) = dS.$$

He argued that when M is found then

$$dz = \frac{p\,dS}{M} + U\,dy,$$

and in this equation z, and therefore U and M, can be treated as functions of y and S. This equation may be integrated if S is held constant, to yield

$$z = \int U\,dy = T + f(S),$$

where T is a function of y and S. So the solution is given as a function of y and S.[3] But still, Euler did not discuss how M the integrating factor, or multiplier as he called it, can be found, except to note that it is sufficient that a solution of the equation $dx + V\,dy = 0$ can be found, and he was unable to give a general account of this aspect of the problem. As he put it:

---

[3]From this it follows that
$$dT = U\,dy + W\,dS,$$
and so, using the fact that $\frac{p}{M} = \frac{\partial z}{\partial S}$, $\frac{p}{M} = W + f'(S)$.

COROLLARY 2 Sect. 148. To this end it is convenient to consider the differential equation $dx + V dy = 0$; for if this can be integrated, likewise thereupon it will be possible to deduce the multiplier M, so that the formula $M(dx + V dy)$ truly becomes the differential of a certain function S, which therefore hence may be found.

Here as always Euler was not interested in fitting the general solution to a set of initial or boundary conditions. Just as with ordinary differential equations, the full solution was a formula with a degree of arbitrariness to it.

Two questions arise: What did Euler mean by saying that a multiplier might not always exist but that it was sufficient that a solution of the ordinary differential equation $dx + V dy = 0$ can be found, and why did he not solve the ordinary differential equation, or at least say that it can be solved? He had, after all, given his solution method for the first-order ordinary differential equation in Volume 1 of the *Institutionum Calculi Integralis*.

To deal with the multiplier issue first, in Problem 22 the partial differential equation is

$$v_y - V v_x = 0.$$

This is an equation for an unknown function $v = v(x, y)$. What can we say about the curves given by $v(x, y) = v_0$, where v_0 is a constant (the level curves of the function v)? Along these curves, $dv = 0$. But we always have $dv = v_x dx + v_y dy$, so, comparing this equation with the given partial differential equation, we find that

$$\frac{dy}{dx} = -\frac{v_x}{v_y} = -\frac{1}{V},$$

so we have the ordinary differential equation

$$dx + V dy = 0.$$

This has as its solutions the curves $v(x, y) = v_0$. If we write

$$dx + V dy = dv$$

then we see that

$$dv = v_x \left(dx + \frac{v_y}{v_x} dy\right) = v_x (dx + V dy),$$

so the integrating factor is v_x, where $v = v(x, y)$ is the solution of the ordinary differential equation

$$dx + V dy = 0.$$

Euler could surely have written this down, and it is not clear why he did not. A partial answer is that he had been committed to the formalism of (inexact) differentials and multipliers for as many as thirty years by now, and saw no reason to change. Another is that he was hung up on the fact that the multiplier is not explicit.

What is all the more curious about his remark about the ordinary differential equation is that in Volume I of the book, published in 1766, Euler had described a method for an approximative solution to any differential equation of the form

$$\frac{dy}{dx} = V(x, y).$$

In Sect. 650, Problem 85, he described how, from an initial point $(x, y) = (a, b)$ one can proceed in small steps of ω in x to $a_1, a_2, \ldots$, setting y successively equal to $b_1 = b + V(a, b)\omega$, then $b_2 = b_1 + V(a_1, b_1)\omega$, etc. The sequence of points (a_j, b_j) lie on a curve that approximates the solution to the differential equation through the initial point (a, b).

He indicated in informal terms that the smaller the steps the more accurate the approximation would be, but he recognised that the further the process was continued the more the errors would add up and that this process would be particularly liable to gross error if V was very large or very small. To investigate the way the error can behave, Euler offered an argument (Sects. 656–667) that brought in more and more terms for the power series for y. The errors grow fastest when some of these terms are large, which can happen when $V(x, y)$ becomes either zero or infinite, and Euler finished his account by providing examples to indicate how to work around this problem.

It is far from certain that Euler thought that his remarks about finding an approximative solution constitute a proof that ordinary differential equations have a solution. He most likely thought that was simply true and never in need of a proof. What he offered was what he said it was: a method of finding an approximate solution that would work if handled with care.

It is much more difficult to understand why he did not regard the ordinary differential equation as solved and therefore the multiplier found. He might have thought that the solution method he had proposed was an infinite sequence of approximations, so it did not provide useful answers. But it is hard to believe that he thought it subject to any significant restrictions.

5.2.1 Second-Order Partial Differential Equations

Euler next turned to the subject of second-order partial differential equations. He began his study of the second-order linear partial differential equation with an explanation of what the first and second partial derivatives are and how they behave under changes of variable (see Part 1, Sect. 2, Chap. 1, Sect. 229, Problem 39). He obtained these equations, which express the partial derivatives of a function z with respect to new variables u and v that are related to the old variables u and v by expressions of the form $u = u(x, y)$, etc.[4]

---

[4]Euler wrote t where we have written v here and throughout.

$$\frac{\partial z}{\partial x} = \frac{\partial v}{\partial x}\frac{\partial z}{\partial v} + \frac{\partial u}{\partial x}\frac{\partial z}{\partial u}, \quad \frac{\partial z}{\partial y} = \frac{\partial v}{\partial y}\frac{\partial z}{\partial v} + \frac{\partial u}{\partial y}\frac{\partial z}{\partial u}$$

$$\frac{\partial^2 z}{\partial x^2} = \frac{\partial^2 v}{\partial x^2}\frac{\partial z}{\partial v} + \frac{\partial^2 u}{\partial x^2}\frac{\partial z}{\partial u} + \left(\frac{\partial v}{\partial x}\right)^2\frac{\partial^2 z}{\partial v^2} + 2\frac{\partial v}{\partial x}\frac{\partial u}{\partial x}\frac{\partial^2 z}{\partial u\partial v} + \left(\frac{\partial u}{\partial x}\right)^2\frac{\partial^2 z}{\partial u^2},$$

$$\frac{\partial^2 z}{\partial x\partial y} = \frac{\partial^2 v}{\partial x\partial y}\frac{\partial z}{\partial v} + \frac{\partial^2 u}{\partial x\partial y}\frac{\partial z}{\partial u} + \frac{\partial v}{\partial x}\frac{\partial v}{\partial y}\frac{\partial^2 z}{\partial v^2} + \frac{\partial v}{\partial x}\frac{\partial u}{\partial y}\frac{\partial^2 z}{\partial u\partial v} + \frac{\partial v}{\partial x}\frac{\partial u}{\partial y}\frac{\partial^2 z}{\partial u\partial v} + \frac{\partial u}{\partial x}\frac{\partial u}{\partial y}\frac{\partial^2 z}{\partial u^2},$$

$$\frac{\partial^2 z}{\partial y^2} = \frac{\partial^2 v}{\partial y^2}\frac{\partial z}{\partial v} + \frac{\partial^2 u}{\partial y^2}\frac{\partial z}{\partial u} + \left(\frac{\partial v}{\partial y}\right)^2\frac{\partial^2 z}{\partial v^2} + 2\frac{\partial v}{\partial y}\frac{\partial u}{\partial y}\frac{\partial^2 z}{\partial u\partial v} + \left(\frac{\partial u}{\partial y}\right)^2\frac{\partial^2 z}{\partial u^2}.$$

In Chap. 2, he then explained how to deal with equations of the form $z_{xx} = P(x, y)$ and showed by integrating twice that the general solution is

$$z = \int \left(\int P dx \right) dx + x F(y) + G(y),$$

where F and G are arbitrary functions. He then discussed what equations can be reduced to one of this form by suitable changes of variable, and explored how to extend the method of changes of variable to equations of the form $z_{xx} = P(x, y, z)z_x + Q(x, y, z)$.

Section 2, Chap. 3 begins in Sect. 296 with the problem of solving the wave equation

$$\frac{\partial^2 z}{\partial y^2} = a^2\frac{\partial^2 z}{\partial y^2},$$

where a is a constant. Euler reduced it in the way just described, by showing that the substitutions

$$t = \alpha x + \beta y, \quad u = \gamma x + \delta y$$

transform the equation to

$$(\beta^2 - a^2\alpha^2)\frac{\partial^2 z}{\partial t^2} + 2(\beta\delta - \alpha\gamma)\frac{\partial^2 z}{\partial t\partial u} + (\delta^2 - a^2\gamma^2)\frac{\partial^2 z}{\partial u^2} = 0.$$

So Euler set

$$\alpha = 1, \ \beta = a, \ \gamma = 1, \text{ and } \delta = -a,$$

thus reducing the equation to

$$\frac{\partial^2 z}{\partial t\partial u} = 0,$$

which he had earlier shown is solved by integrating twice and has the complete solution $z = f(t) + F(u)$, where f and F are arbitrary functions.

Euler then wrote

COROLLARY 1 Sect. 297. Therefore the value z of this is equal to the sum of two arbitrary functions, the one is of $x + ay$, and other of $x - ay$, and both these functions thus can be assumed at will, so that also discontinuous functions are able to be taken in place of these.

COROLLARY 2 Sect. 298. Therefore any two curves described freely by the hand as it pleases are able to be taken according to this usage. Evidently if in one the abscissa is taken as $= x + ay$, and in the other truly the abscissa $= x - ay$, then the sum of the applied lines [i.e. the y -coordinates– Editor's note] will always put in place a suitable value for the function z.

The first corollary is Euler's way of saying first that the solution is a sum of an arbitrary function of $x + ay$ and an arbitrary function of $x - ay$. The second one says that if the coordinates are changed to $u = x + ay$ and $v = x - ay$ then the solution is the sum $f(u) + g(v)$.

Euler then compared his solution method with d'Alembert's, noting that he admitted a much greater range of candidates for the functions f and F than d'Alembert had done.[5] Of more interest in the present context is Euler's Scholium 3 (Sect. 301), where he remarked that our solution has this disadvantage because it leads to an imaginary expression for this equation

$$\frac{\partial^2 z}{\partial y^2} + a^2 \frac{\partial^2 z}{\partial x^2} = 0, \quad (*)$$

evidently

$$z = f(x + ay\sqrt{-1}) + F(x - ay\sqrt{-1}).$$

He then noted that

$$z = \frac{1}{2} f(x + ay\sqrt{-1}) + \frac{1}{2} f(x - ay\sqrt{-1})$$
$$+ \frac{1}{2\sqrt{-1}} F(x + ay\sqrt{-1}) + \frac{1}{2\sqrt{-1}} F(x - ay\sqrt{-1})$$

will always be real. However, although Euler was confident that this reduction to real values would always be possible for curves that he called analytic (i.e. were given as explicit functions of a real variable) or could be represented as series of sines and cosines, he explicitly doubted that this would be possible for arbitrary curves, drawn as he put it by the free motion of the hand. He therefore concluded that this was "a great defect in the calculation, on account of which a great many solutions lose their power".

This means that although Euler had given a recipe for producing a solution to the equation (*) by starting with a real expression such as x^2 or $\sin x$ (and obtaining

---

[5] D'Alembert's arbitrary functions were nonetheless regarded by him as analytic because, in his view, the calculus applied to them. Euler was more and more of the opinion that the initial conditions for the wave equation could be given by much more general curves.

$x^2 - a^2y^2$ or $\sin x \cosh ay$) he could not be sure that all solutions were of this kind.[6] This reflects a deeper lack of knowledge about the passage from complex functions to harmonic functions that only became available after Riemann.

Moreover, the equation (*) is unique among equations of the form

$$A\frac{\partial^2 z}{\partial y^2} + 2B\frac{\partial^2 z}{\partial x \partial y} + C\frac{\partial^2 z}{\partial x^2} + R\frac{\partial z}{\partial y} + S\frac{\partial z}{\partial x} + Tz + V = 0,$$

where A, B, and C are functions of x and y and $B^2 - AC < 0$, in that it has a "twin" with $B^2 - AC > 0$, and so Euler had no way in to analyse any of the others.

He therefore confined his attention to partial differential equations that do not raise this problem, and Chap. 3 steadily establishes that all second-order linear partial differential equations of the form

$$\frac{\partial^2 z}{\partial y^2} - 2P\frac{\partial^2 z}{\partial x \partial y} + (P^2 - Q^2)\frac{\partial^2 z}{\partial x^2} + R\frac{\partial z}{\partial y} + S\frac{\partial z}{\partial x} + Tz + V = 0,$$

where the coefficients are functions of x and y, can be reduced to ones of the form

$$\frac{\partial^2 z}{\partial v \partial u} + P_1\frac{\partial z}{\partial v} + Q_1\frac{\partial z}{\partial u} + R_1 z + S_1 = 0.$$

All the necessary formulae for the changes of variable had been set out in Chap. 1.

His conclusion was that the new variables u and v must satisfy the partial differential equations

$$\frac{\partial v}{\partial y} = (P + Q)\frac{\partial v}{\partial x} \quad \text{and} \quad \frac{\partial u}{\partial y} = (P - Q)\frac{\partial u}{\partial x}.$$

These are of the form that Euler had already shown how to solve.[7]

I add a few remarks about the generality of the solution to the wave equation.[8] D'Alembert always maintained that for a function to be a solution of the equation it had first to be a candidate, and for that reason analytic; he tried to reject the idea that there might be classes of functions to which the calculus did not automatically apply. Challenged by Euler, he gave a geometric argument that rested on the idea that the curvature of the string must vary continuously, which a modern reader could interpret as saying a solution must be twice continuously differentiable as a function of each variable. But Euler was willing to contemplate more general solutions, and so he gave arguments to show that at points where the solution curve is not differentiable

---

[6]He did not say this explicitly but left the result to be deduced from his formulae.

[7]Laplace did the same thing, perhaps independently, in his treatment of partial differential equations in his [174].

[8]See Lützen's [191], 15–23, which also considers Lagrange's ideas in this regard.

there are other solutions that are infinitely close, and so any errors that are introduced at such points will be negligible.

What about the intermediate case

$$\frac{\partial^2 z}{\partial x^2} = \frac{\partial z}{\partial y}?$$

Because only one second-order term appears this is of the form that Euler had treated in Chap. 2. But for this specific equation, his only comments (Sect. 265) were that his methods do not apply and it is "allowed to be understood that the resolution of this has to be thought out with the greatest hardship".

5.3   The Introduction of Characteristics by d'Alembert

D'Alembert was already interested in these questions and had a fruitful insight into the question of the existence of integrating factors after he met Euler in Berlin in 1762, that form a response to Euler's papers of 1763 and 1764. This is his theory of characteristics.[9]

In the fourth volume of his *Opuscules*, d'Alembert argued that given the equation $dx + \alpha dy = 0$, where $\alpha = \alpha(x, y)$, it is possible at each point on the y-axis to draw through it a curve along which $dx + \alpha dy = 0$.[10] In this way, one obtains a family of curves, one through each point of the y-axis, along which $dx + \alpha dy = 0$. Then, if M is the factor that makes the differential $M dx + M \alpha dy$ exact, the integral of this exact differential will be constant along these curves and only vary from curve to curve. So, it is enough to prescribe the values of M arbitrarily along the y-axis for the function M to be known everywhere in the plane. This is the basic method of characteristics: the curves d'Alembert defined are known as the characteristic curves and the method is valid at least locally and for as long as the curves are not tangent to the y-axis.

D'Alembert described the values of M at each point as a height above the (x, y)-plane, so the curves of constant M are curves on constant height, which we might call level curves or contour lines.

He also explained that if M is an integrating factor for a differential $dx + \alpha dy$, so $M(dx + \alpha dy) = du$, say, then so is M times any function of u, because $f(u)M(dx + \alpha dy) = f(u)du$.

He did not connect this insight to the method of changing variables when solving a first-order partial differential equation. Had he done so, he could have said that replacing x by the new variable u reduces the original partial differential equation to one of the form $z_u = 0$, whose solutions are z is an arbitrary function of the other variable y, so the solution is known everywhere once it is known on the y-axis. The

---

[9]For a revision of the mathematics, see Appendix B.

[10]See d'Alembert [56], pp. 225–281, and especially 255–258.

corresponding differential is dy which is exact, as is any differential $f(y)dy$. This suggests that even at this stage the notion of boundary values for a partial differential equation was not clear or established.

D'Alembert was clear that the sought-for function M would be far from unique, and he commented that "it can often happen that M will not be expressible algebraically although it can always be determined geometrically". This was his reply to Euler who, in his ([88], Sect. 20) had said that this was too much to hope for. But D'Alembert was unhappy that his method was far from being explicit, and so he set out a tentative method for finding an expression for M "when this can be done". This method was little more than the hope that there will be a change of variables, obtained by regarding x and y as functions of u and z, that will make the transformed equation tractable because it is now written in variables that separate.

What became known as the characteristic curves are the solutions of the differential equation $dx + \alpha dy = 0$, and the value of $z(x, y)$ along each such curve is determined by the value at an arbitrary given point on it. So, to solve the partial differential equation one finds a curve γ that is not a characteristic curve but crosses all the characteristic curves, and assigns a value to the function z at each point on γ, and then solves the ordinary differential equation (5.7) with those initial values. Under the heading of the method of characteristics, this became a standard technique in the theory of linear and quasi-linear partial differential equations.

5.4   Laplace

In his memoir (1777), Pierre Simon Laplace, a protégé of d'Alembert's, presented what he claimed was the first systematic treatment of linear partial differential equations, one that went beyond the isolated cases treated by d'Alembert and others.[11] He praised d'Alembert as the inventor of the calculus of partial differential equations, and mentioned neither Euler nor Lagrange by name, perhaps because d'Alembert had been helpful to him early in the younger man's career, by securing him a professorship in mathematics at the École Militaire in 1769, the year he turned 20, and d'Alembert and Euler had been rivals until the 1760s.

Laplace began with the first-order equation, which we shall write in a form equivalent to his as

$$\alpha \frac{\partial z}{\partial x} + \beta \frac{\partial z}{\partial y} = V, \tag{5.2}$$

where α and β are functions of x and y and V is a function of x and y if the equation is linear. (Laplace allowed V to be a function of x and y with a linear term in z, a slight generality that we suppress here.) This is the same equation as Euler's equation (5.1), and his solution method was the same as Euler's, except for a little more clarity about

---

[11]On Laplace, see Gillispie [120].

the integrating factor that arises. Like Euler, he said nothing about the solutions being constant along any characteristic curves.

Then, in his ([174], 21–41), Laplace turned to the second-order linear partial differential equation, which he wrote in the form

$$z_{xx} + \alpha z_{xy} + \beta z_{yy} + \gamma z_x + \delta z_y + \lambda z + T = 0, \tag{5.3}$$

where $\alpha, \beta, \gamma, \delta, \lambda$, and T are functions of x and y. He looked for a change of variables $u = u(x, y), v = v(x, y)$ that would put the equation in the form

$$z_{uv} + \text{lower terms} = 0.$$

From the values of the various partial derivatives of z with respect to the new variables, he deduced that the coefficients of z_{uu} and z_{vv} are, respectively,

$$u_x^2 + \alpha u_x u_y + \beta u_y^2, \quad \text{and} \quad v_x^2 + \alpha v_x v_y + \beta v_y^2,$$

so, for the transformed equation to reduce to the required form, the new variables must satisfy the differential equations

$$u_x^2 + \alpha u_x u_y + \beta u_y^2 = 0, \quad v_x^2 + \alpha v_x v_y + \beta v_y^2 = 0. \tag{5.4}$$

He factorised these to get these equations[12]:

$$u_x = u_y(-\alpha/2 + \sqrt{(\alpha/2)^2 - \beta}), \quad v_x = v_y(-\alpha/2 - \sqrt{(\alpha/2)^2 - \beta}), \tag{5.5}$$

which are of the type he had shown how to solve earlier in the memoir, as we described above. Indeed, to solve them, use the fact that $u(x, y) = 0$ implies that $u_x dx + u_y dy = 0$, so in Eqs. (5.4) we may replace u_x/u_y by $-dy/dx$. This leads to the equation

$$(dy)^2 - \alpha dx dy + \beta (dx)^2 = 0,$$

which we write as

$$\left(\frac{dy}{dx}\right)^2 - \alpha \frac{dy}{dx} + \beta = 0.$$

This is a quadratic equation whose solutions are

$$\frac{dy}{dx} = -\alpha/2 + \sqrt{(\alpha/2)^2 - \beta} = \sigma \quad \text{and} \quad \frac{dy}{dx} = -\alpha/2 - \sqrt{(\alpha/2)^2 - \beta} = \tau.$$

The characteristic curves are the solutions of these equations.

---

[12]This corrects a trivial error in Laplace's paper: he wrote α for $\alpha/2$ in the square root.

Along these curves Eqs. (5.4) hold, so the partial differential equation reduces to

$$z_{uv} = 0,$$

which has as its solutions anything of the form $f(u) + g(v)$. So once a characteristic curve is found that crosses these curves, the values of the solution are found, just as in the simple case of the wave equation, by saying that f is constant along with one set of curves and g is constant along with the other.

Laplace then attempted to solve the original Eq. (5.3) in its reduced form so as to obtain its general solution, and to supply extra information to provide particular solutions. He set no store by what we, however, would see as an essential difference between the cases where $\sqrt{(\alpha/2)^2 - \beta}$ is real and where it is complex. In the real case, the change of variables produces new *real* variables and real characteristic curves, but in the complex case, the new variables are complex and the characteristic curves are complex. In this latter circumstance, the only hope for Laplace is that the imaginary parts will somehow vanish at the end. But he made no remark to that effect, and it would seem that he thought that the reduction is always possible. This is remarkable, given that Euler had raised an alarm in the simplest case, and because it breaks an analogy (which perhaps Laplace missed) with the reduction of a curve given by a quadratic equation in two variables to either an ellipse or a hyperbola (or a parabola).[13] No mathematician would have thought that linear transformations could confuse an ellipse with a hyperbola.

This is further evidence that the understanding of partial differential equations in the mid-1770s was often purely formal, and many significant issues remained to be discovered.

We shall just consider the real case. Now there are new variables u and v that, respectively, satisfy the equations

$$\frac{\partial u}{\partial x} - \sigma \frac{\partial u}{\partial y} = 0, \quad \text{and} \quad \frac{\partial u}{\partial x} - \tau \frac{\partial u}{\partial y} = 0$$

and with respect to these new variables the partial differential equation takes the form

$$\frac{\partial^2 z}{\partial u \partial v} + a \frac{\partial z}{\partial u} + b \frac{\partial z}{\partial v} + cz = 0, \tag{5.6}$$

in which a, b, c are functions of u and v. This equation includes, as a particular case, the equation for the transmission of sound, which Euler was to consider in 1776 (as we saw above). Laplace put forward some ingenious but complicated solution methods that we shall not be able to pursue.

---

[13] Strictly speaking, the quadratic must be non-degenerate.

5.4.1   Lagrange's Method

In his [174], Lagrange took the equation

$$z_x = V z_y + Z$$

where V is a function of x and y, and Z is a function of x, y, z, and deduced using the identity $dz = z_x dx + z_y dy$ that

$$dz = (V dx + dy) z_y + Z dx.$$

He then went on Lagrange [174], 83:

> Suppose for a moment that
> $$V dx + dy = 0, \qquad\qquad\qquad (5.7)$$
> then I have an equation in two variables, which I integrate, adding an arbitrary constant α. Now I regard α as a function of x and y determined by this equation; by differentiation I have
> $$V dx + dy = A d\alpha,$$
> A being a function of x, y and α. So, substituting this value in the preceding equation it becomes
> $$dz = A z_y d\alpha + Z dx.$$
> Now, if one replaces y everywhere in this equation by its value in terms of x and α one thence has an equation in x, z, and α, and supposing α constant one will have the equation $dz = Z dx$ between the two variables x and z, which can be integrated with the addition of an arbitrary constant, which will be an arbitrary function of α, thus giving the general solution of the proposed equation at once, because it only remains to replace α by its value in x and y.

This is close to being the first statement of the modern approach, that puts the emphasis on some ordinary differential equations and drops any consideration of differentials and multipliers.[14]

5.5   Exercises

1. Solve the partial differential equations

$$px + qy = 0,$$

$$q = pV,$$

$$q = pV + U,$$

---

[14]Lagrange also showed how to vary the argument when V and Z may involve the derivatives of the unknown function.

and compare your solutions with Euler's.

Questions

1. What sorts of solutions did Euler admit to the partial differential equation

$$\frac{\partial u}{\partial x} = 0 \, ?$$

2. What sorts of functions would Euler or d'Alembert have supposed prescribed values along a transversal to a family of characteristics?

Chapter 6
Lagrange's General Theory of Partial Differential Equations

6.1 Introduction

The first systematic theories of first- and second-order partial differential equations were developed by Lagrange and Monge in the late eighteenth century. Lagrange's approach was not as overwhelmingly algebraic as the bulk of his work might suggest; in particular, he seems to have introduced the idea of envelopes of curves and surfaces into the study of differential equations. Monge's work is, however, the start of the modern geometric theory of partial differential equations, and we shall defer consideration of it to Sect. 8.2 below.

6.2 Clairaut's Paradox

In 1734, Clairaut had raised a paradoxical finding that can be described in modern terms as follows. Let $p = \dfrac{dy}{dx}$, and consider the differential equation

$$y = xp + f(p).$$

Differentiating this gives,

$$p = p + xp' + p'f'(p),$$

or

$$p'(x + f'(p)) = 0.$$

This, combined with the original equation, gives a solution in the parameterised form

$$x = -f'(p), \quad y = xp + f(p).$$

© The Author(s), under exclusive license to Springer Nature Switzerland AG 2021
J. Gray, *Change and Variations*, Springer Undergraduate Mathematics Series,
https://doi.org/10.1007/978-3-030-70575-6_6

What struck Clairaut and others as remarkable was that the solution to a differential equation had been found by further *differentiating* and not by *integrating*, as was to be expected.

When Euler wrote about this some twenty years later (in his E236) he opened by saying

> I propose here to study a paradox in the integral calculus that can appear very strange:

> We sometimes encounter differential equations for which it would seem very difficult to find the integrals by the rules of integral calculus, and which however are easy to find, not by means of an integration, but rather by differentiating the proposed equation again; so a repeated differentiation leads us in these cases to the sought-for integral. It is undoubtedly a very surprising accident, that differentiation can lead us to the same goal, that we are accustomed to find by integration, which is an entirely opposite operation.

Euler went on to connect this paradox to another unexpected result: a differential equation may be solved by an expression that does not arise from the general solution by a choice of the arbitrary constant it contains.[1] He gave this example,

$$x\,dx + y\,dy = dy\sqrt{x^2 + y^2 - a^2}.$$

It has the circle with equation $x^2 + y^2 - a^2 = 0$, which implies $x\,dx + y\,dy = 0$, as a solution, and also the general solution, which is a one-parameter family of parabolas

$$\sqrt{x^2 + y^2 - a^2} = y + c$$

or

$$x^2 - a^2 = 2yc + c^2,$$

but the circle cannot be obtained as one of a family of parabolas.

Euler's explanation of the paradox in this paper did not get close to solving it. He returned to the question in 1763 in the first volume of his *Institutiones Calculi Integralis* (Part 1, Sect. 2, Chap. 4, see Sect. 594 of the English translation), but he looked for an algebraic criterion to resolve it, and that does not get to the heart of the matter.

The envelope of a family of curves is a curve that touches each member of the family. To put that another way, it is a curve that has, at the point where it meets a curve of the family, the same tangent as the curve of the family does. It is a good exercise to check that the parabolas in Fig. 6.1 envelope the circle in this sense.

Let the family of curves by given by the equations $f(x, y, a) = 0$, where a is a parameter that varies from curve to curve. Let the curve $g(x, y) = 0$ be the equation of the envelope. Differentiating the first equation says that, at points on curves in the family, we have $df = 0$, so

$$f_x(x, y, a)dx + f_y(x, y, a)dy + f_a(x, y, a)da = 0.$$

---

[1] See Capobianco, Enea, and Ferraro [29] for a discussion this problem posed for Euler's ideas about the foundations of the calculus.

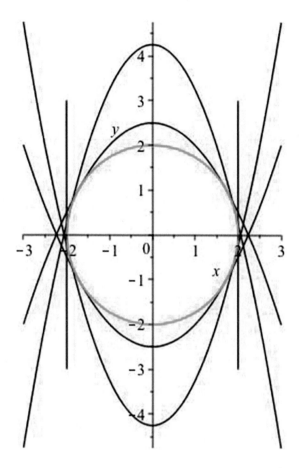

Fig. 6.1 A one-parameter family of parabolas enveloping a circle

If we fix a value of the parameter a then $da = 0$ and so at points on that specific curve in the family

$$f_x(x, y, a)dx + f_y(x, y, a)dy = 0, \quad \text{so} \quad \frac{dy}{dx} = -\frac{f_x(x, y, a)}{f_y(x, y, a)}.$$

Differentiating the second equation says that, at points on curve $g(x, y) = 0$, we have $dg = 0$, so

$$g_x(x, y)dx + g_y(x, y)dy = 0, \quad \text{and} \quad \frac{dy}{dx} = -\frac{g_x(x, y)}{g_y(x, y)}.$$

The slopes of the tangent at the curve with parameter a and the curve $g(x, y) = 0$ will therefore be the same at a point (x, y) where they meet if at that point $f_a(x, y, a) = 0$.

So the envelope is the set of points where both $f(x, y, a) = 0$ and $f_a(x, y, a) = 0$, so it is found by eliminating a between those equations.

One way of looking at envelopes is to imagine the curves $f(x, y, a) = 0$ forming a surface in (x, y, a) space. The equation

$$f_x(x, y, a)dx + f_y(x, y, a)dy + f_a(x, y, a)da = 0$$

can be written as

$$(f_x(x, y, a), f_y(x, y, a), f_a(x, y, a)).(dx, dy, da) = 0,$$

which says that $(f_x(x, y, a), f_y(x, y, a), f_a(x, y, a))$ is a normal to this surface. The further condition that $f_a(x, y, a) = 0$ says that the normals of interest are those that lie in the (x, y)-plane, and this picks out the part of the surface that you see if you look down the a-axis from a long way away.

6.3 Lagrange

A decade later, Lagrange (see Fig. 6.2) made a systematic study of first-order partial differential equations in his three papers, [173–175]. We shall look at his method for solving a general first-order partial differential equation in his (1772); his account of complete, general, and particular solutions from his [173] where he solved Clairaut's problem; briefly at his discussion of problems involving more than two independent variables in his [175]; and finally at his lectures on the subject at the École Polytechnique in 1806.

6.3.1 Lagrange [173]

In his [173], Lagrange considered the general first-order non-linear partial differential equation in two variables x and y for an unknown function $u(x, y)$. This was an ambitious undertaking, given what little had previously been discovered about partial differential equations, and he modelled his approach, naturally enough, on what was known about linear equations and the method of integrating factors.[2]

A partial differential equation is defined by a function F of the five variables x, y, u, p, q that satisfies the equation $F(x, y, u, p, q) = 0$, where, as he wrote $p = \frac{\partial u}{\partial x} = u_x$ and $q = \frac{\partial u}{\partial y} = u_y$, and so, when u is known as a function of x and y, $du = pdx + qdy$. The crucial difficulty with a non-linear equation is that p and q

---

[2]See Engelsman [67].

Fig. 6.2 Joseph Louis Lagrange (1736–1815), artist unknown

no longer occur to the first power but may be squared, multiplied by u, or in other novel combinations.

Lagrange supposed that the partial differential equation can be written in the form $q = q(x, y, u, p)$. The problem, as he now saw it, was to find p in terms of u, x, and y so that the expression $du - pdx - qdy$ is integrable. Before we see why this is true, note that this had been observed by Euler in his ([96], Sect. 128), who remarked that the necessary condition is that, on setting

$$L = \frac{\partial q}{\partial z}, \quad M = -\frac{\partial p}{\partial z}, \quad \text{and } N = \frac{\partial p}{\partial y} - \frac{\partial q}{\partial x},$$

one has

$$Lp + Mq + N = 0, \quad \text{or} \quad p\frac{\partial q}{\partial z} - q\frac{\partial p}{\partial z} + \frac{\partial p}{\partial y} - \frac{\partial q}{\partial x} = 0.$$

However, Euler made no remarks on how this equation could be satisfied, and turned to other aspects of the theory.

Lagrange's (and Euler's) observation is valid because the existence of an integrating factor M such that $M(du - pdx - qdy)$ is exact implies that

$$p\frac{\partial q}{\partial u} - q\frac{\partial p}{\partial u} + \frac{\partial p}{\partial y} - \frac{\partial q}{\partial x} = 0, \tag{6.1}$$

and conversely, a solution p of Eq. (6.1) implies that an integrating factor L for $pdx + qdy$ can be made to supply a suitable M.

To show this, Lagrange argued along these lines. Fix a value of u, then there is an integrating factor L such that $L(pdx + qdy) = dt$, where t is a function of x and y; L is found by solving the ordinary differential equation $pdx + qdy = 0$. Now let u vary. It is clear that $(L + \frac{dt}{du})du - dt = 0$ is integrable. He defined $P = L + \frac{dt}{du}$, and showed that condition that P as a function of t, y, and u is a function of t and u only is (6.1). He then let L' be the integrating factor for $Pdu - dt$, and deduced that $L'L(du - pdx - qdy)$ is exact, and so $LL' = M$.

Let p be a solution of (6.1). Lagrange now showed, by an argument I omit, that it is enough that p contains an arbitrary constant for a complete solution to the original partial differential equation $F(x, y, u, p, q) = 0$ to be found.

So in principle, Lagrange had discovered a general method for solving non-linear first-order partial differential equations in two independent variables, although he had to admit that it could be too complicated to follow even in cases when the solution was already known, for example, for the equation

$$q = pX + Y,$$

where $X = X(x, y)$ and $Y = Y(x, y)$.

True though it is that the various conditions on L, L', and M can be translated into the equations for the characteristics that define the modern solution (see Appendix C), the summary account of Lagrange's work by Eduard von Weber in the German Encyclopedia goes too far in implying that this was known to Lagrange. The use of characteristics is an important later development.[3]

Still less was there progress on partial differential equations in more than two independent variables, despite some inconclusive remarks at the end of the paper. Here, the problem at the time was that the integrating factor method cannot work: there is no good theory of an integrating factor for expressions of the form $udx + vdy + wdz$ because the analogue of Clairaut's equations yields an over-determined system consisting of three equations for the integrating factor.

6.3.2   Lagrange [173]

We now turn to his account in his [173] of the types of solution a partial differential equation may have, and the relationships between what he called complete, general, and particular solutions.

Before looking at it, it will help to look at the account in Courant and Hilbert ([49], Vol. 2, 22–27), which is clearer. Suppose that a first-order partial differential equation $F(x, y, z, p, q) = 0$ has a family of solutions $z = f(x, y, a)$ that depend on a parameter a. If this family has an envelope, then this envelope is also a solution. Geometrically, this is clear: the envelope shares a tangent plane at the point of contact

---

[3] See Weber [264], 338, written in 1900.

with the curve with parameter a, and this plane, therefore, belongs to the family of tangent planes defined by the partial differential equation.

A *complete integral* of the given partial differential equation is one that depends on two independent parameters, so the solution is of the form $z = f(x, y, a, b)$. Given a complete integral, one can impose an entirely arbitrary relation between a and b, say $b = b(a)$, and then the one-parameter family of surfaces $z = f(x, y, a, b(a))$ has an envelope which is again a solution. What makes this valuable is that the new solution is obtained by differentiation (which is easy) and elimination of a parameter (which, however, may not be easy). So a *general solution* is one that depends on one parameter, either by imposing a relation on the parameters such as $b = b(a)$ or is an envelope.

Finally, *singular solutions* of the partial differential equation are obtained from the two-parameter family of solutions as those envelopes that do not arise from a one-parameter family.

Gaspard Monge gave this example in his *Applications d'analyse* [202], Chap. 7. Consider the two-parameter family of surfaces

$$(x - a)^2 + (y - b)^2 + z^2 = 1.$$

They are all spheres of radius 1 with centres in the (x, y)-plane, and it is easy to see that these surfaces all satisfy the partial differential equation

$$z^2(1 + z_x^2 + z_y^2) = 1.$$

We shall see that this makes the surfaces what is called a complete integral of the partial differential equation.

Consider the one-parameter family, where $b = b(a)$—this selects just those spheres with centres on the curve $y = b(x)$ in the (x, y)-plane. The envelope of this family is obtained by eliminating the parameter a from the equations

$$(x - a)^2 + (y - b)^2 + z^2 = 1, \text{ and } x - a + b'(a)(y - b(a)) = 0.$$

It is a tubular surface, called a canal by Monge.

The planes $z = 1$ and $z = -1$ are an envelope of the two-parameter family of spheres, and they satisfy the partial differential equation. They are singular solutions of the partial differential equation, and they are not a tubular surface.

In his [174], Lagrange offered what he believed was the first general analysis of the problematic phenomenon first discovered by Clairaut in 1734. He noted that Euler had discussed it, as by then had d'Alembert, Condorcet, and Laplace (in a paper seen by Lagrange but not yet published). Now Lagrange proposed to offer a new and complete analysis.

He took Euler's example of the differential equation

$$x dx + y dy = dy\sqrt{x^2 + y^2 - a^2},$$

which has the solution

$$\sqrt{x^2 + y^2 - a^2} = y + c$$

or

$$x^2 - 2cy - a^2 - c^2 = 0,$$

where c is an arbitrary constant, and worked his way round to explaining that the circle is the envelope of the family of parabolas.

Lagrange now looked for the a priori reason for the existence of the phenomenon. He found that the differential equation specifies a condition on the solutions that is also met by the envelope of the complete integral, and he expressed this condition first in formal analytic terms and then in geometric terms. He also went on to note what happens when there is no envelope or only a one-point envelope; this explains other aspects of the original question.

In Article V of the paper, Lagrange then extended this analysis to partial differential equations. A partial differential equation in three variables has two independent variables, so a complete integral will have two arbitrary constants, he explained. A general integral has one, and a particular integral none.

6.3.3   Lagrange [175]

The paper Lagrange [175] deals mostly with particular kinds of partial differential equations in which some geometric condition is imposed, and in it Lagrange also generalised some arguments he had used in his [173] to address the question of solving first-order quasi-linear partial differential equations (to use a more recent technical term) with any number of independent variables.[4]

He wrote the partial differential equation in a form that differs only notationally from this:

$$\frac{\partial z}{\partial x} + P_1 \frac{\partial z}{\partial x_1} + \cdots + P_n \frac{\partial z}{\partial x_n} = Z,$$

where z is an unknown function of the $n + 1$ variables $x, x_1, \ldots x_n$ and $P_1, \ldots P_n$, and Z are known functions of $x, x_1, \ldots x_n$, and z.

His method was to form the n ordinary differential equations

$$\frac{dx_1}{dx} = P_1, \ldots, \frac{dx_n}{dx} = P_n.$$

These can be solved, and the solutions involve n arbitrary constants $\alpha_1, \ldots, \alpha_n$. If these constants are regarded as functions of x, and φ is an arbitrary function of $\alpha_2, \ldots, \alpha_n$, then the solution of the partial differential equation is given in the form

---

[4]We shall pick up that story when we look at mechanics and Hamilton–Jacobi theory in Chap. 25.

$$\alpha_1 = \varphi(\alpha_2, \ldots, \alpha_n).$$

To see what this means, it helps to let $n = 2$ and to work in (x, x_1, x_2) space, which is three-dimensional. We now fix values of α_1 and α_2. This gives us a curve in (x, x_1, x_2) space as x varies. If we now let α_2 be a function of α_1 then these curves form a surface. This surface is a solution of the partial differential equation because it is composed of curves that must lie in the solution surface because their tangents satisfy the partial differential equation.

Lagrange did not prove this claim but said that the proof was contained in his paper of 1774. His comment on the method is interesting [175], 625:

> By this method one can therefore integrate in general every first-order partial differential equation in which the differentials only appear in a linear form, whatever the number of variables; at least the integration of these sorts of equations is reduced to that of some ordinary differential equations; but one knows that the art of the integral Calculus of partial differential equations only consists of reducing this calculus to that of ordinary differential equations, and that one regards a partial differential equation as integrated when its integral depends only on that of one or more ordinary differential equations.

6.4 Exercises

1. A ladder slides down a wall which is at right angles to the ground. What curve does it envelope? (Hint: use the angle of the ladder to the vertical as a parameter.)
2. Verify the claims made in connection with Clairaut's paradox. Show that if $f(x, y, a) = x^2 + 2ay - a^2 - b^2 = 0$ then eliminating the parameter a from that equation and the equation $\frac{\partial}{\partial a} f(x, y, a) = 0$ yields the equation $y = a$, from which the result follows.

Questions

1. How important would you say that one-parameter families of curves were in the mathematics of the early eighteenth century?

Chapter 7
The Calculus of Variations

7.1 Introduction

Problems in which the solution is a curve (or perhaps a function) with some maximal or minimal property began to be studied at the end of the seventeenth century, as we saw in Chap. 2. Euler was the first to make a systematic study of problems of this kind, and his book the *Methodus inveniendi* (1744, E 65), which is full of different kinds of examples stimulated the young Lagrange to invent the methods of the calculus of variations. These were not the modern methods, but an inspired, and mostly unexplained, formalism that was very useful but by no means clear. Out of these insights, the so-called Euler–Lagrange equations were discovered, the law of least action was proclaimed, and Lagrangian dynamics was created.

7.2 The Euler–Lagrange Equations Discovered

There are few topics in the history of mathematics that have a clear beginning. Most emerge out of a shifting array of problems, undergo various reformulations, split into branches, merge with others, and acquire new aspects. The calculus of variations went through these preliminary stages so quickly that it may almost be said to have begun with Euler in his *Methodus inveniendi*. Here, Euler set out a general method for finding functions that minimise or maximise a given integral, and are perhaps subject to other constraints.

Euler thought of the calculus in the language of infinitesimals.[1] So he took three points on a curve that is the graphical representation of y a function of x, say (x, y),

---

[1] A good account of Euler's method is provided in Fraser [106]. Importantly, it does not follow the accounts of Carathéodory [31] and Goldstine [121] in interpreting Euler's arguments by bringing them into line with later methods, which obscures some of the points that Lagrange was to criticise. For English translations of some of the work of Euler and Lagrange, see Struik *Source Book* 399–413.

© The Author(s), under exclusive license to Springer Nature Switzerland AG 2021
J. Gray, *Change and Variations*, Springer Undergraduate Mathematics Series,
https://doi.org/10.1007/978-3-030-70575-6_7

(x', y'), and (x'', y'') that are infinitesimal distances apart and varied the curve so that it passed through the infinitesimally nearby point $(x', y' + nv)$ instead. He thought of these changes as if they were finite and then treated them as infinitesimal, not as limits of finite changes.

Euler now supposed that the curve is the one that maximises or minimises a certain integral over an interval of a function Z of x, y, and perhaps some derivatives of y with respect to x. Because the integral is an extreme, the effect of these infinitesimal changes must be zero. Euler let the infinitesimal horizontal distances be dx, the change in y at x be dy, and defined p by the equation $dy = pdx$, with similar changes at $x' = x + dx$ and $x'' = x' + dx$.

He then considered the effect of these changes on the integral. The change is a sum of infinitesimal amounts that reflect the new value of y at x' and the consequent changes in any other quantity that enters the integral, so the changes are $Zdx + Z'dx + \cdots$, where Z is the value of Z at (x, y, p), Z' is the value of Z at (x', y', p'), and so on. But the change in the curve is concentrated in the infinitesimal region around the points (x', y') and $(x', y' + nv)$. So the change in the integral is $Zdx + Z'dx$.

Euler wrote

$$dZ = Mdx + Ndy + Pdp, \quad dZ' = M'dx + N'dy' + P'dp',$$

and proceeded to calculate M, N, P, M', N', P' in terms of the change nv to the curve. He found that

$$dZ = P\frac{nv}{dx}, \quad dZ' = N'nv - P'\frac{nv}{dx}.$$

So the change in the integral is given by

$$(dZ + dZ')dx = nv\left(P + N'dx - P'\right),$$

and this, because the integral is an extremum, is zero.

Euler wrote $P' - P = dP$, replaced N' by N, and obtained $Ndx - dP = 0$ or

$$N - \frac{dP}{dx} = 0 \qquad (7.1)$$

as a necessary condition for the curve to be an extremal of the integral. This is the first occurrence of the Euler–Lagrange equations in the calculus of variations.

The confusion this exposition induces is not simply a matter of the use of infinitesimals. As Fraser notes ([106], 185), Euler has used the d symbol in two ways. First, to denote a fixed infinitesimal separation of the x coordinates, and this was standard practice in the Leibnizian calculus of the day. Second, to denote the change in quantity consequent upon the change in y', and among these we find that $dy' = nv$, $dp = nv/dx$, and $dp' = -nv/dx$ while in this sense $dx = dy = dp'' = 0$. Euler then extended this method to apply to variations subject to constraints, which incidentally had the effect of highlighting the two uses of the d symbol.

The next year Euler received a response to his book that surprised and delighted him. This was a letter written by the 17-year-old Lagrange on 12 August 1755 in which he set out a new method for tackling these problems, which he illustrated with the solution of three examples.

In the letter, Lagrange introduced the symbol δ for the variation in the curve, so, for example, $\delta x = 0$. He wrote $\delta F y$ to denote the change in F, a function of y, and claimed with very little justification that

$$d\delta F y = \delta d F y, \tag{7.2}$$

and so in particular $d\delta y = \delta dy$. His method thereafter is a judicious combination of the new rule in Eq. (7.2) and integration by parts, and among other results Lagrange obtained a better derivation of equation (7.1), which now deserves to be called the Euler–Lagrange equation.

As Fraser notes ([106], 163) Lagrange surely introduced his new symbol δ to sort out Euler's ambiguous use of d, then somehow came to the belief that d and δ commute, and had a good idea (which Euler had missed) of using integration by parts. This is most useful when the integral has fixed end points because the variation is necessarily zero there.

A rich correspondence between the two men followed. Euler did not, at first, appreciate the way Lagrange's δ works. He also queried the crucial step in which the passage from the vanishing of the integral that expresses the variation in the original integral leads to the vanishing of the integrand and thus the Euler–Lagrange equation. Lagrange, in his reply, gave a general argument that remains unconvincing.

Between 1756 and 1760, Lagrange refined his method and focussed it on problems in mechanics, including the brachistochrone problem, and in 1760-1761 he published his own account of the calculus of variations [170]. The argument he set out in this paper is opaque at key stages, and as the calculus of variations developed numerous interpretations were provided that culminated in what today is called the direct method. As Fraser points out, the modern approach is not a reasonable interpretation of what Lagrange did, and the reader is referred to Fraser's paper for the details.

Lagrange argued that to find the minimum or maximum of an integral, say $\int Z$, one does as one does in the calculus and differentiates and equates to zero:

$$\delta \int Z = 0.$$

Here, Z is to be regarded as a function of variables x, y, z and their differences $dx, dy, dz, d^2x, d^2y, d^2z, \ldots$ The equation he said—offering no explanation— could also be written as

$$\int \delta Z = 0.$$

So one writes Z out in full, and "as one sees easily"

$$\delta dx = d\delta x, \quad \delta d^2 x = d^2 \delta x,$$

"and so for the others".

He then suggested that in any given problem, there is always some relationship between $\delta x, \delta y, \delta z, d\delta x, d\delta y, \ldots$.

We can at least look at how Lagrange determined the brachistochrone or curve of quickest descent between two points.[2] He took x, y, and z as a set of three mutually perpendicular axes—with the x axis vertical—so the time of descent is given by

$$\int Z, \quad Z = \frac{ds}{\sqrt{x}}, \quad ds = \sqrt{dx^2 + dy^2 + dz^2}.$$

The integrand Z involves four terms, dx, dy, dz, ds, so according to Lagrange's rules δZ is a sum of these four terms:

$$-\frac{\delta x ds}{2x\sqrt{x}}, \quad \frac{dx\delta dx}{\sqrt{x}ds}, \quad \frac{dy\delta dy}{\sqrt{x}ds}, \quad \frac{dz\delta dz}{\sqrt{x}ds},$$

and all other quantities in the general theory vanish. In his terminology,

$$n = -\frac{ds}{2x\sqrt{x}}, \quad p = \frac{dx}{\sqrt{x}ds}, \quad P = \frac{dy}{\sqrt{x}ds}, \quad \tilde{\omega} = \frac{dz}{\sqrt{x}ds}.$$

To find the curve of quickest descent, one therefore has the equations

$$n - dp = 0, \quad -dP = 0, \quad -d\tilde{\omega} = 0,$$

and so

$$-\frac{ds}{2x\sqrt{x}} - d\frac{dx}{\sqrt{x}ds} = 0, \quad -d\frac{dy}{\sqrt{x}ds} = 0, \quad -d\frac{dz}{\sqrt{x}ds} = 0.$$

For these three equations to represent a unique curve it is necessary, he said, that they reduce to two, which they do because, as he showed, the second and third imply the first.

He then integrated these two equations and obtained

$$\frac{dy}{\sqrt{x}ds} = \frac{1}{\sqrt{a}}, \quad \frac{dz}{\sqrt{x}ds} = \frac{1}{\sqrt{b}},$$

whence

$$\frac{dy}{dz} = \frac{\sqrt{b}}{\sqrt{a}}.$$

---

[2]See Lagrange ([170], 339–341).

This equation shows that the solution curve lies in a vertical plane. He now assumed that the x-axis passes through the curve, which takes care of the constant of integration when integrating both sides of the equation, and gave the equation of the plane as

$$z = y \frac{\sqrt{a}}{\sqrt{b}}.$$

Lagrange then took coordinates x and t in this plane, where $\sqrt{y^2 + z^2} = t$. This gave him z as a function of t and y as a function of t, and finally, on setting $\frac{ab}{a+b} = c$, a differential equation connecting x and t:

$$dt = \frac{\sqrt{x}\,dx}{\sqrt{c - x}},$$

which is "the equation for a cycloid described on a horizontal base by a circle of diameter equal to c".[3]

7.3 Maupertuis and the Principle of Least Action

Abstract though it is, the principle of least action was the occasion for the most unpleasant scientific controversy of the century. Pierre Louis Maupertuis, the head of the Berlin Academy, presented a paper to the Berlin Academy in 1744 in which he claimed to show that light travels in a way that continually minimises a quantity called its action (a concept to be defined below). He published an extended version of the same argument in 1746, which he now applied to the motion of a mechanical system. He also gave the principle a profoundly teleological spin by suggesting that the system evolved to meet a pre-assigned goal, and made it the animating principle of all of nature and the basis of a proof of the existence of God. As he observed, Euler had made a strictly mathematical statement of the principle in his *Methodus inveniendi* (E 65, [74]), but Maupertuis's claim was far grander.

In essence, Maupertuis's claim was that a benign deity had seen to it to produce a world in which everything happened with a minimum of effort, or rather, that a world in which things happened with minimal effort was evidence for a benign deity. Quite why action corresponded to hard work was not clear, and the whole claim was ridiculed by Voltaire in 1759 in one of the great books of the Enlightenment, *Candide*. Voltaire could not accept that we lived in the best of all possible worlds when all too painfully it was the world of the Seven Years War and the Lisbon earthquake of 1755.

Maupertuis was a vain man who courted fame. He was known as 'The Great Flattener' because he had led the successful French expedition to Lapland in the late

---

[3]It is easier to follow this argument on choosing the axes so that everything happens in the plane $z = 0$. See also the derivation in Sect. 7.5.

1730s that verified the flattening of the Earth at its poles and helped to persuade Continental Europe of the merits of the Newtonian calculus and Newton's theory of gravitation.[4] He was friends with Voltaire, and, importantly for this story, the Swiss Academician Samuel König, who Voltaire had persuaded to teach his lover Émilie du Châtelet algebra.[5]

However, in 1751, Samuel König published a paper criticising the principle of least action. Maupertuis took it personally and accused the author of plagiarism. He charged König with forging a letter by Leibniz stating the principle of least action, which would have denied Maupertuis priority. When this failed, he obtained Euler's support and tried to have König driven out of the Academy. It seems that Euler, who was generally a benign man, usually got his way with Maupertuis by humouring him, but felt that on this occasion he had to give something back. Maupertuis's actions enraged Voltaire, who promptly published a pamphlet entitled *Diatribe du docteur Akakia, medicin du Pape*. Emperor Frederick, who took a lordly interest in his Academy in Berlin, publicly supported Maupertuis, and was so enraged by Voltaire's pamphlet that he had it burned by the hangman in public places in Berlin. Matters eventually calmed down, and in 1752 König was given an official censure by the Academy but allowed to remain as an Academician.

What, more precisely, was the principle of least action? Maupertuis expressed it this way:

> Whenever there is a change in nature, the quantity of action necessary for this change is as small as possible. The quantity of action is the product of the mass of the body by its speed and by the distance through which it has moved.

Euler, in the second appendix to his *Methodus inveniendi*, wrote that the path of a body of mass M moving with a speed that it would have acquired by falling through a height v travels a distance ds has a quantity of motion of $Mds\sqrt{v}$ and

> I say that the path the body will describe, by comparison with all the others with the same start and end points, will minimise $\int Mds\sqrt{v}$, or, if M is a constant, $\int ds\sqrt{v}$.

If we say that in falling a height h from rest a body of mass M loses an amount of potential energy equal to Mgh, which is equal to the amount of kinetic energy it gains, $\frac{1}{2}Mu^2$, then $h = \frac{u^2}{2g}$ and Euler is claiming that the path of the particle minimises the integral

$$\int \frac{M}{\sqrt{2g}}uds,$$

which is what Maupertuis said, and which we recognise as the product of the momentum of the body times the distance through which it has moved in an instant. If we also note that $u = \frac{ds}{dt}$, then we can say that the integral is $\frac{M}{\sqrt{2g}} \int u^2 dt$.

---

[4]See Terrall [253] for a detailed account.

[5]Du Châtelet is remembered as the mathematician who translated Newton's *Principia* into French. For an account of the difficulties this involved, and the merits of her extensive commentaries, see Zinsser [278].

This agrees with the definition of the action in use today, which, in problems involving particle motion, is the difference of the kinetic and potential energies in the motion. If the kinetic energy is represented by T and the potential energy by V, then the action is $T - V$, and in problems about motion under gravity, we have that $T = -V$, so $T - V = 2T$.

The crucial point is that whatever deductions follow from the new principle have to agree with those that follow Newton's laws. So the correct formula for the action is not a new quantity; it must be a disguised version of one already known, and the minimising principle can only be a disguised version of Newton's laws. Or, of course, it could be the other way round: the action principle is fundamental, Newton's laws follow from it, and we just happened to have discovered them first, which, given the origin of mechanics in astronomy was surely inevitable. Either way, the philosophical implications Maupertuis drew were ultimately spun out of a misunderstanding.

Before we dismiss them, however, it is worth observing that they are curious. It is not difficult to imagine a particle feeling a force at every instant and responding to it. It is much harder to imagine a particle considering every possible path between two points before deciding which one to take. In the first case, one can imagine the particle needs no help deciding what to do, but in the second case, one is tempted to imagine it appealing to an all-knowing higher authority, which is the essence of Maupertuis's theological argument.

That said, we are left with the problem of showing the equivalence of the principle of least action and Newton's laws; however, a valid modern derivation requires more theory than we have at present.[6]

There is also a strong pragmatic reason for preferring the principle of least action. It can be much easier to use in problems in dynamics because it fits very well with the framework of generalised coordinates, to which we now turn.

7.4 Euler's Later Approach

In his paper (E420, [97]), Euler returned to the calculus of variations and derived its fundamental equations in a way that pointed the way to all future treatments. Given the problem of finding an extremal of an integral involving function $y(x)$, Euler supposed there was a one-parameter family of functions $y(x, t)$, such as $y(x, t) = y(x) + tV(x)$, so that $y(x)$ can be approximated arbitrarily closely. He then considered the variation in t:

$$\delta y = \frac{\partial y}{\partial t} dt$$

evaluated at $t = 0$, and argued that at an extremal this variation should vanish.

---

[6]It is nicely explained in the Notes for the Harvard course Mechanics 151, see (http://www.people. fas.harvard.edu/~djmorin/chap6.pdf).

The differential of a function $Z(x, y)$ is, as he wrote, $dZ = M dx + N dy$, where $M = \frac{\partial Z}{\partial x}$ and if the variation in y is all that is considered then $dx = 0$ and $dy = \frac{\partial y}{\partial t} dt$, so one can write $\frac{\partial Z}{\partial t} = N \frac{\partial y}{\partial t}$.

He wrote higher derivatives as

$$y = \frac{\partial p}{\partial x}, \quad q = \frac{\partial p}{\partial x} = \frac{\partial^2 y}{\partial x^2}, \quad r = \frac{\partial q}{\partial x} = \frac{\partial^2 p}{\partial x^2} = \frac{\partial^3 y}{\partial x^3}, \quad \ldots,$$

and

$$\frac{\partial p}{\partial t} = \frac{\partial^2 y}{\partial x \partial t}, \quad \frac{\partial q}{\partial t} = \frac{\partial^3 y}{\partial^2 \partial t}, \quad \frac{\partial r}{\partial t} = \frac{\partial^4 y}{\partial x^3 \partial t} \ldots.$$

So now, if

$$dZ = M dx + N dy + P dp + Q dq + R dr + \cdots$$

then, holding x fixed, so $dx = 0$, the infinitesimal variation in y produces

$$dy = \frac{\partial y}{\partial t} dt, \ dp = \frac{\partial p}{\partial t} dt = \frac{\partial^2 y}{\partial x \partial t} dt, \ dq = \frac{\partial^3 y}{\partial^2 x \partial t} dt, \ dr = \frac{\partial^4 y}{\partial^3 x \partial t} dt, \ldots.$$

So, the variation in Z is given by

$$\frac{\partial Z}{\partial t} dt = N \frac{\partial y}{\partial t} dt + P \frac{\partial^2 y}{\partial x \partial t} dt + Q \frac{\partial^3 y}{\partial^2 x \partial t} dt + R \frac{\partial^4 y}{\partial^3 x \partial t} dt + \cdots.$$

The variation of an integral goes like this:

$$\delta \int Z dx = \int \delta Z dx = \int \frac{\partial Z}{\partial t} dt dx = dt \int \frac{\partial Z}{\partial t} dx.$$

Applying this to the expansion of $\frac{\partial Z}{\partial t}$, Euler deduced that variation of the integral is given by the power series

$$dt \left(\int \left(N \frac{\partial y}{\partial t} dx + P \frac{\partial^2 y}{\partial x \partial t} dx + Q \frac{\partial^3 y}{\partial x^2 \partial t} dx + \cdots \right) \right).$$

He then integrated by parts, restricted attention to variations that vanish at the boundary, and deduced that for an extremal

$$dt \int dx \frac{\partial y}{\partial t} \left(N - \frac{dP}{dx} + \frac{d^2 Q}{dx^2} - \cdots \right) = 0,$$

and therefore that the Euler–Lagrange equation holds:

$$N - \frac{dP}{dx} + \frac{d^2 Q}{dx^2} - \cdots = 0.$$

He discussed the geometric meaning of these assumptions in a section at the end of the paper, and remarked that $\delta y = 0$ means that $y(x, t)$ and $y(x)$ agree at the boundary points, $d\delta y = 0$ means that they have parallel tangents at the boundary, and so on.

Euler also showed how to extend the subject to find the equations governing the extremal of integrals of functions of two variables.

7.5 Brachistochrone and the Calculus of Variations

The Euler–Lagrange equation for an integrand $F(x, y, y')$ is

$$\frac{d}{dx} F_{y'} - F_y = 0.$$

Remember that this involves the total differential with respect to x, not the partial derivative, so we have for any differentiable function G

$$\frac{d}{dx} G(x, y, y') = G_x + G_y y' + G_{y'} y''.$$

Now for the brachistochrone, or curve of quickest descent. The problem is to find the curve along which a frictionless mass point sliding along the curve will descend under gravity (acting in the y direction) from $A = (x_0, 0)$ to $B = (x_1, y_1)$ in the shortest time, given that its initial velocity is zero. When the point has descended a distance y its vertical velocity will be $\sqrt{2gy}$. At that moment, in an instant of time dt it moves a distance ds, where

$$ds^2 = (dx^2 + dy^2) = \left(1 + \left(\frac{dy}{dx}\right)^2\right) dx^2 = (1 + y'^2)dx^2,$$

so at each instant of time

$$dt = \sqrt{\frac{1 + y'^2}{2gy}} dx.$$

So the time of descent is

$$T = \int_{x_0}^{x_1} \sqrt{\frac{1 + y'^2}{2gy}} dx.$$

We write $F(x, y, y') = \sqrt{\dfrac{1 + y'^2}{2gy}}$.

To compute the Euler–Lagrange equation, set $h = (1 + y'^2)^{1/2}$ and $(2g)^{-1/2} = a$, so

$$F(x, y, y') = ahy^{-1/2}.$$

Now calculate $h_{y'}$, $\frac{d}{dx}h$, $\frac{d}{dx}h_{y'}$, $F_{y'}$, and F_y, and verify that the Euler–Lagrange equation simplifies to

$$y'^2 y'' y + \frac{1}{2}h^2 y'^2 = h^2 y'' y + \frac{1}{2}h^4,$$

and therefore to

$$(y'^2 - h^2)y'' y = \frac{1}{2}h^2(h^2 - y'^2).$$

All is not lost, because $h^2 - y'^2 = 1$, so the Euler–Lagrange equation becomes

$$-y'' y = \frac{1}{2}h^2 = \frac{1}{2}(1 + y'^2).$$

Verify that the equation

$$-y'' y = \frac{1}{2}(1 + y'^2)$$

is obtained by differentiating

$$y(1 + y'^2) = k,$$

where k is a constant, and deduce that

$$y' = \sqrt{\frac{k - y}{y}}.$$

Set $y = k \sin^2(z)$ and deduce that the Euler–Lagrange equation has become

$$2k \sin^2 z\, dz = dx,$$

and so

$$x = \frac{k}{2}(2z - \sin 2z).$$

So x and y have been expressed in terms of a parameter z, and the brachistochrone is found to be a cycloid obtained by a point on the circumference of a wheel of radius k that rolls on the x-axis.

7.6   Generalised Coordinates

This material is here for later use, in Chaps. 24 and 25.

Lagrange introduced the method of generalised coordinates in mechanics in a memoir of 1788, when he returned to a problem he had looked at in 1764 on the libration of the Moon.[7] The virtue of his method was that it enabled him to choose new variables with which to analyse libration, and this led to new differential equations that led to solutions that were easier to interpret.

Equally importantly, in those intervening years, as Fraser [105] discusses, Lagrange moved away from relying on the principle of least action as a fundamental principle in physics. This principle, as Maupertuis and even Euler had described it, had a metaphysical aspect that Lagrange increasingly disliked. He much preferred to put his trust in formal, algebraic arguments. By 1788, when Lagrange published his *Méchanique Analitique*, he disparaged the use of such phrases as "least action"

> as if these vague and arbitrary denominations comprise the essence of the laws of mechanics and can by some secret virtue establish in final causes the simple results of the known laws of mechanics.

He might have picked up this attitude from his mentor, d'Alembert. Or, as Fraser speculates, he might also have come to realise that his approach, which grew out of d'Alembert's, was more general, which it is because it does not require the forces in a problem to be given by a potential function. The calculus of variations remained fundamental to Lagrange's approach, not because it led to the formulation of a physical problem as the stationary value of an integral but because it led to a formulation in terms of differential equations.

The manner in which Lagrange presented his new theory of mechanics is not easy to read or to describe, and in the absence of a thorough modern account that makes it easy to read him carefully and accurately I have chosen to follow Lützen ([192], 640–642) and Pulte [231] and to indicate the outlines of his achievements while suppressing the details of his methods.[8]

We shall suppose that what is at issue is the motion of n particles, and that the jth particle has mass m_j and coordinates (x_j, y_j, z_j) with respect to some Cartesian frame of reference. Lagrange introduced new variables $q_1, q_2, \ldots, q_N$, $N = 3n$, and supposed that every x_j, y_j, and z_j is a function of the new variables $q_1, q_2, \ldots, q_N$.

It is clear that in principle any system of equations that expresses the motion of the n particles can be written as a system of equations in the new variables. Specifically, consider the kinetic energy of the system,

---

[7]Libration is the slow oscillation in the motion of the Moon that enables us to see a little more than half of its surface (about 59%). It is largely a result of the elliptical orbit of the Moon. d'Alembert had published theoretical papers on it in 1761, and Cassini and Mayer had published observational accounts.

[8]For the original treatment, see Lagrange *Mécanique analytique* [178], reprinted in Lagrange *Oeuvres* 11, Part 2, Sect. IV, especially pp. 334 and 336.

$$T = \frac{1}{2} \sum_{j=1}^{n} m_j (\dot{x}_j^2 + \dot{y}_j^2 + \dot{z}_j^2).$$

Assume also that the forces are conservative, which means that they are given by the gradient of a potential function $U(x_1, \ldots, z_n)$. Then T will be a function of the variables $q_1, \ldots, q_N, \dot{q}_1, \ldots \dot{q}_N$, and U will be a function of the variables $q_1, \ldots, q_N$.

What Lagrange showed was that the equations of motion can be written as

$$\frac{d}{dt} \frac{\partial T}{\partial \dot{q}_j} = \frac{\partial T}{\partial q_j} - \frac{\partial U}{\partial q_j}.$$

Later writers introduced the 'Lagrangian' $L = T - U$ in terms of which the equations of motion take the form

$$\frac{d}{dt} \frac{\partial L}{\partial \dot{q}_j} = \frac{\partial L}{\partial q_j}.$$

Lagrange was at least clear about the advantages of his method. Even the simplest mechanical problem expressed in rectangular Cartesian coordinates is tiresome to convert to a coordinate system better adapted to the problem, as, for example, spherical coordinates often are. The new coordinates are entirely general, so the equations are coordinate-free, and symmetries of the problem turn up as simpler systems of equations.

> Because these equations can have various forms that are less simple or more simple, and above all easier to integrate, it is not a matter of indifference in which form they are presented at the start; and it is perhaps one of the principal advantages of our method that it always provides the equations for each problem in the most simple form relative to the variables that it employs, and puts one in a position to judge in advance which are the variables to use that will most simplify the integration. Here, for this purpose, are some general principles that one will see applied in what follows to the solution of different problems. (Lagrange, *Oeuvres* Vol. 11, 336–337.)

7.7   Exercises

1. Prove that the shortest curve joining two points in the plane is the straight line between them.
2. Show that the curve in the (x, y)-plane joining the points $(-a, b)$ and (a, b) that produces the surface of least area when rotated around the x-axis is the catenary $y = \alpha \cosh x$ for a suitable value of α. This is equivalent to minimising the integral $\int_{-a}^{a} y(1 + y'^2)^{1/2} dx$.

Questions

1. If Newton's laws and the principle of least action give the same answers to the same problems in dynamics, what is the difference between them?
2. Find and enjoy articles on Jakob Steiner and the isoperimetric problem.
3. The physicist Richard Feynman greatly appreciated the principle of least action and made it the basis of his theory of quantum mechanics. Find out what you can about his views and arguments—he provided numerous accessible accounts.

Chapter 8
Monge and Solutions to Partial Differential Equations

8.1 Introduction

Lagrange's account of the solution of first-order partial differential equations is, at its core, formal and algebraic and not easy to understand. The account that has become the basis of all further elementary accounts was published in 1809 by Monge, in his *Application de l'analyse à la géométrie*—see the Addition, pp. 367–414—and it brings out his remarkable geometrical gifts. Here, we look briefly at its origins in Monge's earlier work, and how he tried to extend these ideas to second-order partial differential equations.

8.2 Monge and First-Order Partial Differential Equation

Monge (see Fig. 8.1) gave two accounts of how to solve partial differential equations in two memoirs in the *Histoire de l'Académie Royale des Sciences* for 1784 (published in 1787). The first eliminates arbitrary constants and arbitrary functions from an equation by repeated differentiation to show that the result is a partial differential equation. The second paper then reverses this idea and shows how arbitrary functions arise in the solutions of partial differential equations.

In §3 of his (1787b), Monge took up the quasi-linear first-order partial differential equation:

$$Mp + Nq + L = 0,$$

where M, N, L are functions of x, y, and the unknown function z, and $p = z_x$ and $q = z_y$.

He argued that the partial differential equation cannot be solved for p and q, yet, paradoxically, it seems that it can be. For, using the equation $dz = pdx + qdy$, one immediately obtains these two equations for p and q:

© The Author(s), under exclusive license to Springer Nature Switzerland AG 2021
J. Gray, *Change and Variations*, Springer Undergraduate Mathematics Series,
https://doi.org/10.1007/978-3-030-70575-6_8

Fig. 8.1 Gaspard Monge
(1746–1818), artist unknown

$$Mdz + Ldx = q(Mdy - Ndx),$$

$$Ndz + Ldy = p(Ndx - Mdy).$$

Monge's way out of this apparent impossibility was to argue that these equations are to be understood as a set of restrictions on any function z that satisfies the partial differential equation. They say that the system of differential equations

$$Mdy - Ndx = 0, \quad Mdz + Ldx = 0, \quad Ndz + Ldy = 0$$

can be solved simultaneously (any two imply the third) and their solution, for any arbitrary point (x_0, y_0, z_0), defines a curve through that point that lies in the solution surface. In other words, all solution surfaces through that point have this curve in common.

To obtain the solution of the partial differential equation Monge then said that if two of these equations, or two of their consequences, can be integrated explicitly, say to provide equations of the form

$$f(x, y, z) = a \quad \text{and} \quad g(x, y, x) = b,$$

where a and b are constants, then the complete integral of the partial differential equation will be of the form

$$f(x, y, z) = \varphi(g(x, y, z)),$$

where φ is an arbitrary function.

8.2.1 A Comparison with the Modern Account

We write the first-order quasi-linear partial differential equation in the form

$$ap + bq = c,$$

where

$$a = a(x, y, z), \quad b = b(x, y, z), \quad c = c(x, y, z),$$

and, as usual, $p = z_x, q = z_y$.

The solution to this partial differential equation will be a surface with an equation of the form $z = z(x, y)$, and any tangent to such a surface at the point (x, y, z) is of the form (dx, dy, dz), which is normal to $(p, q, -1)$, so the curve

$$(a(x(t), y(t), z(t)), \quad b(x(t), y(t), z(t)), \quad c(x(t), y(t), z(t))),$$

where $x(t)$, $y(t)$, $z(t)$ are functions of a variable t, is everywhere tangent to this surface, and so the curves that satisfy

$$dx : dy : dz = a : b : c$$

or

$$\frac{dx}{dt} = a, \quad \frac{dy}{dt} = b, \quad \frac{dz}{dt} = c$$

lie in the surface, and indeed fill it out. They are called the *characteristic curves* and they project down onto curves in the (x, y) plane that were called the characteristic curves before.

It is a simple, instructive, and reassuring exercise to connect these equations to the equations that Monge derived in 1784 (write everything in terms of $\frac{d}{dt}$).

This suggests that we think of directional vectors at each point of space: at the point (x, y, z), there is the vector $(a(x, y, z), b(x, y, z), c(x, y, z))$. So if we are given a curve Γ in space in the form $(x(s), y(s), z(s))$ say for $0 \le s \le 1$, then the characteristic curves through the curve Γ, which we assume are not tangent to Γ, will form the solution surface, and this is indeed the case—I omit the proof, but note that there is something to prove.

It is helpful to give an example. We take the partial differential equation

$$zp + q = 1,$$

with initial conditions

$$x = s, \quad y = \cos s, \quad , z = \sin s, \quad 0 \le s \le 1.$$

The transversality condition to be met is that along this curve

$$a\frac{dy}{ds} - b\frac{dx}{ds} \neq 0,$$

and we have

$$-\sin^2 s - \cos s \neq 0,$$

which is correct for $0 \leq s \leq \pi/2$ (indeed, until $\cos s = 1/2(1 - \sqrt{5})$).

The solution of the partial differential equation is then given by solving the system of ordinary differential equations

$$\frac{dx}{dt} = z, \quad \frac{dy}{dt} = 1, \quad \frac{dz}{dt} = 1,$$

for which

$$x = \frac{1}{2}t^2 + \alpha t + \beta, \quad y = t + \gamma, z = t + \alpha,$$

and finding the solution that meets the initial conditions, which are that when $t = 0$

$$\beta = s, \gamma = \cos s, \alpha = \sin s.$$

So, the solution of the partial differential equation that meets the initial conditions is

$$x = \frac{1}{2}t^2 + t\sin s + s, \quad y = t + \cos s, \quad z = t + \sin s.$$

8.2.2   The General First-Order Case

Before we return to Monge, it is instructive to see how far one can go pursuing his method but on the general first-order partial differential equation.

We write the partial differential equation in the form

$$F(x, y, z, p, q) = 0.$$

The aim is to get equations for dx, dy, and dz. We shall also need equations for dp and dq, which we did not need before because p and q were presented linearly then.

We can differentiate this equation with respect to p and obtain

$$\frac{dF}{dp} = 0 = F_p + F_q\frac{dq}{dp}. \tag{8.1}$$

As with the quasi-linear case, it is helpful to think of a surface with an equation $z = z(x, y)$ that satisfies this partial differential equation. Let (x_0, y_0, z_0) be a point on this surface, then the tangent plane to the surface at that point satisfies the equation

$$z - z_0 = p(x - x_0) + q(y - y_0),$$

If we differentiate the equation for the tangent plane with respect to p, we obtain

$$0 = x - x_0 + (y - y_0)\frac{dq}{dp},$$

and by thinking of (x, y) as infinitesimally close to (x_0, y_0) we write this as

$$dx + dy\frac{dq}{dp} = 0. \tag{8.2}$$

We eliminate $\frac{dq}{dp}$ from Eqs. (8.1) and (8.2) and deduce that

$$\frac{dx}{F_p} = \frac{dy}{F_q}.$$

A nice piece of elementary algebra tells us that therefore

$$\frac{dx}{F_p} = \frac{dy}{F_q} = \frac{dz}{pF_p + qF_q}, \tag{8.3}$$

or, equivalently, that

$$\frac{dx}{dt} = F_p, \quad \frac{dy}{dt} = F_q, \quad \frac{dz}{dt} = pF_p + qF_q.$$

These are equations for curves that lie in the solution surface, but they are not enough to determine the solution surface given some initial data, because the equations still contain p and q. We can of course assume that locally without much loss of generality that we can solve the partial differential equation for q and obtain

$$q = q(x_0, y_0, z_0, p).$$

This is an equation for all the planes at the point (x_0, y_0, z_0) that envelope a cone that touches the solution surface at that point. It became known as the Monge cone.

However, purely formally, we have

$$\frac{dp}{dt} = p_x\frac{dx}{dt} + p_y\frac{dy}{dt} = p_xF_p + p_yF_q,$$

$$\frac{dq}{dt} = q_x\frac{dx}{dt} + q_y\frac{dy}{dt} = q_xF_p + q_yF_q,$$

and from the partial differential equation we obtain by differentiating with respect to x and y

$$F_x + F_z p + F_p p_x + F_q q_x = 0,$$

$$F_y + F_z q + F_p p_y + F_q q_y = 0.$$

So, the previous two equations can be written as

$$\frac{dp}{dt} = -F_x - F_z p,$$

$$\frac{dq}{dt} = -F_y - F_z q.$$

These equations, and the three before, give a set of five that reduce the solution of a first-order partial differential equation $F(x, y, z, p, q) = 0$ to little more than algebra:

$$\frac{dx}{dt} = F_p \tag{8.4}$$

$$\frac{dy}{dt} = F_q \tag{8.5}$$

$$\frac{dz}{dt} = p F_p + q F_q \tag{8.6}$$

$$\frac{dp}{dt} = -F_x - p F_z \tag{8.7}$$

$$\frac{dq}{dt} = -F_y - q F_z. \tag{8.8}$$

Geometrically, these equations determine a curve that lies in the solution surface and a family of tangent planes that are tangent not only to the curve but—this is the contribution of the equations for $\frac{dp}{dt}$ and $\frac{dq}{dt}$—also to the surface. They define what came to be called a characteristic strip.

The solution of the partial differential equation then proceeds much as before. An initial curve in space is given, and if it is crossed by a family of characteristic strips that are never tangent to it then these strips determine a unique solution, at least in a neighbourhood of the initial curve.

For a statement of the existence and uniqueness theorem for the first-order partial differential equation

$$F(x, y, z, p, q) = 0$$

and a sketch of its proof, see the Appendix (Chap. C).

8.3 Monge on General First-Order Equation

Monge reworked and extended this analysis in his [202]. His argument demonstrates his acute geometrical vision, but for that reason, it is not easy to follow. The comparison with Cauchy's later, analytic method (see Sect. 31.1) is instructive both mathematically and historically.

Monge first satisfied himself that any problem involving the envelope of a one-parameter family of surfaces is a problem involving a first-order partial differential equation:

$$F(x, y, z, p, q) = 0.$$

To do so, he began with a one-parameter family of surfaces

$$G(x, y, z, \alpha, \beta) = 0,$$

where β is a function of α, say $\beta = \varphi(\alpha)$, and considered what he called the envelope of these surfaces.

If S_α is a surface in the family, and S is the surface they envelope, then S_α touches S along a curve C_α that Monge called a characteristic because it characterises the contact of S_α and S. We are to think of the surface S_α moving, changing its shape as it goes, and sweeping out a surface S. At each moment, S_α and S touch along C_α, so we can also think of C_α as sweeping out the surface S.

Monge considered the tangent plane to the envelope S at a point P and noted that there are distinguished directions at the point. The tangent plane can roll in many ways, but two stand out: along the characteristic, and about the tangent to the characteristic (thought of as an axis). In the first case, the tangent plane stays on the same characteristic; in the second case, it pushes in a direction that leaves the characteristic and defines a new curve that Monge called a trajectory.

Thus motivated, he considered a partial differential equation of the above form. Any solution of it will be a surface and locally therefore of the form $z = z(x, y)$. Differentiating F gives

$$X dx + Y dy + Z dz + P dp + Q dq = 0,$$

and because one always has

$$dz = p dx + q dy \tag{8.9}$$

Monge deduced that

$$(X + pZ)dx + (Y + qZ)dy + P dp + Q dq = 0. \tag{8.10}$$

He then looked at the characteristics with parameters α and $\alpha + d\alpha$, which amounts to letting

$$dp = 0, \quad dq = 0,$$

and deduced that

$$(X + pZ)dx + (Y + qZ)dy = 0. \tag{8.11}$$

which defines the projection of a trajectory on the (x, y)-plane. This equation is equally well obtained by stipulating that $Pdp + Qdq = 0$, so any surface satisfying this condition will touch the curve.

Varying the tangent plane by rolling it along the envelope generates a developable surface, and this allowed Monge to differentiate $dx = pdx + qdy$ keeping dx, dy, dz fixed. In this way Monge obtained the equation

$$dpdx + dqdy = 0$$

for the line common to two 'neighbouring' tangent planes. If in particular the tangent plane rolls along the characteristic, and therefore rotates about trajectory, then the value of $\frac{dy}{dx}$ is given by Eq. (8.11) and so

$$(X + pZ)dq - (Y + qZ)dp = 0. \tag{8.12}$$

Similarly, if the moving plane rolls along a trajectory, then

$$Pdy - Qdx = 0, \tag{8.13}$$

which is an equation for the characteristic.

So, Monge concluded, the Eqs. (8.9), (8.10), (8.12), and (8.13), belong to the characteristic.

Monge then observed that these four equations involve the differentials dp, dq, dx, dy, dz and that one can deduce ten equations each of which involves only two differentials. He listed these equations on p. 380 but, as he pointed out, only four are independent:

$$\frac{dx}{P} = \frac{dy}{Q} = \frac{dz}{Pp + Qp} = -\frac{dp}{X + pZ} = -\frac{dx}{Y + qZ}. \tag{8.14}$$

When it came to solving these equations, Monge dealt first with what he called the linear case

$$Pp + Qq = L,$$

where P, Q, and L involves only x, y, and z. In this case, the equations that are necessary are only

$$Pdy - Qdx = 0, \quad Pdz = Ldx = 0, \quad Qdx - Ldy = 0,$$

which describe the projections of the characteristics on the three coordinate planes.

He noted that if these differential equations are easy to solve then one should do so, but if they are more intractable, then because they define a curve, one can regard

x, y, and z as all being functions of a single variable, that one might as well take to be z. In which case, eliminating y between these equations leads to a second-order ordinary differential equation for x as a function of z. He then showed how, in this case, the envelope can be regarded as swept out by a curve that moves and changes its form in space in a way that is prescribed by the partial differential equation.

Monge then turned to the general case. His method was to see what can be done by regarding everything possible as a function of p. Four of the ten equations he had listed before involve either $\frac{dx}{dp}$, $\frac{dy}{dp}$, $\frac{dz}{dp}$, or $\frac{dq}{dp}$ in an equation with x, y, z, p, q. Systematic elimination produces a third-order ordinary differential equation for q as a function of p that involves only p and q but not x, y, or z.

Monge then, in a way, I shall not describe, showed how to produce from this equation the solution to the partial differential equation. But he admitted that this process could lead to long analytic difficulties and that it would be more useful to turn to interesting special cases, with which he proceeded to conclude his account.

I omit Monge's lengthy description of the most general case, because, as he put it himself (p. 409):

> The geometrical considerations on which we have based the study of the equations for the characteristics are familiar to students at the École Polytechnique, but they can be hard going for other readers, and we shall therefore derive the same equations by a purely analytical process. We shall begin with the case of linear equations, and then pass to the general case.

Monge also discussed the difficulties that arise in the many cases in which the equations for the characteristics are not immediately integrable and then turned his attention to partial differential equations that are reducible to the above (quasi-) linear form. Among these was a class of developable surfaces with equation

$$F(z - px - qy, p, q) = 0$$

that, Monge remarked,

> includes a great many of those that M. Lagrange has treated in his beautiful work on particular integrals, printed in the *Mémoires de l'Académie de Berlin* for the year 1774.

In fact, Monge's equations are enough to solve the equation $F(x, y, z, p, q) = 0$. Given a point (x_0, y_0, z_0) and p_0, q_0 such that

$$F(x_0, y_0, z_0, p_0, q_0) = 0$$

the equations determine a curve through the point and a set of planes, one for each point on the curve, along which the planes are tangent to a surface $z = z(x, y)$ that satisfies the partial differential equation. This curve is a characteristic curve, and the curve and these tangent planes define a characteristic strip. If the initial point lies on a curve that is transversal to a family of characteristics, then the family of planes envelope a surface $z = z(x, y)$ that satisfies the partial differential equation.

8.4   Monge on Second-Order Partial Differential Equation

In his ([201], Sect. 8) Monge extended these methods to the study of various second-order partial differential equations, which he wrote in the form

$$Ar + Bs + Ct + D = 0,$$

where A, B, C, D are arbitrary functions of x, y, z, p, q. He now argued that the partial differential equation cannot yield expressions for r, s, and t, but the use of the equations[1]

$$dp = rdx + sdy \quad \text{and} \quad dq = sdx + tdy$$

seemingly leads to these expressions for them:

$$Bdpdy + Cdqdy - Cdpdx + Ddy^2 = -r(Ady^2 - Bdxdy + Cdx^2),$$

$$Adpdy + Cdqdx + Ddxdy = s(Ady^2 - Bdxdy + Cdx^2),$$

$$Adpdx - Adqdy + Bdqdx + Ddx^2 = -t(Ady^2 - Bdxdy + Cdx^2).$$

These equations cannot hold in general, so if they hold simultaneously it must be along certain curves in the solution surface. If these equations

$$Ady^2 - Bdxdy + Cdx^2 = 0,$$

$$Bdpdy + Cdqdy - Cdpdx + Ddy^2 = 0,$$

$$Adpdy + Cdqdx + Ddxdy = 0,$$

$$Adpdx - Adqdy + Bdqdx + Ddx^2 = 0,$$

hold simultaneously (any two imply the other two), and if two of these equations have the solutions $v = a$ and $u = b$, where a and b are arbitrary constants of integration, then the general solution of the partial differential equation is $v = \varphi(u)$, where φ is an arbitrary function.

He gave some examples of how his method works in practice. In §12 he supposed that A, B, C, D are constants. The equation

$$Ady^2 - Bdxdy + Cdx^2 = 0$$

gives rise to equations

---

[1] Recall that $r = \frac{\partial^2 z}{\partial x^2}$, $s = \frac{\partial^2 z}{\partial x \partial y}$, and $t = \frac{\partial^2 z}{\partial y^2}$.

$$dy - kdx = 0 \quad \text{and} \quad dy - k'dx = 0,$$

and so

$$y - kx = a \quad \text{and} \quad y - k'x = a',$$

where k and k' are the roots of the equation $Ak^2 - Bk + C = 0$. The equation

$$Adpdy + Cdqdx + Ddxdy = 0$$

then becomes

$$Akdp + Cdq + Dkdx,$$

which implies

$$Akp + Cq + Dkx = b.$$

Monge deduced that a first integral of the partial differential equation is

$$Akp + Cq + Dkx = \varphi'(y - kx),$$

and another is

$$Ak'p + Cq + Dk'x = \psi'(y - k'x),$$

where the functions φ and ψ are arbitrary, so he deduced that the general solution of the original partial differential equation is

$$Az + \frac{1}{2}Dx^2 = \varphi(y - kx) + \psi(y - k'x).$$

Monge did not distinguish the cases when k, k' are real and when they are complex conjugate (or even when they are equal), and this indifference to concerns raised by Euler forty years earlier may be down to Monge's optimism that the problem Euler had pointed to could be solved more easily than was to turn out to be the case.

Monge's next example (§13) was the partial differential equation for surfaces generated by a line moving in space while remaining parallel to the (x, y)-plane:

$$q^2r - 2pqs + p^2t = 0.$$

His method leads to the equations

$$q^2dy^2 + 2pqdxdy + p^2dx^2 = 0 \quad \text{and}$$

$$q^2dpdy + p^2dydx.$$

The first leads to

$$(qdy + pdx)^2 = 0,$$

which implies that $z = a$, a constant. The second then implies that

$$pdq - qdp = 0$$

and so $q = bp$ where b is a constant, and after a little work the solution of the partial differential equation is found to be

$$x + y\varphi(z) = \psi(z),$$

a result already known. (Exercise 1 below asks you to derive the partial differential equation generated by a line moving as described.)

8.5 Lagrange at the École Polytechnique, 1806

In his lectures in 1806 at the still-new École Polytechnique Lagrange gave another account, which was very close to the one Monge had given some years before. He began Lecture 20 by observing that there are three immediate consequences of an equation $F(x, y, z) = 0$, which are obtained by finding the total differential of F and taking the partial derivatives with respect to x and y:

$$F_x dx + F_y dy + F_z dz = 0,$$

$$F_x + F_z z_x = 0,$$

$$F_y + F_z z_y = 0.$$

Lagrange then showed that these are useful when looking for a solution of a first-order partial differential equation, by giving two examples. In the first

$$z_x + M z_y = N,$$

where M and N are constants. Lagrange introduced the 'primitive' equation

$$z - Nx = \varphi(y - Mx),$$

and differentiated it partially with respect to x and y. The resulting two equations imply that $z = Nx + \varphi(y - Mx)$ is a solution of the partial differential equation.

In the second example, M and N are now regarded as functions of x, y, z. If the solution (or primitive equation) is of the form $F(x, y, z) = 0$ then partial differentiation of F with respect to x and y implies that

$$dF = F_x + M F_y + N F_z = 0,$$

and by the first of the three consequences above, regarding x, y, z as functions of a variable t, the partial differential equation becomes

$$dF = (dy - Mdx)F_y + (dz - Ndx)F_z = 0.$$

So, said Lagrange, the solution of the partial differential equation is found by solving (the fourth and fifth consequences)

$$(dy - Mdx) = 0 \text{ and } (dz - Ndx) = 0.$$

This is exactly what we have already seen in Monge's treatment of the quasi-linear equation.

The solution of these ordinary differential equations introduces two arbitrary constants a and b, and the result of eliminating the variable z from them will be a second-order equation in x and y in which one can set $dx = 1$ if one wants to regard y as a function of x, or $dy = 1$ if one wants to regard x as a function of y. Then z can be found as a function of x and y.

The answer is now given as an implicit function $F(x, y, z) = 0$ in which the constants a and b appear, which Lagrange wrote as $\Phi(a, b)$, highlighting that a and b are functions of each other. Moreover, the function F involves only these two constants (other than the ones in M and N), and so eliminating two of the variables x, y, z will eliminate the third (because $dF = 0$ and so F cannot be a function of the remaining variable). The primitive equations yield expressions for a and b as functions of x, y, z, say $a = P(x, y, z)$ and $b = Q(x, y, z)$, so the primitive equation becomes $\Phi(P, Q) = 0$.

Then he returned to the case of two independent variables and, as it were, ran the above argument backwards to show how first-order partial differential equations arise by eliminating the two parameters a and b from an equation of the form

$$F(x, y, z, a, b) = 0,$$

which he called the complete primitive equation (he required that a single differentiation cannot eliminate both parameters at the same time). This equation, he showed, leads to a more general one involving an arbitrary function, on supposing that $b = \varphi(a)$ and letting a satisfy

$$\frac{\partial}{\partial a} F(a, \varphi(a)) = 0.$$

Finally, the singular primitive equation is obtained by letting a and b satisfy

$$\frac{\partial}{\partial a} F = 0, \quad \frac{\partial}{\partial b} F = 0.$$

He gave this example. The partial differential equation

$$z = xz_x + yz_y$$

has the complete primitive equation

$$z = ax + by.$$

To obtain the primitive general equation, he set $b = \varphi(a)$ and took derivatives with respect to a only, thus finding

$$z = ax + y\varphi(a), \quad x + y\varphi'(a) = 0, \tag{8.15}$$

from which it was necessary to eliminate a. Because φ was arbitrary, this led to an infinity of different complete primitives, each with two arbitrary constants. If, for example,

$$\varphi(a) = A - \frac{a^2}{4B},$$

the procedure just outlined results in

$$a = \frac{2Bx}{y} \text{ and } z = Ay + \frac{Bx^2}{2y}$$

as a new form for the primitive equation.

In this way, he remarked, one can find as many complete primitives as one likes, but the general primitive equation is never among them. In the present case, it is impossible to find a function $\varphi(a)$ such that the resulting primitive equation is $z = Ax + By$. I omit the proof.

Lagrange then discussed the theory of envelopes and its role in the theory of partial differential equations and remarked (p. 348)

> One can see, in the writings of Monge, the theory of the generation of these surfaces and the equations that can represent them developed to its full extent and with particular and ingenious considerations that belong to them.

He then returned to the solution method for first-order partial differential equations that he had discussed 34 years earlier, in his [173], and noted that some difficulties remained to be resolved that, moreover, he had not been able to treat in the *Théorie des fonctions*.[2]

Lagrange's method for tackling first-order partial differential equations is frequently presented in textbooks as the Lagrange–Charpit method or even as Charpit's method. Many historians say that we know almost nothing about Charpit, and report

---

[2]Lagrange also considered linear equations in more than two independent variables in this paper. This was the territory that no one else had mastered, and he indicated that his approach generalised, but it cannot be treated here.

only that Lacroix said that Paul Charpit submitted a memoir in 1784 on the general solution of first-order partial differential equations, but it was never published and that Charpit died that year (28 December 1784).[3] However, as Grattan-Guinness (1990, 151) explained, the manuscript is not lost; indeed two copies survive. Lagrange saw the manuscript in 1793, and the Lagrange–Charpit method is essentially his more rigorous account of what Charpit wrote, which Lagrange presented in his *Leçons sur le calcul des fonctions* ([177], Lecture 20).[4]

8.5.1 Lacroix's Traité *(1798)*

Monge's ideas formed the basis of the rather superficial account given by Lacroix, the great textbook writer of the period, in his *Traité* of 1798, but this indicates the difficult nature of the subject for even the best students of the time. He dealt with what we would call the quasi-linear first-order case and some examples of second-order equations. For the linear second-order partial differential equation

$$az_{xx} + bz_{xy} + cz_{yy} = V(x, y)$$

with constant coefficients a, b, c he factorised the equation for the characteristics and proceeded formally to a solution without caring whether the roots of the equation

$$am^2 - bm + c = 0$$

were real or complex, but only that they were distinct.

He noted that many equations escaped this analysis, notably the equation

$$z_{xx} = z_y,$$

but that in this case the equation could be solved by an infinite series of sums of terms of the form

$$e^{m_j x + m_j^2 y}.$$

Lacroix then commented on some remarks about the generality of the solution, noting that Laplace had been of the opinion that the equation could not be solved

---

[3] On Paul Charpit de Ville Coer and his manuscript, see Grattan-Guinness and Engelsman [122]. It seems that Charpit came from Strasbourg to Paris in 1782, and became an assistant to Monge, who taught him solid geometry. He read an extract of his paper to the Académie des Sciences on 30 June 1784, but nothing was done with it. Laplace acquired a copy, and he passed it on to Lagrange (who had been in Berlin in 1784) in 1793. Lagrange in due course sent it to Charpit's friend Arbogast, who made a copy that is now in Florence. In 1798 Lacroix described Charpit's paper in his *Traité du calcul différentiel et du calcul intégral* ([168], 496–497, 513–516). Curiously, his copy of the manuscript is shorter than Arbogast's and seems less reliable.

[4] Lagrange's *Leçons* are in vol. 10 of his *Oeuvres*. Charpit's name is not mentioned.

by an arbitrary function, but the Italian mathematician Pietro Paoli had disagreed. Evidently, Lacroix had little idea of the power inherent in series of that kind. Such was the situation when Fourier took up the problem, as we shall see in Chap. 10.

8.6   Exercises

1. Explain why a line in space that is parallel to the (x, y)-plane has an equation of the form $y = m(z)x + z$ (ignoring lines parallel to the x-axis). Eliminate $m(z)$ by differentiating twice, and deduce that the equation of the surface generated by a line moving in space while remaining parallel to the (x, y)-plane is

$$q^2 r - 2pqs + p^2 t = 0.$$

2. What is the Monge cone for a quasi-linear first-order partial differential equation?
3. Follow through Monge's analysis for the first-order partial differential equations considered by Euler.
4. Follow through Monge's analysis for the wave equation and Laplace's equation.

Questions

1. What does the near silence about the heat equation, even as a partial differential equation without any physical interpretation, say about the solution methods available around 1800?
2. To what extent does Monge's geometrical analysis help you? How do you find it compares with the more formal account in Sect. 8.2.2 above?
3. How fair is it to say that Lagrange's account of first- and second-order partial differential equations in two independent variables is Monge's without the geometry? What advantages and disadvantages are there in the two approaches?

Chapter 9
Revision

9.1 Revision and Assessment 1

This chapter is given over to revision and discussion of the first assignment, see H.2 in Appendix H.

9.1.1 Comments

The Assessment asked students to reflect on what they had studied either by imagining becoming a student of mechanics and dynamics around 1770, or more generally, on what is involved around 1770 in the study of partial differential equations.

They were to do so by writing a letter (some years as an English professor writing to a student, some years as the student writing back to their former professor). I asked for a letter, not a history, to help them get into the way of seeing things through the protagonists' eyes. No one can see the future, and historians shouldn't try. And in fact, generally speaking, answers that began as a letter made it easier for the writer to engage with the developments historically.

No comparison between the mechanics of Newton and Euler should diminish Newton's remarkable achievements in his *Principia Mathematica*. His major work is a theory of celestial mechanics that delivered a remarkably accurate theory of the motion of the planets and their satellites, in which to a high level of detail, only the motion of the Moon around the Earth remained unaccounted for.

Euler managed no comparable single achievement in this field, but he produced theories of motion for rigid bodies of any kind (Newton could only handle spheres, which he showed could be treated as points), and of fluids and gases. There is also his account of the vibrating string.

Whereas Newton's method was largely geometrical but intermittently invoked calculus-type series of arguments, starting from a set of three laws of motion, Euler always started with infinitesimal pieces of bodies and worked towards equations

© The Author(s), under exclusive license to Springer Nature Switzerland AG 2021
J. Gray, *Change and Variations*, Springer Undergraduate Mathematics Series,
https://doi.org/10.1007/978-3-030-70575-6_9

of motion expressed in calculus terms as differential equations. He reformulated Newton's laws of motion as equations of motion, and his calculus of variations was a completely novel approach to geometrical and mechanical questions, one rapidly improved by Lagrange. The central importance of differential equations, and the means of solving them, unite all of Euler's work in this area.

Perhaps nothing compares with the invention of (single-variable) calculus and the creation of celestial mechanics, but if anything does it might be Euler's breadth of applications coupled with the generality of his (several-variable) calculus. But Newton's ingenious methods were not what anyone could follow, and Euler's were.[1]

The origins of partial differential equations are the wave equation and its investigations by d'Alembert, Euler, Daniel Bernoulli, and Lagrange. This led to the idea that its solutions are any functions of the form $f(x + ct) + g(x - ct)$, but that led to deep questions about what is an arbitrary function.

First-order linear partial differential equations were first studied by treating them like ordinary differential equations, so a good answer would go into a little mathematical detail about them because later methods were different. Note, for example, the advance in the theory of characteristics made by d'Alembert.

The theory of second-order linear partial differential equations was much less developed. Success with the wave equation came with Euler's worry about (what came to be called) the Laplace equation.

There are other aspects worth mentioning: fluids, the propagation of sound, and, perhaps, the calculus of variations.

---

[1] A good source is an essay by Maronne and Panza entitled Euler: Reader of Newton. See https://hal.archives-ouvertes.fr/hal-00415933.

Chapter 10
The Heat Equation

10.1 Introduction

Fourier series are infinite series of sines or cosines that are used to represent a function. They had been used by Euler and others in the eighteenth century in the study of the vibrating string and in celestial mechanics, but Fourier's name is rightly attached to them because of the great generality and utility he envisaged for them and the success he put them to in the study of heat diffusion. His strong claims for them were also to prove a valuable challenge to mathematicians who came after him and demanded more rigour in analysis.

10.2 Fourier and His Series

Joseph Fourier (Fig. 10.1) was born in France in 1768, and for much of his life he was caught up in the political transformation of France. When he was an assistant lecturer at the École Polytechnique in 1795 he came to the attention of Gaspard Monge, who became a prominent supporter of Napoleon, and who selected Fourier for the French expedition to Egypt in 1798. When that ended in defeat Fourier returned to France in 1801, but Napoleon, impressed by his organisational talents, made him the prefect of Governor of the Department of Isère, and, still impressed, a Baron in 1808.

The defeat of Napoleon led to a difficult period in Fourier's life, as the new regime came down hard on those it could accuse of having supported either the revolution or Napoleon, but he recovered and with the support of Laplace he became the permanent secretary of the Académie des Sciences in 1822 and was elected to the Académie Française in 1827. In 1830, he died from complications of an illness caught in Egypt.

© The Author(s), under exclusive license to Springer Nature Switzerland AG 2021
J. Gray, *Change and Variations*, Springer Undergraduate Mathematics Series,
https://doi.org/10.1007/978-3-030-70575-6_10

Fig. 10.1 Joseph Fourier
(1768–1830) by Amédée
Félix Barthélemy Geille,
after Jules Boilly, c. 1823.
*Portraits et Histoire des
Hommes Utiles, Collection
de Cinquante Portraits,*
Société Montyon et Franklin,
1839–1840

His book, *Théorie analytique de la chaleur* (*The analytical theory of heat*) came out in 1822.[1] It is devoted to the topic of heat diffusion.[2] By this time, it was reasonably well known that heat was a state of the hot material, not a fluid substance permeating the material in an amount proportional to its temperature. Not much else was understood, however, and accordingly, Fourier made as few assumptions as possible about the nature of heat. Rather, he concentrated on formulating the way heat passes from one part of the body to an adjacent part in a very short interval of time. He argued that it was enough to suppose that the amount of heat that passes is proportional to the duration of the time interval, the infinitesimal temperature difference between adjacent parts, and a certain function of the distance between the parts. He was able to examine homogeneous bodies of simple shapes with simple temperature distributions on their boundaries, say when the boundaries are kept at fixed temperatures. He also tested his results experimentally by heating simple shapes and measuring the temperature at various points and various times and found good agreement with his theoretical predictions.[3]

One of his examples was a window—of an unusual shape, being infinite from side to side and top to bottom, but of finite thickness—kept at a fixed temperature on each side and warmer inside than out. In this case, the temperature drops linearly with the distance from the warmer side. Another of his examples was that of an oven.

The problems that Fourier considered have two aspects. One concerns the flow of heat in the body, and he showed that that is described by a differential equation. The other concerned the temperature on the boundaries of the body, and although these

---

[1] It was written in several stages and has a complicated publication history that Fourier described in its Preliminary Discourse. In 1812, a version of his theory won a prize of the Institut de France in Paris Academy, but Lacroix and Laplace were unable to overcome the objections of Lagrange and let Fourier's account be published. Lagrange accepted that the correct equation for heat diffusion had been found, but not the generality of the solutions. For a rich introduction to Fourier and his work, see Grattan-Guinness and Ravetz [123].

[2] See the translation by A. Freeman in the Internet Archive.

[3] See Grattan-Guinness and Ravetz ([123], 421–440) for Fourier's account of 1807.

can be arbitrary they are easiest to handle if the boundary is made of simple shapes. Only if the *boundary conditions*, as these specifications are called, are simple can explicit solutions be found.

We now turn to one of his examples to see how he formulated the mathematical equation that describes the flow of heat.

Fourier considered semi-infinite bars with either semi-circular or square cross sections, in which one end is kept hot and the rest uniformly cool. The problem is to find how hot the bar becomes when it reaches a steady state. He argued that at each point in the interior of the bar, heat—measured by the temperature, v, a function of x, y, z—passes through in each of the $x-$, $y-$, and z-directions. In §98, he considered an infinitesimal cube in the interior of the bar and stated that the amount that enters the face with sides dx and dy is $-Kdxdy\frac{\partial v}{\partial z}$ evaluated at that face, where K is a quantity determined by the nature of the body, and what leaves the opposite face is $-Kdxdy\frac{\partial v}{\partial z}$ evaluated at that face.[4] The minus sign arises because heat flows from a hot body to a cold one.

What left at the second face was found by replacing z by $z + dz$, and so the difference between what enters and what leaves is the difference in the values of $-Kdydx\left(\frac{\partial v}{\partial z}\right)$ at the two faces, which works out to be $Kdxdydz\,\frac{\partial^2 v}{\partial z^2}$. Because the temperature is in a steady state, the sum of these quantities taken over the three pairs of opposite faces of a cube is zero, and the resulting equation is

$$\frac{\partial^2 v}{\partial x^2} + \frac{\partial^2 v}{\partial y^2} + \frac{\partial^2 v}{\partial z^2} = 0. \tag{10.1}$$

The distribution of heat in the body is described by the solution of this differential equation that satisfies the stated boundary conditions.

A similar argument allowed Fourier to derive the equation for the distribution of heat in a body that is not in a steady state, so the temperature $v = v(x, y, z, t)$ is now a function both of position and time. He now argued (§142) that, by regarding the body as made up of little cubes, the differential equation (10.2) will hold for the flow of heat in any body. As before, the amount of heat leaving a cube in the z-direction is $-Kdxdydz\frac{\partial^2 v}{\partial z^2}$, but now the sum over the pairs of opposite faces is proportional to the rate of change of temperature, which is $\frac{\partial v}{\partial t}$. The result (§128) is the *heat equation*:

$$K\left(\frac{\partial^2 v}{\partial x^2} + \frac{\partial^2 v}{\partial y^2} + \frac{\partial^2 v}{\partial z^2}\right) = \frac{\partial v}{\partial t}. \tag{10.2}$$

As before, solutions of this partial differential equation are required that satisfy some given boundary conditions (and, if v is a function of t, initial conditions). These conditions, however, generally forced him to suppose that the body has a simple shape, such as a cuboid, or one of a limited range of other shapes that can be handled by finding suitable coordinate transformations.

---

[4]I set side Fourier's remarks about what physical properties affect K.

To show how this could be done, Fourier had first to show how to find a general solution to the partial differential equation, and then show how to fit the general solution to the boundary conditions. He began (§166) with the simplest two-dimensional case, a semi-infinite strip of a given width, π in suitable units, located between $-\pi/2$ and $+\pi/2$, included between two parallel infinite sides at a given temperature 0 in some units which at its base has a temperature of 1. He chose coordinates (here relabelled to be more familiar) in which y measures the height above the base and x measures the distance of a point from the mid-line of the strip. The differential equation of the steady-state distribution, which now involves only two variables, is

$$K\left(\frac{\partial^2 v}{\partial x^2} + \frac{\partial^2 v}{\partial y^2}\right) = 0, \tag{10.3}$$

It is likely he was inspired by existing treatments of the wave equation. At all events, he looked for a solution of the form

$$v(x, y) = f(x)g(y).$$

This forces

$$g''(y)/g(y) = -f''(x)/f(x), \tag{10.4}$$

so both sides must be constant, say m^2, and the solutions are of the form

$$f(x) = \cos mx, \quad g(y) = e^{-my}. \tag{10.5}$$

The boundary conditions force m to be positive, otherwise e^{my} would become infinitely great, and if the solution is to vanish for $x = \pm\pi/2$ for all y, then m must be an odd integer.

This led Fourier to contemplate solutions of the form, as he wrote (§169),

$$ae^{-y}\cos x + be^{-3y}\cos 3x + ce^{-5y}\cos 5x + de^{-7y}\cos 7x + etc. \tag{10.6}$$

subject to the boundary condition at the base that

$$1 = a\cos x + b\cos 3x + c\cos 5x + d\cos 7x + etc. \tag{10.7}$$

The infinitely many arbitrary constants are now to be determined from the boundary conditions. Fourier gave two methods. The first is an impressive tour de force, but the second one is much easier and has been used ever since, although with much more attention to the conditions under which it is valid.

Fourier argued as follows. We have $f(x) = 1$, $-\pi/2 \le x \le \pi/2$. The corresponding Fourier series is $\sum_{n=1} a_n \cos nx$, so

$$1 = \sum_{n=1}^{\infty} a_n \cos nx,$$

so, multiplying with sides by $\cos jx$ and integrating

$$\int_{-\pi/2}^{\pi/2} \cos jx\, dx = \int_{-\pi/2}^{\pi/2} \left(\sum_{n=1}^{\infty} a_n \cos nx \cos jx \right) dx.$$

The left-hand side is

$$\frac{1}{j} \sin jx \big|_{-\pi/2}^{\pi/2} = \frac{2}{j} \sin(j\pi/2).$$

The quantity $\sin(j\pi/2)$ is zero when j is even, it is $+1$ when j is of the form $4k+1$, and it is -1 when j is of the form $4k-1$. So the left hand side is zero if j is even, it is $\frac{2}{j}$ if j is of the form $4k+1$, and it is $-\frac{2}{j}$ if j is of the form $4k-1$.

The right-hand side is

$$a_j \frac{\pi}{2},$$

so a_j is is zero if j is even, it is $\frac{4}{j\pi}$ if j is of the form $4k+1$, and it is $-\frac{4}{j\pi}$ if j is of the form $4k-1$.

This gives the result

$$1 = \frac{4}{\pi} \left(\cos x - \frac{1}{3} \cos 3x + \frac{1}{5} \cos 5x + \cdots \right),$$

as Fourier said.

It gave him that for all values of y between $-\pi/2$ and $+\pi/2$

$$e^{-y} \cos x - \frac{1}{3} e^{-3y} \cos 3x + \frac{1}{5} e^{-5y} \cos 5x - \frac{1}{7} e^{-7y} \cos 7x + etc., \qquad (10.8)$$

where on the boundary[5]

$$1 = \frac{4}{\pi} \left(\cos x - \frac{1}{3} \cos 3x + \frac{1}{5} \cos 5x - \cdots \right). \qquad (10.9)$$

Here is a graph of the sum of the first 105 terms of the series for the function $F(x) = \pm\pi/4$ (Fig. 10.2):

$$\cos(x) - \frac{1}{3} \cos(3x) + \frac{1}{5} \cos(5x) - \frac{1}{7} \cos(7x) + \cdots.$$

Note how small the difference between the sum of the series and the sum of its first 105 terms has become.

---

[5]Note that we have a series of continuous (indeed, analytic) functions that defines a function that is plainly not continuous.

Fig. 10.2 The sum of the first 105 terms of the Fourier series for $F(x) = \pm\pi/4$

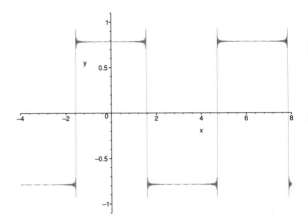

To reach these conclusions, he had observed in §220 that when $j \neq k$ (the switch from cosine to sine is irrelevant here)

$$\int_0^\pi \sin jx \sin kx\,dx = \frac{1}{2}\left(\frac{1}{k-j}\sin(k-j)x - \frac{1}{k+j}\sin(k+j)x\right)\bigg|_0^\pi = 0,$$
(10.10)

and that the integral is $\pi/2$ when $j = k$. He accordingly deduced that the coefficients of the series can be found by multiplying the series by $\sin jx$, for each value of j, and integrating. For, if

$$f(x) = \frac{1}{2}a_0 + \sum_{n=1}^\infty b_n \sin nx,$$

then multiplying both sides by $\sin jx$ and integrating gives

$$\int_0^\pi f(x)\sin jx\,dx = \int_0^\pi \frac{1}{2}a_0 \sin jx\,dx + \int_0^\pi \sum_{n=1}^\infty b_n \sin nx \sin jx\,dx.$$

This, he simply assumed, is equal to

$$\int_0^\pi \frac{1}{2}a_0 \sin jx\,dx + \sum_{n=1}^\infty \int_0^\pi b_n \sin nx \sin jx\,dx.$$

In this expression, all the terms $\int_0^\pi b_n \sin nx \sin jx\,dx$ vanish except for the one in which $n = j$, and this one is equal to $\frac{\pi}{2}$. So

$$\int_{-\pi}^\pi f(x)\sin jx\,dx = \pi b_j,$$

and therefore,

$$b_j = \frac{1}{\pi} \int_{-\pi}^{\pi} f(x) \sin jx \, dx.$$

Similar results apply to series of cosines, to series of sines and cosines, and to series obtained when the period is different (such as π or 1). For example, changing the range of integration from $(-\pi/2, \pi/2)$ to $(-\pi, \pi)$, which is much more convenient for later use—when $j \neq k$

$$\int_{-\pi}^{\pi} \sin jx \sin kx \, dx = \frac{1}{2} \left(\frac{1}{k-j} \sin(k-j)x - \frac{1}{k+j} \sin(k+j)x \right) \Big|_{-\pi}^{\pi} = 0,$$
(10.11)

and that the integral is π when $j = k$.

He went on to claim that any function f defined on the interval $[-\pi, \pi]$ can be written as an infinite series of sines and cosines in any of these forms, depending on how they are to be continued outside the interval on which they have been defined:

- $f(x) = \frac{1}{2}a_0 + \sum\limits_{n=1}^{\infty} a_n \cos nx + b_n \sin nx$ (mixed series)
- $f(x) = \frac{1}{2}a_0 + \sum\limits_{n=1}^{\infty} a_n \cos nx$ (cosine series, which works only for even functions)
- $f(x) = \frac{1}{2}a_0 + \sum\limits_{n=1}^{\infty} b_n \sin nx$ (sine series, which works only for odd functions)

The coefficients of the mixed series are given by the formulae $a_0 = \frac{1}{\pi} \int_{-\pi}^{\pi} f(x) dx$ and

$$a_k = \frac{1}{\pi} \int_{-\pi}^{\pi} f(x) \cos kx \, dx \, , \qquad b_k = \frac{1}{\pi} \int_{-\pi}^{\pi} f(x) \sin kx \, dx.$$
(10.12)

The result is that the heat equation can be solved, for bodies with simple shapes, by the method of Fourier series, and the answer is written as an infinite series of either sines, cosines, or both. The solution of the partial differential equation appears as an infinite series with largely arbitrary coefficients. The boundary conditions allow the coefficients to be determined, and a unique solution is exhibited.

Or rather, we should say, Fourier claimed that this could be done. Very quickly his claim became a challenge to mathematicians to prove that it is correct—and this was to become a long and fascinating story that we can only begin to tell here.

Fourier's ideas generated quite some discussion in print. Siméon Denis Poisson rightly complained that Fourier's methods for finding the coefficients in a Fourier series[6] "has not in fact been demonstrated in a precise and rigorous manner". A decade later he objected (again correctly) that the fundamental assumption that an arbitrary function can be expanded as an infinite series of sines and cosines had not been proved, and Charles-François Sturm joined in, remarking that "Fourier and

---

[6]Poisson ([226], 46), quoted in Bottazzini ([21], 188).

other geometers seem to have misunderstood the importance and the difficulty of this problem, which they have confused with that of determining the coefficients".[7]

10.2.1   Dirichlet on the Convergence of Fourier Series

Peter Gustav Lejeune Dirichlet, who had got to know Fourier personally during a long stay in Paris, took up the subject of the convergence of Fourier series in the late 1820s, by which time it was a topic of some discussion, although Dirichlet said that he knew of no other attempt on the problem than Cauchy's, which he found to be flawed.[8] In his opinion it was remarkable that a Fourier series expansion of an arbitrary function converges (which suggests that he doubted neither the existence nor the convergence of the series), and so he proposed to establish the convergence of a Fourier series directly, and to show that the series and the function agree. He succeeded, by a very fine application of Cauchy's own $\varepsilon - \delta$ analysis and the theory of convergence, in showing that the Fourier series representation of a function that is piecewise continuous and piecewise monotonic on an interval converges and agrees with the function except at the point where the function jumps. At points where

$$\lim_{x\to a-} f(x) = \alpha \neq \beta = \lim_{x\to a+} f(x)$$

the Fourier series takes the value $\frac{1}{2}(\alpha + \beta)$.

Dirichlet's rigorous argument made it clear that to prove Fourier's claim in any greater degree of generality would be hard work, and indeed later generations of mathematicians would discover that Fourier's claim is in fact false in general and applies only to functions that do not oscillate wildly.

10.2.2   Fourier Integrals

Fourier also considered how heat diffused in an *infinite* bar under two distinct conditions. In the first, a part of the bar is raised to a temperature given by a function $F(x)$, the rest being at temperature zero. In the second, one end of the bar is kept at a constant temperature.

He began (§345) by supposing that bar is an infinite line, modelled by the positive real axis $\{x : 0 \leq x\}$, and the heated region is the interval $[0, 1]$, the temperature being given by a function $F(x)$ on $[-1, 1]$. He then supposed that the whole line is considered, and the data extended to the negative real axis by defining $F(-x) = F(x)$. The problem is to be solved by a function $v(x, t)$.

---

[7] See Poisson ([227], 186) and Sturm ([250], 400).
[8] See Dirichlet ([63] and, for a historical account in keeping with the present book, [127]).

The equation to be solved is

$$\frac{\partial v}{\partial t} = k \frac{\partial^2 v}{\partial x^2} - h v,$$

where h and k are constants determined by the physical properties of the wire. He set $v = e^{-ht} u$ and reduced the equation to

$$\frac{\partial u}{\partial t} = k \frac{\partial^2 u}{\partial x^2}.$$

The equation is solved, for example, by

$$u = a \cos(qx) e^{-kq^2 t},$$

where a and q are arbitrary constants and therefore, Fourier supposed, by an infinite sum of such expressions:

$$u = \sum_j a_j \cos(q_j x) e^{-k q_j^2 t}.$$

By supposing the successive qs vary only a little, this led him to look for a solution of the form

$$u(x, t) = \int_0^\infty f(q) \cos(qx) e^{-kq^2 t} dq,$$

where $f(q)$ is an as-yet unknown function of q that can be found from the initial data, because

$$u(x, 0) = F(x).$$

This led him to this equation for $f(q)$:

$$F(x) = \int_0^\infty f(q) \cos(qx) dq,$$

which he called (§346) "a remarkable problem whose solution demands attentive examination".[9]

He solved this equation by going back to thinking of the solution as an infinite series and using his old method for integrating in a way that picked up a single term of the series. This led him to the solution

$$f(q) = \frac{2}{\pi} \int_0^\infty F(x) \cos qx dx.$$

---

[9]We recognise this as the introduction of "Fourier transforms", whose properties Fourier did indeed begin to study.

Therefore, the solution to the heat diffusion problem in this case is

$$u(x,t) = \frac{2}{\pi} \int_0^\infty \left(\int_0^\infty F(x)\cos qx dx \right) \cos(qx) e^{-kq^2 t} dq.$$

Fourier then (§348) considered the special case when $F(x) = 1$ when $-1 \leq x \leq 1$ and $F(x) = 0$ otherwise. The x-integral becomes

$$\frac{2}{\pi} \int_0^1 \cos qx dx = \frac{2}{\pi} \frac{\sin q}{q}.$$

Here, Fourier remarked that the discontinuous function $F(x)$ has been expressed by a definite integral.

To deal with a semi-infinite bar heated at one end, Fourier found it convenient (§351) to think of an infinite bar heated in the middle by a function $F(x)$ for which $F(-x) = -F(x)$. In this case, he found the solution to be

$$u(x,t) = \frac{2}{\pi} \int_0^\infty \left(\int_0^\infty F(\alpha)\sin(q\alpha)d\alpha \right) e^{-kq^2 t} \sin(qx)dq.$$

At the risk of being somewhat arbitrary, I add this remark by Fourier (§358):

We might deduce also from the transformation of series into integrals the properties of the two expressions

$$\frac{2}{\pi} \int_0^\infty \frac{\cos qx dq}{1+q^2}, \quad \text{and} \quad \frac{2}{\pi} \int_0^\infty \frac{\sin qx q dq}{1+q^2};$$

the first (Art. 350) is equivalent to e^{-x} when x is positive, and to e^x when x is negative. The second is equivalent to e^{-x} when x is positive, and to $-e^x$ when x is negative, so the two integrals have the same value, when x is positive, and have values of contrary sign when x is negative.

10.3   The Analysis of Fourier Integrals

Fourier was as over-confident here as he had been when dealing with infinite series— or perhaps we should say that rising standards of rigour were to catch up with him. Rather than trace the history of analyses of this aspect of his work, I shall leap to a satisfactory solution from just over a century later by the American mathematician A.G. Webster that indicates what had to be done.[10]

We begin with the Fourier series for a function defined on the interval $[-l, l]$:

$$f(x) = \frac{1}{2}a_0 + \sum_{j=1}^\infty \left(a_j \cos \frac{j\pi x}{l} + b_j \sin \frac{j\pi x}{l} \right).$$

---

[10]See Webster ([265], 153–156).

where

$$a_j = \frac{1}{l}\int_{-l}^{l} f(s)\cos\frac{j\pi s}{l}ds, \quad b_j = \frac{1}{l}\int_{-l}^{l} f(s)\sin\frac{j\pi s}{l}ds,$$

and we suppose that the function f is such that these last two integrals converge. The question is: what happens as $l \to \infty$?

Note first that if $f(s)$ is absolutely integrable on the whole real line then a_j and b_j both tend to zero as $l \to \infty$.

We define

$$S(l, m, x) = \int_{-l}^{l} f(s)\left(-\frac{1}{2} + \sum_{j=0}^{j=m}\cos\frac{j\pi}{l}(s - x)\right)ds$$

and consider its value as both l and m tend to infinity. This value depends on the order in which we take these limits.

Now, a detailed argument (here omitted, see Webster [265], 154) leads to the conclusion that if the function f is continuous at x then

$$f(x) = \lim_{p \to \infty} \frac{1}{\pi}\int_{-\infty}^{\infty}\left(\int_{0}^{p}\cos(t(s - x))dt\right)f(s)ds,$$

where $p = \pi m/l$. So we require that $m/l \to \infty$, and so we let $m \to \infty$ first.

It then seems natural to suppose that

$$\lim_{p \to \infty} \frac{1}{\pi}\int_{-\infty}^{\infty}\left(\int_{0}^{p}\cos(t(s - x))dt\right)f(s)ds = \frac{1}{\pi}\int_{-\infty}^{\infty}\left(\int_{0}^{\infty}\cos(t(s - x))dt\right)f(s)ds.$$

Indeed, this is the form in which Fourier gave it. But this, to quote Webster (p. 155) "makes no sense" because the integral

$$\int_{0}^{p}\cos(t(s - x))dt = \frac{\sin(p(s - x))}{s - x}$$

does not tend to a limit as $p \to \infty$ but instead oscillates.

Instead, as Webster pointed out (see pp. 155–156), it is possible to switch the order of integration—I omit the argument—and to deduce that the correct value of the Fourier integral is

$$\frac{1}{\pi}\int_{0}^{\infty}\left(\int_{-\infty}^{\infty}f(s)\cos(t(s - x))ds\right)dt. \tag{10.13}$$

Webster at this point quoted Kronecker (*Vorlesungen über die Theorie der einfachen und vielfachen Integrale*, 81) to indicate the continuing importance of Fourier's result:

This so-called Fourier double-integral made at its discovery a tremendous impression on the mathematical world. It was shown for the first time how an almost arbitrary function, satisfying only the limitations mentioned, fits itself into mathematical forms. The formula (10.13) maintains its correctness, as was shown by P. du Bois-Reymond, for various fluctuating functions, inserted instead of the cosine.

10.4   Stokes and Laplace Transform

In the 1860s, as we shall see more fully in Chap. 20, Thomson and Stokes used the heat equation to successfully describe the transmission of electricity down a wire. In the course of that work, Stokes used a Fourier integral approach which we shall now examine.[11]

Stokes treated the problem as being one of heat diffusion down a semi-infinite wire ($x \geq 0$) with an arbitrary initial distribution of heat (or electricity) along with it. The case of heat (or electricity) concentrated at the end point $x = 0$ is of particular interest because it corresponds to an impulse at that end.

He wrote the equation for the temperature $v(x, t)$ as

$$\frac{\partial v}{\partial t} = \frac{\partial^2 v}{\partial x^2},$$

with the conditions that

$$v(x, 0) = 0, \quad v(0, t) = f(t).$$

He then looked for a solution in the form

$$v(x, t) = \int_0^\infty u(\alpha, t) \sin \alpha x \, d\alpha. \tag{10.14}$$

First, differentiation with respect to t under the integral sign gives

$$\frac{\partial v}{\partial t} = \int_0^\infty \frac{\partial u(\alpha, t)}{\partial t} \sin \alpha x \, d\alpha.$$

But, he observed, differentiation under the integral sign with respect to x does not work, because v does not vanish when $x = 0$, and it is necessary to add the term

$$\frac{2}{\pi} \alpha v(0, t) = \frac{2}{\pi} \alpha f(t),$$

---

[11] It is well worth consulting https://www.math.ubc.ca/~feldman/m267/pdeft.pdf for an account of how the Fourier transform can be applied to partial differential equations, including the wave equation and the telegraphist's equation. The same method applied to the heat equation is described in http://web.math.ucsb.edu/~helena/teaching/math124b/heat.pdf . See also Appendix D.

as he had explained in a paper he had published earlier.
  It follows that

$$\frac{\partial^2 v}{\partial x^2} = \int_0^\infty \left(\frac{2}{\pi} \alpha f(t) - \alpha^2 u \right) \sin \alpha x \, d\alpha.$$

Hence,

$$\frac{\partial v}{\partial t} - \frac{\partial^2 v}{\partial x^2} = \int_0^\infty \left(\frac{\partial u(\alpha, t)}{\partial t} - \frac{2}{\pi} \alpha f(t) + \alpha^2 u \right) \sin \alpha x \, d\alpha,$$

and so

$$\frac{\partial u(\alpha, t)}{\partial t} = \frac{2}{\pi} \alpha f(t) - \alpha^2 u.$$

This is an ordinary differential equation whose solution was known. It is

$$u(\alpha, t) = \frac{2}{\pi} e^{-\alpha^2 t} \int^t \alpha f(t') e^{+\alpha^2 t'} dt',$$

where the constant of integration has been expressed as an arbitrary lower end point
of the integral.
  The initial condition $v(x, 0) = 0$ implies that

$$u(\alpha, t) = \frac{2}{\pi} \alpha \int_0^t f(t') e^{+\alpha^2(t-t')} dt'.$$

So Stokes wrote that the temperature v is given by

$$v = \frac{2}{\pi} \int_0^\infty \int_0^t f(t') \alpha e^{+\alpha^2(t-t')} \sin \alpha x \, d\alpha \, dt'.$$

Equation 10.14 implies that this means

$$v = \frac{2}{\pi} \int_0^\infty \left(e^{-\alpha^2 t} \int^t \alpha f(t') e^{+\alpha^2 t'} dt' \right) \sin \alpha x \, dx.$$

He now switched the order of integration, so that the first integral to be evaluated
is the α integral, which is of the form

$$\int_0^\infty e^{-a\alpha^2} \sin b\alpha \, \alpha \, d\alpha.$$

This is a derivative of a known integral:

$$\int_0^\infty e^{-a\alpha^2} \cos b\alpha \, d\alpha = \frac{1}{2} \left(\frac{\pi}{a} \right)^{1/2} e^{-b^2/4a},$$

so

$$\int_0^\infty e^{-a\alpha^2}\sin b\alpha\, \alpha\, d\alpha = -\frac{d}{db}\frac{1}{2}\left(\frac{\pi}{a}\right)^{1/2}e^{-b^2/4a}$$

$$= \frac{b\pi^{1/2}}{4a^{3/2}}e^{-b^2/4a}.$$

Whence, as Stokes said, writing $t - t'$ for a and x for b,

$$v(x,t) = \frac{x}{2\pi^{1/2}}\int_0^t (t-t')^{-3/2}e^{x^2}4(t-t')f(t')dt'.$$

Three comments are in order. First, switching the order of integration is impermissible, although the conclusion remains correct.

Second, as we shall see in Chap. 20, the x^2 term is critical.

Third, the known integral is an example of a Laplace transform, which we now proceed briefly to discuss.

Ingenuity with integrals was a necessary skill of the eighteenth-century mathematician, and Euler and others discovered many clever results without being encumbered by rigour. One of Laplace's contributions in this line was the study of integrals of the form

$$\int_0^\infty e^{-st}f(t)dt,$$

which are today called the Laplace transform of the function $f(t)$.

When the function $f(t)$ is a power of t the Laplace transform can be found recursively, starting with $f(t) = t$. When $f(t)$ is an exponential it is elementary to find the transform, and in this way the Laplace transforms of the trigonometric and hyperbolic functions are found. More complicated integrals, such as the one Stokes used, require ingenuity and were the stock in trade of mathematicians of the day (as they still are for many kinds of engineer).

The Laplace integral

$$\int_{-\infty}^\infty e^{-x^2}dx = \sqrt{\pi},$$

which implies that

$$\int_{-\infty}^\infty e^{-\alpha x^2}dx = \sqrt{\pi/\alpha},$$

is one of the pleasures of contour integration in elementary complex function theory. I leave that for you to find, but this anecdote about it is irresistible.[12]

---

[12]From Thompson [254], 1139 quoted in Lützen [192], 146. The author here is Sylvanus P. Thompson, in his biography of Thomson.

When Thomson was young he went to Paris where he formed a high opinion of Joseph Liouville. He asked him about this integral, and Liouville immediately evaluated it. This mightily impressed Thomson, and

Once when lecturing he [Thomson] used the word "mathematician" and then interrupting himself asked the class: "Do you know what a mathematician is?". Stepping to the blackboard he wrote upon it:

$$\int_{-\infty}^{\infty} e^{-x^2} dx = \sqrt{\pi}.$$

Then, putting his finger on what he had written, he turned to his class and said: "A mathematician is one to whom that is as obvious as that twice two makes four is to you. Liouville was a mathematician." Then he resumed his lecture.

10.5 Exercises

1. Find Fourier series expansions of some very simple functions, such as $f(x) = 1$ and $f(x) = x$ on a suitable interval.
2. Confirm Liouville's claim that

$$\int_{-\infty}^{\infty} e^{-x^2} dx = \sqrt{\pi}.$$

3. Fourier's first method for finding Fourier coefficients invoked Wallis's series for π as an infinite product. Find out what this is and how Fourier used it.

Questions

1. Euler said that the heat equation could only be solved with great effort. Fourier tackled it with methods drawn from the theory of the vibrating string. What does that say about methods for solving partial differential equations around 1800?
2. The eighteenth-century debate about arbitrary solutions to the wave equation and their representation by infinite series of sines and cosines was inconclusive, but the nineteenth-century debate was hugely productive. Why do you think this was?

Chapter 11
Gauss and the Hypergeometric Equation

11.1 Introduction

The hypergeometric equation is arguably the richest example of a linear ordinary differential equation with polynomial functions as coefficients. It has deep roots in the study of elliptic integrals, and its study throughout the nineteenth century was to be promoted by Gauss, Riemann, and others.[1]

11.2 Elliptic Integrals

The simplest, paradigmatic, elliptic integral is

$$u = \int_0^v \frac{dt}{\sqrt{1 - t^4}}. \tag{11.1}$$

It measures arc length along the lemniscate $r^2 = \cos 2\theta$, which is a curve in the shape of a figure eight.

The Italian mathematician Count Fagnano had found some interesting results about this integral in 1714, and after Fagnano submitted his life's work, the *Produzioni* [101] to the Berlin Academy they were sent to Euler, who greatly extended them in the early 1750s.[2] Even so, it was becoming an embarrassment that the integral could not be expanded as a function of its upper endpoint, except as a power series that revealed no significant properties of the integral.

---

[1] For a much fuller account, see Gray [124] and Bottazzini and Gray [22].

[2] The first paper, E252, was presented to the Berlin Academy in January 1752, and the second, E251, was presented to the St. Petersburg Academy in April 1753. Both were published for the first time in 1761, which says something about the turbulent conditions of the time.

© The Author(s), under exclusive license to Springer Nature Switzerland AG 2021   129
J. Gray, *Change and Variations*, Springer Undergraduate Mathematics Series,
https://doi.org/10.1007/978-3-030-70575-6_11

Euler's work on many topics was a great stimulus to Adrien-Marie Legendre, and he took up the challenge of including elliptic integrals in an extended theory of functions. He concentrated on the integrals

$$F(x) = \int_0^x \frac{dt}{\Delta} \quad \text{and} \quad E(x) = \int_0^x \Delta dt,$$

where $\Delta = \sqrt{(1-t^2)(1-c^2t^2)}$, which he called integrals of the first and second kinds, respectively. He called the parameter c the modulus and required it to be real. He was also interested in the corresponding complete integrals $\int_0^1 \frac{dt}{\Delta}$ and $\int_0^1 \Delta dt$, which he denoted F^1 and E^1, respectively, and as $F^1(c)$ and $E^1(c)$ when he wanted to think of them as functions of the modulus.

Legendre's three-volume *Exercises de calcul intégral* [184] draws together his life's work on the subject. Among the many results it contains was one that showed that the complete elliptic integrals satisfy linear differential equations as functions of the modulus[3]:

$$(1-c^2)\frac{d^2F^1}{dc^2} + \frac{1-3c^2}{c}\frac{dF^1}{dc} - F^1 = 0 \qquad (11.2)$$

$$(1-c^2)\frac{d^2E^1}{dc^2} + \frac{1-c^2}{c}\frac{dE^1}{dc} + E^1 = 0. \qquad (11.3)$$

He solved these equations by the method of undetermined coefficients and so obtained power series expansions for the complete integrals $F^1(c)$ and $E^1(c)$. He also established a strikingly attractive result connecting complete integrals of the first two kinds with complementary moduli (c and $b = \sqrt{1-c^2}$):

$$\frac{\pi}{2} = F^1(c)E^1(b) + F^1(b)E^1(c) - F^1(b)F^1(c). \qquad (11.4)$$

Legendre was keen to show that his new functions would be useful. He discussed at length how to calculate table of values for them, and then investigated three problems in detail: the rotation of a solid about a fixed point; the motion (either in a plane or in space) of a body attracted to two fixed bodies; and the gravitational attraction due to an homogeneous ellipsoid. In the first volume of his *Traité* (1828), he further investigated motion under central forces, the surface area of oblique cones, the surface area of ellipsoids, and the problem of determining geodesics on an ellipsoid. But for all this work, these integrals did not reveal their most fundamental properties to him, as we shall now see.

---

[3]It is sometimes said that Euler studied the first of these in 1750, supposedly in the paper E154, which is indeed about the rectification of the ellipse. But this is not quite true; there Euler studied the similar but different differential equation $(1-p^2)\frac{d^2q}{dp^2} - \frac{1+p^2}{p}\frac{dq}{dp} + q = 0$.

Fig. 11.1 Carl Friedrich
Gauss (1777–1855) by
Christian Albrecht Jensen,
1840

11.3 Gauss

Gauss (Fig. 11.1) was arguably the first mathematician to leave the circumscribed
eighteenth-century domain of functions given by explicit expressions and moves with
ease into the large class of functions known only indirectly through some prescribed
property.

This move confronts all who take it with the question: when is a function "known"?
One answer is to develop a theory of functions in terms of some characteristic traits
that can be used to mark certain functions out as having particular properties: they
are periodic, or they have no zeros, for example. A second answer sidesteps the
question and regards the inter-relation of functions given in power series as itself
the answer, which is what Gauss did in his study of the hypergeometric series. Most
mathematicians adopted a mixture of the two approaches depending on their own
success with a given problem. Later, Weierstrass and his followers in the Berlin school
based their theory of functions on the study of series. On the other hand, Riemann
and later workers, chiefly Klein and Poincaré, sought more geometric answers.

Gauss was a brilliantly gifted mathematician born at an unusual time. In 1801, the
year Gauss became famous at the age of 24, Lagrange was 64, Laplace 51, Legendre
48, and Monge 54. Contact with them, and the younger generation of Cauchy and
Fourier, would have been difficult for Gauss because of the Napoleonic war, and
perhaps distasteful, given his conservative disposition. His teachers, Pfaff (then 35)
and Kaestner (81), were not of the first rank, and nor were his contemporaries Bartels
and Farkas Bolyai. By the time the next generation of young mathematicians emerged
(Jacobi and Abel, for example) Gauss had become confirmed in a lifelong avoidance
of mathematicians, and was closer to German astronomers, notably Bessel, in whose
subject he worked increasingly.

Gauss left several of best discoveries unpublished, and they only became known with the publication of his collected works after his death in 1855. His sympathy for the work of János Bolyai and Lobachevskii, when it was revealed, helped change attitudes to non-Euclidean geometry, and his work on elliptic functions and his work on the hypergeometric equation and the hypergeometric serieswere also revelatory. To introduce it, we must make a digression and consider what is called the arithmetico-geometric mean (agm).

Gauss discovered the agm for himself when he was 15. It is defined as follows for positive numbers a_0 and b_0. Set $a_1 = \frac{1}{2}(a_0 + b_0)$, their arithmetic mean, and $b_1 = \sqrt{a_0 b_0}$, their geometric mean. The iteration of this process, defining

$$a_{n+1} = \frac{1}{2}(a_n + b_n) \text{ and } b_{n+1} = \sqrt{a_n b_n},$$

produces two sequences (a_n) and (b_n) that converge to the same limit, α, called the agm of a_0 and b_0. Convergence follows from the inequality $a_{n+1} - b_{n+1} < \frac{1}{2}(a_n - b_n)$.

Gauss denoted the agm of a and b by $M(a, b)$. Plainly $M(\lambda a, \lambda b) = \lambda M(a, b)$ and Gauss considered various functions of the form $M(1, x)$. For example,

$$M(1, 1 + x) = M(1 + \frac{x}{2}, \sqrt{1 + x}),$$

so setting $x = 2t + t^2$ he obtained power series expansions with undetermined coefficients for M in terms of x and then in terms of t, from which the coefficients could be calculated. They display no particular pattern, but various manipulations led Gauss to this dramatic series for a reciprocal of M:

$$y = M(1 + x, 1 - x)^{-1} = 1 + \frac{1}{4}x^2 + \frac{9}{64}x^4 + \frac{25}{256}x^6 + \cdots =$$

$$1 + \left(\frac{1}{2}\right)^2 x^2 + \left(\frac{1.3}{2.4}\right)^2 x^4 + \left(\frac{1.3.5}{2.4.6}\right)^2 x^6 + \cdots .$$

As a function of x, y satisfies the differential equation

$$(x^3 - x)\frac{d^2 y}{dx^2} + (3x^2 - 1)\frac{dy}{dx} + xy = 0,$$

which is Legendre's equation (11.2). Gauss also found another, linearly independent, solution $M(1, x)^{-1}$.

The substitution $x^2 = z$ turns the equation for $M(1 + x, 1 - x)^{-1}$ into this example of the hypergeometric equation

$$x(1 - x)\frac{d^2 y}{dx^2} + (1 - 2x)\frac{dy}{dx} - \frac{1}{4}y = 0$$

as we shall now explain. It is another form of Legendre's equation.

11.3.1 The Hypergeometric Equation

Gauss published only his study of the hypergeometric series, [114] in 1812. The second part, on the hypergeometric equation [115], which is the differential equation satisfied by the hypergeometric series, was found among the extensive *Nachlass* and follows on from the first in numbered paragraphs (Sects. 38–57).

The published paper is not remarkable by Gauss's standards, although it considers x as a complex variable, and contains the earliest rigorous argument for the convergence of a power series and a study of the behaviour of the function at a point on the boundary of the circle of convergence, as well as a thorough examination of continued fraction expansions for certain quotients of hypergeometric functions. Part two is given over to finding several solutions of the hypergeometric equation and the relationships between them, and is of more interest to us here.

In the first part, Gauss observed that the series

$$F(\alpha, \beta, \gamma, x) = 1 + \frac{\alpha\beta}{1.\gamma}x + \frac{\alpha(\alpha+1)\beta(\beta+1)}{1.2.\gamma(\gamma+1)}x^2 + \cdots,$$

where α, β, and γ are real numbers, is a polynomial if either $\alpha - 1$ or $\beta - 1$ is a negative integer, and is not defined at all if γ is a negative integer or zero (this case he excluded). In all other cases, the ratio test shows that the series is convergent for $x = a + bi$ whenever $a^2 + b^2 < 1$. It is striking that Gauss was willing to introduce a new function as a complex-valued functions of a complex variable.

He gave, following Pfaff [209], a list of functions which can be represented by means of hypergeometric functions. For example,

$$e^t = \lim_{k \to \infty} F(1, k, 1, \frac{t}{k});$$

the trigonometric functions can now be obtained. Gauss then introduced the idea of contiguous functions (Sect. 1, Sect. 7): $F(\alpha, \beta, \gamma, x)$ is contiguous to any of the six functions $F(\alpha \pm 1, \beta \pm 1, \gamma \pm 1, x)$ obtained from it by increasing or decreasing one coefficient by 1. He obtained 15 equations connecting $F(\alpha, \beta, \gamma, x)$ with each of the 15 pairs of its different contiguous functions by systematically permuting the αs, βs, γs, etc. and comparing coefficients. As an illustration, here is the first of these equations:

$$(\gamma - 2\alpha - (\beta - \alpha)x)F(\alpha, \beta, \gamma, x) + \alpha(1 - x)F(\alpha + 1, \beta, \gamma, x) - (\gamma - \alpha)F(\alpha - 1, \beta, \gamma, x) = 0.$$

As Felix Klein was to remark ([157], 16) these establish that any three contiguous functions satisfy a linear relationship with rational functions for coefficients. As a result, there are linear relationships over the rational functions between any three functions of the form $F(\alpha \pm m, \beta \pm n, \gamma \pm p, x)$, where m, n, and p are integers. Gauss then showed that how to use contiguous functions to provide continued fraction expansions of quotients of hypergeometric functions, e.g.

$$\frac{F(\alpha, \beta + 1, \gamma + 1, x)}{F(\alpha, \beta, \gamma, x)},$$

and hence for several familiar elementary functions. Observe, as Gauss did at the start of the second paper, that $F(\alpha, \beta, \gamma, x)$ and $\dfrac{dF(\alpha, \beta, \gamma, x)}{dx}$ are contiguous in the obvious generalised sense, the relationship between them being, essentially, the differential equation itself. This is because

$$\frac{d}{dx} F(\alpha, \beta, \gamma, x) = \frac{\alpha\beta}{\gamma} F(\alpha + 1, \beta + 1, \gamma + 1, x).$$

In the third and final sections of the published paper, Gauss considered the question of the value of $F(\alpha, \beta, \gamma, 1)$, i.e. of $\lim_{x\to 1} F(\alpha, \beta, \gamma, x)$, for real α, β, γ. He then defined

$$\Pi(k, z) = \frac{1.2.\ldots.k.k^z}{(z+1)(z+2)\ldots(z+k)},$$

where k is a positive integer, and $\Pi(z) = \lim_{k\to\infty} \Pi(k, z)$, which may be called (Gauss's) factorial function and is his version of the Gamma function. The limit certainly exists for $Re(z) > 0$ and Π satisfies the functional equation $\Pi(z + 1) = (z + 1)\Pi(z)$ with $\Pi(0) = 1$, from which it follows that $\Pi(n) = n!$ for positive integral n. Π is infinite at all negative integers.[4] The factorial function enabled Gauss to obtain many results that earlier mathematicians had obtained only with great effort. As Gauss puts it: "Whence many relations, which the illustrious Euler could only get with difficulty, fall out at once".

Gauss began the second and unpublished part of the paper, "Determinatio series nostrae per Aequationem Differentialem Secundi Ordinis", by observing that $P = F(\alpha, \beta, \gamma, x)$ is a solution of the hypergeometric equation:

$$z(1 - z)\frac{d^2 w}{dx^2} + (\gamma - (\alpha + \beta + 1)x)\frac{dw}{dx} - \alpha\beta w = 0.$$

To find a second linearly independent solution he set $1 - y = x$, when the equation becomes the first equation with γ replaced by $\alpha + \beta + 1 - \gamma$. It, therefore, has a solution $F(\alpha, \beta, \alpha + \beta + 1 - \gamma, 1 - x)$, and the differential equation, in general, has solutions of the form

$$M F(\alpha, \beta, \gamma, x) + N F(\alpha, \beta, \alpha + \beta + 1 - \gamma, 1 - x), \tag{11.5}$$

where M and N are constants.

Other solutions may arise which do not at first appear to be of this type, but, he remarked, any three solutions must satisfy a linear relationship with constant

---

[4]In the usual notation from Legendre ([184], Vol. II), $\Pi(z) = \Gamma(z + 1)$.

coefficients. This fact was of most use to him when transforming the differential equation by means of a change of variable.

The substitutions he considered are of two types: the following transformations of x:

$$x = 1 - y, \ x = \frac{1}{y}, \ x = \frac{y}{y-1}, \ x = \frac{y-1}{y},$$

and these transformations of P:

$$P = x^\mu P', \ P = (1-x)^\mu P'$$

for particular values of μ. These gave him several solutions to the original equation in terms of functions like $F(-,-,-,x)$ and $F(-,-,-,1-x)$ etc.—where the blanks stand for expressions in α, β, and γ—possibly multiplied by powers of x and $1 - x$, and also some linear identities between triples of such solutions.

The paper concludes with a discussion of certain special cases that can arise when α, β, and γ are not independent, for example, when $\beta = \alpha + 1 - \gamma$, and the quadratic change of variable $x = 4y - y^2$ can be made.

Gauss made a very interesting observation at this point. The equation has as one solution in this case:

$$F(\alpha, \beta, \alpha + \beta + \frac{1}{2}, 4y - 4y^2) = F(2\alpha, 2\beta, \alpha + \beta + \frac{1}{2}, y).$$

If, he said, y is replaced by $1 - y$ this produces

$$F(\alpha, \beta, \alpha + \beta + \frac{1}{2}, 4y - 4y^2) = F(2\alpha, 2\beta, \alpha + \beta + \frac{1}{2}, 1 - y),$$

as we can see by looking at the basis exhibited in Eq. (11.5) above, and we are led to the seeming paradox

$$F(2\alpha, 2\beta, \alpha + \beta + \frac{1}{2}, y) = F(2\alpha, 2\beta, \alpha + \beta + \frac{1}{2}, 1 - y).$$

"which equation is certainly false" (Sect. 55).

To resolve the paradox he distinguished between the sign F when it stood for a function that satisfies the hypergeometric equation, and when the sign F stood for the sum of an infinite series. The sum is only defined within its circle of convergence, but the function is to be understood for all values of its fourth term that have been obtained by continuous change, whether real or imaginary, provided the values 0 and 1 are avoided. However, this "function" may be many-valued, and it is in this case.

This being so, he argued that one would no more be misled than one would infer from $\arcsin \frac{1}{2} = 30°$ and $\arcsin \frac{1}{2} = 150°$ that $30° = 150°$, the reason being that a (many-valued) function such as $\arcsin$ may have different values even though its variable has taken the same value, whereas a series may not.

Gauss here confronted the question of analytically continuing a function outside its circle of convergence. It was his view that the solutions of the differential equation exist everywhere but at 0, 1, (and ∞, although he avoided the expression), whereas their representation in power series is a local question. However, if the function is a many-valued function then the series expression may not be recaptured if the variable is taken continuously along some path and restored to its original value, and neglect of this fact can lead to absurd expressions like the one Gauss produced.

Because he here talked of continuous change in the variable in the complex number plane, one may thus infer that Gauss here was truly discussing analytic continuation, and not merely the plurality of series solutions at a given point.

In these papers, Gauss introduced a large class of functions of a complex variable that were defined by the hypergeometric equation and were capable of various expressions in series. The main direction of his research was in studying relationships between the series, which in turn provided information about the nature of the functions under consideration.

The second part of Gauss's paper on the hypergeometric series raises two main types of question. First, it would be useful to have a systematic account of the solutions obtained by the various substitutions, and of the nature of the substitutions themselves. Second, it would be instructive to connect the hypergeometric functions with the newer functions in analysis, especially in complex analysis, such as the elliptic functions. It is striking that Kummer in his paper [167] set himself both these tasks and resolved them while, moreover, observing Gauss's restrictions where the work would otherwise be too difficult (for example, by considering only real coefficients).

11.4   Kummer and His 24 Solutions

Ernst Eduard Kummer had studied Mathematics at the University of Halle, and in 1836, when he published his paper on the hypergeometric equation, he was 26 and a Lecturer at the Liegnitz Gymnasium. Although he never attended a lecture by Dirichlet, he considered him to have been his real teacher, which is an indication of Dirichlet's great influence on mathematics in Germany, an influence that was then extended to Kummer's best student at Liegnitz, Leopold Kronecker. Kummer, with Kronecker and Weierstrass, went on to dominate the Berlin school of mathematics from 1856 until Kummer retired in 1883. His students regarded him as a gifted teacher and organiser of seminars, and he was diligent in his concern for them. He was also a man of great charm, and he had a great appetite for administration, being Dean of the University of Berlin twice, Rector once, and Perpetual Secretary of the physics-mathematics section of the Berlin Academy from 1863 to 1878.

At the start of his long paper (1836), Kummer remarked of Gauss's paper that

> But this work is only the first part of a greater work as yet unpublished, and wants comparison of hypergeometric series in which the last element x is different. This will therefore be the principal purpose of the present work; the numerical application of the discovered formulae

will preferably be made to elliptic transcendents, to which in great part the general series corresponds.

The hypergeometric equation has three singular points, $x = 0, x = 1, x = \infty$, and in a neighbourhood of these points one expects the solution to be of the form of a hypergeometric series in either x, $1 - x$, or $1/x$, respectively, possibly multiplied by some power of x, $1 - x$, or $1/x$. This was one reason for Kummer to look for appropriate changes of variables in the study of the hypergeometric equation. Another was the need to find hypergeometric series that yield independent solutions of the hypergeometric equation other than the canonical hypergeometric series, because the differential equation has a basis of two independent solutions.

Kummer investigated what happens to the hypergeometric equation under changes of variable, and found that the only changes of variable that can be made (unless there are special relations between the coefficients α, β, γ) are the ones that Gauss considered—including the identity transformation $x = z$ and the one Gauss had not written down, $x = 1/(1 - z)$.

The details of his argument allowed him to deduce more. For example, if α is replaced by $\gamma - \alpha$ and β is replaced by $\gamma - \beta$ this produces a new solution:

$$y = (1 - x)^{\gamma - \alpha - \beta} F(\gamma - \alpha, \gamma - \beta, \gamma, x).$$

Other similar changes produce the solutions

$$y = x^{1-\gamma} F(\gamma - \alpha, \gamma - \beta, \gamma, x),$$

and

$$y = x^{1-\gamma}(1 - x)^{\gamma - \alpha - \beta} F(1 - \alpha, 1 - \beta, 2 - \gamma, x).$$

These solutions can also be checked directly, by long but straightforward calculations.

In this way, he was led to his finest achievement in this paper (Kummer 1836 [167], 52–53): an enumeration of a family of 24 solutions to the hypergeometric equation that between them form what can be considered as the complete solution to the equation (see Fig. 11.2).[5] In Kummer's work, the variable is real, and he regarded the 24 solutions as the best way to represent solutions valid near the singular points in a variety of convenient forms. Later writers, Riemann and Schwarz, allowed the variable to be complex, in which case Kummer's 24 solutions provide not only sets of bases for the solutions everywhere, but their inter-relations (which he also described) give a description of their analytic continuation on the complex sphere. However, this interpretation was almost completely lacking in Kummer's work (despite the remarks of some later mathematicians and historians such as Klein ([158], 267) and Biermann ([16], 523)).[6]

---

[5] See also his *Collected Papers* Vol. II, 88, 89.

[6] Kummer concluded the paper with an unremarkable study of what happens when x is allowed to be complex but α, β, and γ stay real that has no bearing on the issue of analytic continuation.

We could check these solutions by introducing a new variable, say $z = x/(x-1)$, writing the hypergeometric equation in terms of z, and plugging in the new solution. If we do so, it is important to note that the transformed equation may no longer look like the hypergeometric equation.

As you can see in the table he provided, Kummer expressed the solutions of the hypergeometric equation in terms of a hypergeometric series possibly multiplied by a power of x or $1 - x$, where the variable in the hypergeometric series is one of

$$x, \ 1 - x, \ \frac{1}{x}, \ \frac{1}{1-x}, \ \frac{x}{x-1}, \ \text{or} \ \frac{x-1}{x}.$$

The various solutions are valid on several different domains of continuity, which for Kummer were intervals on the real line separated by the points $x = 0$ and $x = 1$ and which Gauss would have understood as discs and half-planes in the complex plane.

If we let x be a complex variable, then it is easy to see that, because $F(\alpha, \beta, \gamma, x)$ converges for $|x| < 1$, which is the disc D_0 centre 0 and radius 1, the series in the variables

$$1 - x, \ \frac{1}{x}, \ \frac{1}{1-x}, \ \frac{x}{x-1}, \ \text{and} \ \frac{x-1}{x}$$

converge, respectively, in the domains

- $|1 - x| < 1$, which is the disc D_1 disc centre 1 and radius 1;
- outside the disc D_0;
- outside the disc D_1;
- in the half-plane of complex numbers with real part less than $\frac{1}{2}$;
- in the half-plane of complex numbers with real part greater than $\frac{1}{2}$.

Because some of these domains overlap, Kummer also established the linear relations that exist between any three of the 24 solutions that converge on a common neighbourhood. Some are simple equalities, for example, between expressions (1), (2), (17), and (18) – (see [167], 54–55)

$$F(\alpha, \beta, \gamma, x) = (1 - x)^{\gamma - \alpha - \beta} F(\gamma - \alpha, \gamma - \beta, \gamma, x)$$

$$= (1 - x)^{-\alpha} F(\alpha, \gamma - \beta, \gamma, x/(x-1)).$$

In fact, there are six different families of four equal solutions thus:

1, 2, 17, 18; 3, 4, 19, 20; 5, 6, 21, 22; 7, 8, 23, 24; 9, 12, 13, 15; and 10, 11, 14, 16.

So, to find all the linear relations between the 24 solutions is enough to consider the six different ones 1, 3, 5, 7, 13, 14. Of these, 5 and 7 converge or diverge exactly when 13 and 14 diverge or converge, respectively (Kummer here restricted x to be real, but the observation is valid for complex x). The problem is thus reduced to finding the relations between the following triples:

$$1, 3, 5; \quad 1, 3, 7; \quad 1, 3, 13; \quad 1, 3, 14; \quad 1, 5, 7,$$

and after some work he listed the relationship that arise. They all arise from evaluating $F(-, -, -, x)$ at $x = 0$ or $x = 1$.

What Kummer's solutions also show, on letting x be complex, is that a solution of the hypergeometric equation is analytic everywhere except at the three points $0, 1, \infty$ and that in a neighbourhood of any of those points the solution becomes analytic on being multiplied by a suitable power of x or $1 - x$. In the language of later writers, this says that the hypergeometric equation is an ordinary differential equation with three regular singular points.

We shall see in Chap. 14 how Riemann deepened that insight.

11.5 The Method of Undetermined Coefficients

We can solve the hypergeometric equation

$$(x - x^2)y''(x) + (c - (a + b + 1)x)y'(x) - aby = 0$$

in a neighbourhood of the origin by the method of undetermined coefficients, and in this way realise that the solutions of the hypergeometric equation that are displayed in Fig. 11.2 are correct for various changes of variable.

To do this, we write

$$y = x^k(a_0 + a_1 x + a_2 x^2 + \cdots + a_n x^n + \cdots) = x^k f(x), \quad a_0 \neq 0.$$

This gives

$$y'(x) = kx^{k-1}(a_0 + a_1 x + a_2 x^2 + \cdots + a_n x^n + \cdots) + x^k(a_1 + a^2 x + \cdots + a_{n+1}x^n + \cdots),$$

$$y''(x) = k(k-1)x^{k-2}(a_0 + a_1 x + a_2 x^2 + \cdots + a_n x^n + \cdots) + 2kx^{k-1}(a_1 + a^2 x + \cdots + a_{n+1}x^n + \cdots)$$

$$+ x^k(2a_2 + 6a_3 x + \cdots + (n+2)(n+1)a_{n+2}x^n + \cdots.$$

So

$$(x - x^2)y''(x) + (c - (a + b + 1)x)y'(x) - aby = k(k-1)x^{k-1} + ckx^{k-1} + higher terms,$$

so

$$k(k - 1) + ck = 0,$$

and so

$$k = 0 \quad \text{or} \quad k = 1 - c.$$

Fig. 11.2 Kummer's 24
solutions, from Kummer
([167], 52–53)

1) $F(\alpha, \beta, \gamma, x)$,

2) $(1-x)^{\gamma-\alpha-\beta} F(\gamma-\alpha, \gamma-\beta, \gamma, x)$,

3) $x^{1-\gamma} F(\alpha-\gamma+1, \beta-\gamma+1, 2-\gamma, x)$,

4) $x^{1-\gamma} (1-x)^{\gamma-\alpha-\beta} F(1-\alpha, 1-\beta, 2-\gamma, x)$,

5) $F(\alpha, \beta, \alpha+\beta-\gamma+1, 1-x)$,

6) $x^{1-\gamma} F(\alpha-\gamma+1, \beta-\gamma+1, \alpha+\beta-\gamma+1, 1-x)$,

7) $(1-x)^{\gamma-\alpha-\beta} F(\gamma-\alpha, \gamma-\beta, \gamma-\alpha-\beta+1, 1-x)$,

8) $x^{1-\gamma} (1-x)^{\gamma-\alpha-\beta} F(1-\alpha, 1-\beta, \gamma-\alpha-\beta+1, 1-x)$,

9) $x^{-\alpha} F\left(\alpha, \alpha-\gamma+1, \alpha-\beta+1, \dfrac{1}{x}\right)$,

10) $x^{-\beta} F\left(\beta, \beta-\gamma+1, \beta-\alpha+1, \dfrac{1}{x}\right)$,

11) $x^{\alpha-\gamma} (1-x)^{\gamma-\alpha-\beta} F\left(1-\alpha, \gamma-\alpha, \beta-\alpha+1, \dfrac{1}{x}\right)$,

12) $x^{\beta-\gamma} (1-x)^{\gamma-\alpha-\beta} F\left(1-\beta, \gamma-\beta, \alpha-\beta+1, \dfrac{1}{x}\right)$,

13) $(1-x)^{-\alpha} F\left(\alpha, \gamma-\beta, \alpha-\beta+1, \dfrac{1}{1-x}\right)$,

14) $(1-x)^{-\beta} F\left(\beta, \gamma-\alpha, \beta-\alpha+1, \dfrac{1}{1-x}\right)$,

15) $x^{1-\gamma} (1-x)^{\gamma-\alpha-1} F\left(\alpha-\gamma+1, 1-\beta, \alpha-\beta+1, \dfrac{1}{1-x}\right)$,

16) $x^{1-\gamma} (1-x)^{\gamma-\beta-1} F\left(\beta-\gamma+1, 1-\alpha, \beta-\alpha+1, \dfrac{1}{1-x}\right)$,

17) $(1-x)^{-\alpha} F\left(\alpha, \gamma-\beta, \gamma, \dfrac{x}{x-1}\right)$,

18) $(1-x)^{-\beta} F\left(\beta, \gamma-\alpha, \gamma, \dfrac{x}{x-1}\right)$,

19) $x^{1-\gamma} (1-x)^{\gamma-\alpha-1} F\left(\alpha-\gamma+1, 1-\beta, 2-\gamma, \dfrac{x}{x-1}\right)$,

20) $x^{1-\gamma} (1-x)^{\gamma-\beta-1} F\left(\beta-\gamma+1, 1-\alpha, 2-\gamma, \dfrac{x}{x-1}\right)$,

21) $x^{-\alpha} F\left(\alpha, \alpha-\gamma+1, \alpha+\beta-\gamma+1, \dfrac{x-1}{x}\right)$,

22) $x^{-\beta} F\left(\beta, \beta-\gamma+1, \alpha+\beta-\gamma+1, \dfrac{x-1}{x}\right)$,

23) $x^{\alpha-\gamma} (1-x)^{\gamma-\alpha-\beta} F\left(1-\alpha, \gamma-\alpha, \gamma-\alpha-\beta+1, \dfrac{x-1}{x}\right)$,

24) $x^{\beta-\gamma} (1-x)^{\gamma-\alpha-\beta} F\left(1-\beta, \gamma-\beta, \gamma-\alpha-\beta+1, \dfrac{x-1}{x}\right)$.

In the case when $k = 0$, we find by looking at the coefficient of x that $-aba_0 + ca_1 = 0$, so $a_1 = \dfrac{ab}{c} a_0$. Likewise, on tidying up, we find that

$$a_2 = \frac{(a+1)(b+1)}{2(c+1)},$$

and by looking carefully at the coefficient of x^n that

$$(-ab - n(a+b+1) - n(n-1))a_n + (n+1)(c+n)a_{n+1} = 0.$$

So

$$(ab + na + nb + n^2)a_n = (n+1)(c+n)a_{n+1},$$

and so

$$a_{n+1} = \frac{(a+n)(b+n)}{(n+1)(c+n)}a_n.$$

So, in this case, the solution of the hypergeometric equation is the hypergeometric series

$$y(x) = a_0 \left(1 + \frac{ab}{1.c}x + \frac{a(a+1)b(b+1)}{1.2.c(c+1)}x^2 + \cdots \right).$$

It is an interesting exercise to determine the domains of convergence of these series, which relates to the changes of variable used by Gauss, Kummer, and later Riemann. The contiguous equations mentioned by Gauss can be used to find a second, linearly independent solution to the hypergeometric equation and convergent on the same domain, as Gauss indicated.

11.6 Exercises

1. Derive Eq. (11.2)

$$(1 - c^2)\frac{d^2 F^1}{dc^2} + \frac{1 - 3c^2}{c}\frac{d F^1}{dc} - F^1 = 0$$

and (11.3)

$$(1 - c^2)\frac{d^2 E^1}{dc^2} + \frac{1 - c^2}{c}\frac{d E^1}{dc} + E^1 = 0.$$

2. Calculate the arithmetico-geometric means of several pairs of numbers, including this one that Gauss did, the arithmetico-geometric mean of 1 and $\sqrt{2}$.

Questions

1. Find out what Gauss found significant about the arithmetico-geometric mean of 1 and $\sqrt{2}$ (it is nr. 98 in his mathematical diary).
2. Why do you think Gauss published his investigations into the hypergeometric series but not into the hypergeometric equation? What other subjects did Gauss investigate but not publish?

Chapter 12
Existence Theorems

12.1 Introduction

It was surely inevitable that the confident methods of the eighteenth century for solving ordinary differential equations, which were more formal than rigorous, would be critically examined by Augustin Louis Cauchy (Fig. 12.1), and they were. Curiously, however, although he appreciated the difference between real and complex methods he chose to advertise his complex power series methods and seemingly forgot that he had been the first to give a good account of ordinary differential equations in line with the insights of his own $\varepsilon - \delta$ analysis.

Cauchy also built on the work of Monge and Lagrange to provide existence theorems for partial differential equations, but in this case his use of real analysis for the first-order case and analytic functions for the general case reflects a fundamental difference in what could be established. This is discussed in Chap. 17.

12.2 Cauchy and Ordinary Differential Equations

Cauchy's interest in analysis was not confined to providing improved foundations for the subject. He applied it in many domains of mathematics and in so doing greatly extended it. As one example of this, we discuss his work on the topic of showing that ordinary differential equations have solutions.

He tackled this question twice in his life. The first time was as early as 1820 or 1821, when he was lecturing at the École Polytechnique, and the second some 15 years later, when he was in self-imposed exile in Prague. As we shall see, for many years the second method eclipsed the first, for reasons that tell us a lot about the mathematical community of the time and its priorities.

In 1820, advanced mathematics students learned how to solve differential equations from Lacroix's *Traité du calcul différentiel et du calcul intégral*. This was a

© The Author(s), under exclusive license to Springer Nature Switzerland AG 2021
J. Gray, *Change and Variations*, Springer Undergraduate Mathematics Series,
https://doi.org/10.1007/978-3-030-70575-6_12

Fig. 12.1 Augustin Louis
Cauchy (1789–1856), artist
unknown, after J. Rollet,
1840

deliberate ragbag of methods. Cauchy broke with this approach, and with the views
of the Directorship of the École, by tacking a question that Lacroix had ignored:
Does a first-order ordinary differential equation have a solution?

It might seem surprising that Lacroix had not dealt with this question himself,
but there are several reasons why he did not. Insofar as differential equations arise
in mathematical physics, it can seem obvious that they have solutions—the only
problem is how to find them. Lacroix was not much interested in rigour, and it is
always possible to check that a suggested solution is correct. Even the theoretical
formulation of a differential equation, as a curve about which some information about
the behaviour of the tangents is given (or a function about which some information
about the behaviour of its derivatives is given) seems to imply that there *is* a function
that has the prescribed property.

But rigour in these matters greatly concerned Cauchy. Christian Gilain, the modern
editor of Cauchy's work in the 1820s on the topic, commented that Cauchy's existence
theorem is not a discovery inserted in an otherwise classical text, but ([39], xxi)

> it has a very important place not only in the great number of pages devoted to the proof and
> its examples, but in its role in the whole organisation of the course. It is truly a work in the
> foundations in the general theory of differential equations …

The lecture notes of a course on ordinary differential equations that Cauchy gave
were lost, and were not republished in the 31 volumes of his *Oeuvres*. However, a
set of 13 lectures, in the original edition printed by the École Polytechnique, was

found by the Gilain in the Bibliothèque Nationale in Paris and published in 1981. Unfortunately, it is apparently impossible to date them precisely; Cauchy seems to have had the idea by 1820 or 1821, and probably gave the lectures in 1823 or 1824.

In his opening five lectures, Cauchy largely followed tradition and dealt with explicit solution methods. Lecture 6 prepared the way for Lecture 7, in which Cauchy observed that explicit solutions to a differential equation of the form

$$\frac{dy}{dx} = f(x, y) \tag{12.1}$$

yield a family of solutions, so one could also find the solution that took a particular value y_0 at a particular point x_0.

Cauchy now proposed that the important problem was to establish for every differential equation the existence of a solution that took a particular value y_0 at a particular point x_0. The existence of a general solution would then follow by allowing x_0 and y_0 to vary. This is not a pedantic point. Lacroix, like others before him, had regarded a solution as a formula and the methods for finding a solution were largely formal— that is what was meant by having a general solution. Cauchy inverted the process, and asked if there is any solution at all that would have the required properties in the light of his revised standards for analysis as a whole.

In Lecture 7, he sketched a method for proving the existence of solutions to the differential equation

$$\frac{dy}{dx} = f(x, y).$$

For the first time, we get restrictions on the function f—without which one suspects nothing can be proved. He showed that

Theorem 12.1 *If the function f is continuous and bounded by $\pm A$ as a function of x and y on a neighbourhood of the point (x_0, y_0) in the plane, and the partial derivative $\dfrac{\partial f}{\partial y}$ is also continuous and bounded in that neighbourhood, then there is a solution to the equation.*

He argued the solution function is very close to the collection of points (x_j, y_j) given by

$$y_j - y_{j-1} = (x_j - x_{j-1})(f(x_{j-1}, y_{j-1}))$$

when the points x_j are close together. This is what you would expect, because when $x_j - x_{j-1}$ is small the quotient

$$\frac{y_j - y_{j-1}}{x_j - x_{j-1}}$$

should be very close to the value of $\frac{dy}{dx}$ and so to the function f at the point (x_{j-1}, y_{j-1}). Then the stated conditions imply, by a limiting argument, that as the points x_j get closer and closer together, the points (x_j, y_j) lie on a curve that is the graph of the integral $\int f(x, y)dx$, which is the solution of the differential equation and for which $y(x_0) = y_0$, so the theorem is proved.

To be more precise, let us follow Cauchy and fix some notation. In Lecture 7, Cauchy considered the ordinary differential equation

$$\frac{dy}{dx} = f(x, y)$$

for $x \in [x_0, X]$ on the assumptions that f is continuous and bounded, $|f(x, y)| \leq A$, and f_y is continuous and bounded, $|f_y(x, y)| \leq C$. He supposed the initial value of y was y_0 and he set $X - x_0 = H$.

He took a sequence of points $x_0 < x_1, < \ldots < x_n = X$ and defined a sequence of points $y_0, y_1, \ldots, y_n = Y$ by the equations

$$y_1 - y_0 = (x_1 - x_0)f(x_0, y_0),$$

$$y_j - y_{j-1} = (x_j - x_{j-1})f(x_{j-1}, y_{j-1}),$$

$$Y - y_{n-1} = (X - x_{n-1})f(x_{n-1}, y_{n-1}).$$

He then took it for granted (validly) that

$$Y = F(x_0, x_1, \ldots, x_j, \ldots x_{n-1}, X, y_0)$$

is a continuous function of all its variables, and proved that as a result

$$Y = y_0 + (x - x_0)f(x_0 + \Theta H, y_0 + \Theta A H) \tag{12.2}$$

for some $0 < \theta < 1$, and $0 < \Theta < 1$. (He did not use the continuity of f_y here.)

The proof has three small steps. First, summing up the equations above gives $Y - y_0$ on the left-hand side, and on the right-hand side a term that can be rewritten as $X - x_0$ times some average value of the various $f(x_{j-1}, y_{j-1})$ that Cauchy wrote as θA by what we would call the intermediate value theorem, for some $0 < \theta < 1$.

Second, similarly, for every j,

$$|y_j - y_{j-1}| \leq (X - x_0)A = HA$$

and so may be replaced by an average, $y_0 + \Theta H A$, for some $0 < \theta < 1$.

Third, therefore, the various values of the $f(x_{j-1}, y_{j-1})$ are all of the form $f(x_0 + \theta H, y_0 + \Theta A H)$. A further averaging argument (or, we would say, use of the intermediate value theorem) gave the required result.

Cauchy then investigated the dependence of Y on the initial value y_0, and now he needed the conditions on f_y. He proved the theorem that if the initial point is y_0', where $y_0' - y_0 = \beta_0$, then

$$|Y(y_0') - Y(y_0)| = \Theta|\beta_0|e^{CH}$$

for some Θ, $0 < \Theta \le 1$. This says that the solution (if it exists) is unique.

The proof began by looking at y_1' and defining β_1 by the equation $y_1' - y_1 = \beta_1$. This gives

$$|\beta_1| - |\beta_0| = (x_1 - x_0)(f(x_0, y_0') - f(x_0, y_0)).$$

The term in f can be written (again by the intermediate value theorem) as

$$f_y(x_0, y_0 + \theta_0(y_0' - y_0))(y_0' - y_0) = |\beta_0|f_y(x_0, y_0 + \theta_0(y_0' - y_0)),$$

for some $0 < \theta_0 < 1$, so

$$|\beta_1| = |\beta_0|(1 + (x_1 - x_0)f_y(x_0, y_0 + \theta_0(y_0' - y_0)).$$

By a further use of the intermediate value theorem this becomes

$$|\beta_1| = |\beta_0|(1 + (x_1 - x_0)\Theta_0 C),$$

for (yet another) $0 < \Theta_0 < 1$. A similar argument applies at every step.

To combine the equations that result, Cauchy noted that

$$1 + \Theta_0 C(x_1 - x_0) < 1 + C(x_1 - x_0) < e^{C(x_1 - x_0)}.$$

Therefore

$$|\beta_n| < |\beta_0|e^{C(x_1 - x_0)}e^{C(x_2 - x_1)} \cdots e^{C(x_n - x_{n-1})} = |\beta_0|e^{C(X - x_0)} = |\beta_0|e^{CH},$$

and the result follows.

Cauchy then checked that the theorems are true independently of the choice of the x_j, and could then state the theorem that if $y = F(x)$ is the limiting value of the function $Y = F(x_0, x_1, \ldots, x_j, \ldots x_{n-1}, X, y_0)$ as n increases indefinitely and the distances $x_j - x_{j-1}$ decrease indefinitely, then $y = y(x)$ satisfies the differential equation $\frac{dy}{dx} = f(x, y)$ with the initial condition $y(x_0) = y_0$.

The proof is little more than using the intermediate value theorem to show that y is a continuous function of x, to bound the variation in y as a function of x, and to deduce that in the limit as the intervals between (otherwise arbitrary) x-values tend to zero the function y is a differentiable function of x and $\frac{dy}{dx} = f(x, y)$ in the region considered.

In Lecture 8, he strengthened the theorem to show that the stated conditions on f and $\frac{\partial f}{\partial y}$ imply that there is a solution to the differential equation on some interval $x_0 \le x \le a$. He showed that the differential equation has a solution for values of x between x_0 and $x_0 + a$ and values of y between $y_0 - Aa$ and $y_0 + Aa$, where a is a quantity determined by the function f and A is greater than the absolute value of

$f(x, y)$ in the interval considered (so A depends on a). Curiously, Cauchy did not prove that the constant a always has a non-zero value but only assumed it.

As in Lecture 7, Cauchy took a sequence of points $x_0, x_1, \ldots, x_n$ and the corresponding points $y_0, y_1, \ldots, y_n$ such that

$$y_{j+1} - y_j = (x_{j+1} - x_j) f(x_j, y_j),$$

and for all j, $|y_{j+1} - y_j| < Aa$. Adding these equations gave him

$$y_m - y_0 = \sum_{j=0}^{m-1} (x_{j+1} - x_j) f(x_j, y_j),$$

and the term on the right-hand side is equal to $(x_m - x_0) f^*$, where f^* is an "average" of the $f(x_j, y_j)$s. By assumption, for every j,

$$|f(x_j, y_j)| < A, \quad 0 \le j \le m - 1,$$

and so their average is also less than A. Therefore, if

$$y_0 - Aa \le y_j \le y_0 + Aa, \quad 0 \le j \le m - 1$$

then also $y_0 - Aa \le y_m \le y_0$, and so each y_j is such that $|y_j - y_0| < Aa$.

It follows that for $x_0 < x < x_0 + a$ and $y_0 - Aa < y < y_0 + Aa$ the functions $f(x, y)$ and $f_y(x, y)$ are continuous and bounded; and $|f(x, y)| < A$.

Therefore, the limit as the x_j get closer and closer together exists and $y - y_0 = \int_{x_0}^{x_0+a} f(x, y) dx$ is a solution of the differential equation that satisfies the given initial conditions.

Cauchy did not stop with this local result. He sought conditions which would allow him to prolong the solutions out of the neighbourhood in which they have been shown to exist. He found that this could only be done indefinitely if certain necessary conditions were met, and gave examples to show that some differential equations have solutions that only exist for a limited range of the variable x. We can note this simple one: the differential equation $\dfrac{dy}{dx} = 1 + y^2$, for which $x = 0$ implies $y = 0$, has the solution $y = \tan x$, but this is only defined on the interval $(-\pi/2, \pi/2)$.

We should also note that these theorems give sufficient conditions for a solution to exist, but they are not necessary.

Before we proceed, it will help to look at two examples. We start with the simple case

$$\frac{dy}{dx} = -\frac{x}{y},$$

with the initial conditions that at $x = 0$ we have $y = a$.

Suppose we use the theorems to look for a solution for which when $x = 0$ we have $y = 1$. The theorems tell us that there is a unique solution valid in some neighbourhood of the point $(x, y) = (0, 1)$.

But if we use the theorems to look for a solution for which when $x = 1$ we have $y = 0$ then the theorems tell us nothing, because in any neighbourhood of the point $(1, 0)$ $\frac{x}{y}$ is unbounded.

These results correspond to the fact that the solution of the differential equation in each case is $x^2 + y^2 = 1$, and in the first case the initial conditions imply that the unique solution in this case is $y = (1 - x^2)^{1/2}$, whereas there is no single-valued solution to the differential equation in any neighbourhood of the point $(1, 0)$. This also brings up the important point that for Cauchy a solution to a differential equation may be given only implicitly, in the form $f(x, y) = 0$, and not explicitly (as $y = y(x)$).

Now consider the first-order ordinary differential equation

$$\frac{dy}{dx} = \frac{1}{x + y}.$$

It looks as if something could go "wrong" when $x + y = 0$; let us see what happens.

We can solve this equation explicitly by writing

$$x + y = z, \quad \text{so } dx + dy = dz,$$

when the equation becomes

$$\frac{dz}{dx} - 1 = \frac{1}{z}$$

or

$$\frac{dz}{dx} = \frac{1 + z}{z}.$$

We separate the variables and it becomes

$$dx = \frac{1 + z - 1}{1 + z} dz = dz - \frac{dz}{1 + z},$$

so

$$x = z - \log a(1 + z),$$

where a is an arbitrary positive constant. This equation, in the original variables, says

$$x = x + y - \log a(1 + x + y),$$

or

$$y = \log a(1 + x + y),$$

Fig. 12.2 The graph of
$e^y = 2(1 + x + y)$

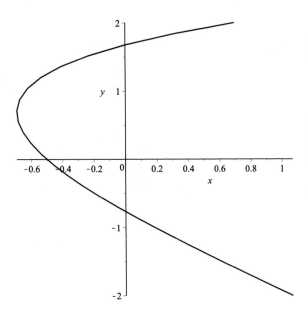

or

$$e^y = a(1 + x + y).$$

This says that when $x + y = 0$ that $y = \log a$, which, unexpectedly, is a constant. What has happened is illustrated in Fig. 12.2, where $a = 2$. It is clear that y cannot be a single-valued function of x in a neighbourhood of the point where $-x = y = \log 2$.

The figure confirms what the "infinite" value of $\dfrac{dy}{dx}$ already said: the implicitly defined function that is the solution of the differential equation cannot be written as a function y of x in a neighbourhood of the "bad" point.

It might seem that Cauchy paid a high price for rigour. Whereas earlier mathematicians offered general formulae, he could only provide solutions in a neighbourhood of an initial point where the differential equation was well behaved. This is, of course, entirely of a piece with his theory of functions, in which, for example, power series may converge only for a limited range of the variable.

These lectures by Cauchy provide the first proof of the existence of solutions (locally, at least) to a first-order differential equation. The historian of mathematics has a particular interest in them because they illuminate the frailty of our knowledge of even the recent past. Unlike much of Cauchy's work, they are not contained in the 31 volumes of his *Oeuvres complètes*. They are not even listed by the editors of those works as being among the texts omitted from the collected works. It was known that they had existed only because Cauchy mentioned them in a resumé of 1835, and because his friend the Abbé Moigno gave an account of them in his book of 1841. Only in 1979 did Gilain track down a printed copy of the first 13 lectures in the archives of the Institut de France.

The disappearance of these lectures is hard to explain, because Cauchy was an energetic publisher and republisher of his own results. A crucial factor is likely to have been Cauchy's discovery of a different proof in 1835, one he did publish (twice) and which did catch on.

By then, Cauchy was in Prague, officially as a tutor to the Bourbon Dauphin (who, unsurprisingly, had no interest in learning what Cauchy had to say). In the introduction to his Prague paper, Cauchy stated that the integration of differential equations by series was "illusory, so long as one did not provide any means of assuring that the series so obtained were convergent".[1] This emphasis on series solutions is a sign of an important departure from this earlier presentation, although he did claim that his new method shared the advantages of the earlier one. But now he worked from the start with systems of first-order ordinary differential equations satisfying given initial conditions. Following what he called Hamilton's "wonderful paper" of 1834 on the differential equations of dynamics (see Sect. 24.3), he reduced the system of ordinary differential equations to a single first-order linear partial differential equation. He then showed how to find enough particular solutions of the partial differential equation to find the general solution of the original ordinary differential equation. He then used his rigorous methods of analysis to show that the particular integrals could be expanded in convergent power series.

What is curious about this paper is that the theoretical passages treat the independent variable x as a *complex* variable, but in the examples he gave to illustrate the theory the variable x is regarded as *real*. In their comments on this, the historians Bottazzini and Gray remark that[2]:

> Apparently, even as late as this Cauchy did not seem to recognize the deep difference between the real and the complex case of his existence theorems, and the ambiguity between the real and complex ran through his entire paper, so much so that in his concluding remark he proudly stated that his new theorems could "easily" be extended to the solution of differential equations in which the variables and functions involved become imaginary. Precisely for this reason he preferred his second existence theorem based on the calculus of limits, which had transformed the integration of differential equations into a rigorous theory.

So in this paper, and indeed in a number of papers he went on to write that drew on this paper, the condition on the functions that enter the differential equation is that they are complex analytic. This was to turn out to be a very much stricter condition than being infinitely differentiable, and very much stronger than the modest conditions on differential equations of a real variable that Cauchy had assumed in 1821. The only reasons Cauchy can have preferred his Prague paper to his earlier account are that either he did not appreciate the difference in the conditions, or he thought that the more interesting case was functions of a complex variable anyway. The former reason is plausible for the period. The latter one is also plausible and it fits to the intermittent but deep interest Cauchy had in establishing a rigorous theory of functions of a complex variable.

---

[1] See Cauchy ([35], 400).
[2] See Bottazzini and Gray ([22], 162).

12.2.1   Later Developments

Cauchy's early emphasis on the local theory of real ordinary differential equations
was taken up and polished by Rudolf Lipschitz and Émile Picard to give a sound
account, using different methods, of the existence of solutions to ordinary differential
equations. For a brief look of the creation of the modern theory of real ordinary
differential equations, see Appendix G. Picard's account is the final part of his study
[211] of partial differential equations, which we shall look at in Chap. 28.

12.3   Exercises

1. Show that Cauchy's theorems 7 and 8 do not apply to the differential equation
   $y' = y^{1/2}$. What happens in this case?
2. Show that Cauchy's theorems 7 and 8 do not apply to the differential equation
   $y' = \frac{y}{x}$. What happens in this case?

Questions

1. Cauchy gave these lectures around the time he gave the famous courses in which
   his new approach to analysis was put on record for the first time. Which of the
   ideas introduced in theorems 7 and 8 do you think need further investigation
   before they could be said to be rigorous?
2. Ask yourself this question again when you have seen Cauchy's lectures from 1819
   on first-order partial differential equations in Chap. 17.

Chapter 13
Riemann and Complex Function Theory

13.1 Introduction

Complex function theory was relatively new in the mid-nineteenth century.[1] After decades of intermittent interest, Cauchy began to draw his insights together in the late 1840s so that a younger generation could appreciate them, and more-or-less independently Riemann (Fig. 13.1) began to develop his theory, starting in 1851. This chapter looks specifically at Riemann's approach to the subject, which greatly transformed it through a balance of intuitive and often geometrical ideas and a profound connection to the theory of harmonic functions, although at a cost in lack of rigour that some were to feel was too high to pay.

13.2 Complex Function Theory

In this section, some of the more important results will be indicated. The more intuitive ones will be explained; others, which require a more careful explanation, are standard in any book on complex function theory, and were given various accounts in the nineteenth century.

Riemann had begun his [234] with an insight that it had taken Cauchy years to get clear: a function f of two real variables x and y that is a complex-valued function of a single complex variable $z = x + iy$ is best understood (and had always been informally understood) to be a function on a domain in the complex plane on which it is complex differentiable. That is, if and only if the limit of the quotient

$$\frac{f(z + dz) - f(z)}{dz}$$

---

[1] See Bottazzini and Gray [22] for a comprehensive history.

© The Author(s), under exclusive license to Springer Nature Switzerland AG 2021 J. Gray, *Change and Variations*, Springer Undergraduate Mathematics Series, https://doi.org/10.1007/978-3-030-70575-6_13

exists as dz tends to zero and does not depend on the direction of dz—or, to put that point another way, if it does not depend on the components dx and dy.

By letting $dz = dx$ and then letting $dz = idy$ and equating the results, we obtain the partial differential equation

$$\frac{\partial f}{\partial x} + i\frac{\partial f}{\partial y} = 0,$$

from which, on writing $f(x, y) = u(x, y) + iv(x, y)$, the familiar Cauchy–Riemann equations are obtained:

$$u_x = v_y; \ u_y = -v_x.$$

They form a system of two coupled first-order partial differential equations, and it is interesting to note that although Riemann was willing to base his new theory of complex analytic functions on them Weierstrass later was not—an indication of Weierstrass's preference for power series, to be sure, but also an indication of how opinions stood on the subject of partial differential equations in the mid- to late nineteenth century.[2]

Riemann appreciated, but Cauchy did not, that these equations imply that at points where the derivative of f does not vanish f are conformal (angle preserving), because the differential can then be written as

$$\begin{pmatrix} u_x & u_y \\ v_x & v_y \end{pmatrix} = \begin{pmatrix} u_x & u_y \\ -u_y & u_x \end{pmatrix} = g(z) \begin{pmatrix} \cos t & -\sin t \\ \sin t & \cos t \end{pmatrix},$$

for a suitable function $g(z)$ and a parameter t.

---

[2]For more on this, see Bottazzini and Gray ([22], Chap. 6). The connection with Dirichlet's principle is explored in Sect. 19.2.

Riemann also noted that if you differentiate these equations you obtain the equations

$$u_{xx} + u_{yy} = 0, \quad v_{xx} + v_{yy} = 0,$$

so u and v, the real and imaginary parts of f, are harmonic functions. He might well have learned of the conformal nature of complex analytic functions from Gauss, who had observed this fact in 1822, and he surely learned to appreciate harmonic functions from his mentor Dirichlet, so much so that much of Riemann's theory of analytic functions is derived from a study of harmonic functions.

It follows from the work of Cauchy that a function that is complex differentiable is infinitely differentiable and can be written as a power series in a neighbourhood of a point where it takes a finite value.[3] Riemann knew this, but generally preferred to treat complex functions geometrically.

13.3 The Riemann Mapping Theorem

Riemann believed that the Dirichlet problem (see Sect. 18.3) had a solution. In other words, given a simply connected region T and a continuous function defined on the boundary of T, the function has a unique continuous extension to a harmonic function u defined on the interior of T. It followed that there was a unique complex differentiable function $u + iv$ once the conjugate function was specified at a point.[4]

He gave a proof of this claim that was inadequate, because the theoretical understanding of the Dirichlet problem was too poor, and the claim divided mathematicians for a generation. Some found it almost certainly true and worth using until a rigorous proof came along, and at the other extreme some found it beyond hope. (It was eventually established under very general conditions.)

Riemann used it as the basis of an argument for a much stronger and deeper result that became called the Riemann mapping theorem[5]:

Theorem 13.2 *The Riemann mapping theorem: Any two simply connected regions are not only topologically equivalent but analytically equivalent.*

(Riemann assumed that such regions are bounded by curves homeomorphic to circles, which need not be the case, but topology was being pulled into existence by this and other papers of his.) His proof that any two such regions, with boundaries as described, are in fact topologically equivalent was already novel. That they are analytically equivalent means that there is essentially only one simply connected domain for the purposes of complex function theory, and that can be taken as the unit disc.

---

[3] For the complicated history of this result, known today as Cauchy's integral theorem, see Bottazzini and Gray [22].

[4] See Riemann ([234], Sect. 19).

[5] Riemann only had in mind domains whose boundaries are topological circles; questions about boundaries were only investigated much later.

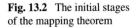 **Fig. 13.2** The initial stages
of the mapping theorem

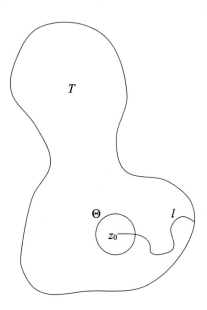

This was immediately seen to be a powerful result, but Riemann's argument for it (in Sect. 21) was naive, and attracted repeated attempts to prove it, some of which we shall meet below.

His argument, somewhat over-simplified, went as follows (see Fig. 13.2). Pick a point z_0 in the interior of T, and let Θ be a small disc centre z_0 that lies entirely in T. Consider the function $\log(z - z_0)$ on Θ cut along a radius ℓ, so that the branch of the log function jumps by $-2\pi i$ as it crosses ℓ clockwise. Extend ℓ to a curve (also called ℓ) that does not cross itself and reaches to the boundary of T. Extend the function to a continuous complex function $f(z)$ on the whole of T that is purely imaginary on the boundary of T, and likewise jumps by $-2\pi i$ as it crosses ℓ. Therefore, the imaginary part of this function goes from 0 to $2\pi i$ as z completes a circuit of the boundary of T clockwise. Show that the integral of $f(z)$ around the boundary of Θ is zero, and over all of T is finite. Deduce that it is possible to define what is called a Green's function (see Sect. 18.2): a function $g(z)$ that is infinite at z_0 and has zero real part on the boundary of T.

On ℓ the value of the real part of $h(z) = f(z) + g(z)$ goes from $-\infty$ at z_0 to 0 where ℓ meets the boundary of T. Let $-\infty < a < 0$, and look at $C_a = \{z \in T \mid Re(h(z)) = a\}$ (see Fig. 13.3). Riemann claimed that this can only be a single loop that does not cross itself, because T is simply connected. So T is filled out by these loops, which are the level curves of the real part of $h(z)$, and they do not intersect each other. The function $e^{h(z)}$ maps the boundary of T, which is the outermost of these loops, onto the unit circle, and all the other loops onto concentric circles; the point z_0 is mapped to the centre of the circle.

Therefore, T is analytically equivalent to the unit disc, and therefore any two simply connected regions are analytically equivalent.

Fig. 13.3 An indication of
the final stages of the
mapping theorem

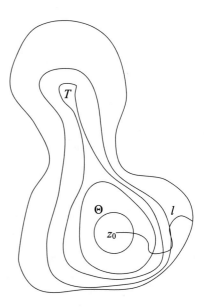

13.4   A Look Ahead

Chapter 14 discusses how Riemann tackled the hypergeometric equation. His analysis focussed on the fact that the coefficients of the equation are infinite at precisely three points: $z = 0, 1, \infty$, and therefore very likely the solutions, which will otherwise be finite, will be infinite at those points. In fact, as Gauss had shown, the solutions take the form of a power of the variable z times a convergent power series, or a power of $1 - z$ times a convergent power series. So Riemann, who knew Gauss's papers [114, 115] very well, could see in advance that the solutions were likely to be branched at $z = 0, 1, \infty$ and would usually be what were called in the nineteenth century many-valued functions.

Riemann could also see, as Kummer had done, that Gauss was hinting that the transformations $z \to 1/z$ and $z \to 1 - z$ were particularly relevant, and this is connected to what are called Möbius transformations, which have the special property of mapping triangles whose sides are arcs of circles to other circular-arc triangles in an angle-preserving way.

It was also becoming clear to mathematicians by 1850 that although a power series is convergent only inside some circular disc (which might be the whole plane of complex numbers) the complex analytic function it defines might be defined on a much larger region (think of $1 + z + z^2 + \cdots$ and $(1 - z)^{-1}$). For Weierstrass, in particular, this invited the question of how a complex function could be defined on a family of overlapping discs, and this led to a theory of analytic continuation.

Finally (for now), it was discovered that if a (non-constant) complex function is defined everywhere including ∞ then it would have to take the value ∞ somewhere (this can be handled with elementary limiting arguments). For example, the simplest functions on the complex plane with a point at infinity are of the form

$$z \to \frac{az + b}{cz + d},$$

where a, b, c, d are constants and $ad - bc \neq 0$ (the Möbius transformations). They take the value $1/c$ when $z = \infty$ and the value ∞ when $z = -d/c$. Discovery of this fact about complex functions is disputed between Hermite and Liouville, to whom it more properly belongs, and Cauchy[6] It was also known to Riemann, who may perhaps have discovered it for himself.

For convenience, a brief account of each of these topics will be found in Appendix E.

The results we need about Möbius transformations are simple, and concern their effect on straight lines and circles. As shown in Appendix F, a Möbius transformation maps a straight line either to a straight line or a circle, and it also maps a circle to either a straight line or a circle. Informally in this context mathematicians used to think of a straight line as a circle passing through ∞. With that convention in place, we can say that a Möbius transformation which maps a circle through three given points to three other given points maps the circle through the first triple of points to the circle through the second triple.

13.5   Exercises

1. Describe the image of the upper half-plane under the two-valued "map" $z \mapsto z^{1/2}$. Extend your account to describe the image of the whole plane.
2. Find all the Möbius transformations mapping the upper half-plane to itself.
3. Find a Möbius transformation mapping the upper half-plane to the unit disc centred on the origin, and interpret it as an inversion.
4. Hence, find all the Möbius transformations mapping the disc to itself.

Questions

1. One of the great divides in mathematics is between a real-valued function of a real variable being differentiable, and a complex-valued function of a complex variable being complex differentiable. Operationally, it is hidden behind the formalism, which looks like nothing more than switching x with z. For as long as it seemed natural to mathematicians to believe that functions were differentiable, even infinitely differentiable, this distinction was even harder to see. What signs have you seen of it already, whether appreciated or overlooked by the mathematicians you are considering?

---

[6]See Lützen ([192], Chap. XIII).

Chapter 14
Riemann and the Hypergeometric Equation

14.1 Introduction

Kummer's 24 solutions to the hypergeometric equation in the form of a hypergeo-metric series in x, $1 - x$, $\frac{1}{x}$, $1 - \frac{1}{x}$, $\frac{x}{x-1}$, or $\frac{1}{1-x}$, possibly multiplied by some powers of x and/or $(1 - x)$, presents each solution in a form that restricted it to a certain domain and then gave the relationship between overlapping solutions. In his [235], Riemann proposed to apply his new, geometric methods to study the hypergeometric equation as an equation for functions of a complex variable, noting correctly but dramatically that these methods were essentially applicable to all linear differential equations with algebraic coefficients.

   Like Gauss, Riemann took the independent variable to be complex, and a possibly unexpected result of doing this is that what look like two independent complex solutions of the hypergeometric equation generally appear as two branches of the same many-valued function. Indeed, a typical way that branched or many-valued functions arise is in the solution of ordinary differential equations.

   The importance of Riemann's work is both general and specific: general in that it opened up the theory of ordinary differential equations to complex functions of a complex variable, and specific in that it showed that equations like the hypergeometric equation can be recovered completely from a knowledge of the behaviour of their solutions under analytic continuation.

14.1.1 Ordinary Differential Equations and Many-Valued Functions

First, we recall some elementary facts about the solution of the hypergeometric equation. We shall suppose that in the hypergeometric equation

© The Author(s), under exclusive license to Springer Nature Switzerland AG 2021
J. Gray, *Change and Variations*, Springer Undergraduate Mathematics Series,
https://doi.org/10.1007/978-3-030-70575-6_14

$$x(1-x)\frac{d^2w}{dx^2} + (\gamma - (\alpha + \beta + 1)x)\frac{dw}{dx} - \alpha\beta w = 0$$

the variable x is complex. The standard method for solving this equation, and any linear ordinary differential equation, in a neighbourhood of the origin is to substitute a power series of the form

$$x^\lambda (a_0 + a_1 x + a_2 x^2 + \cdots)$$

and to look at the coefficient of the lowest power of x. This will be an equation for λ, and in this case the equation is

$$\lambda(\lambda - 1 + \gamma) = 0,$$

so

$$\lambda = 0 \text{ or } 1 - \gamma.$$

The recurrence relation for the coefficients then gives solutions in the form

$$y = F(\alpha, \beta, \gamma, x) \text{ and } x^{1-\gamma} F(\alpha - \gamma + 1, \beta - \gamma + 1, 2 - \gamma, x).$$

These solutions form a basis of solutions for the differential equation in a neighbourhood of the origin. For use below let us call these solutions y_{01} and y_{02}.

14.1.1.1   Result 1

We shall now show that, certain exceptional cases aside, the solutions of the hypergeometric equation in the complex case are all branches of the same complex function.

By what we said earlier, if the variable x goes on a small circle around the point $x = 0$ the solution y_{01} returns to its original value, and the solution y_{02} returns multiplied by $e^{2\pi i(1-\gamma)}$. This does not allow us to infer that the two solutions are branches of the same function, so we must look further afield. For convenience, and before we proceed, we write this information as a matrix equation (writing $\tilde{y}$ for the value of y at $e^{2\pi i z}$)

$$\begin{pmatrix} \tilde{y}_{01} \\ \tilde{y}_{02} \end{pmatrix} = \begin{pmatrix} 1 & 0 \\ 0 & e^{2\pi i(1-\gamma)} \end{pmatrix} \begin{pmatrix} y_{01} \\ y_{02} \end{pmatrix}. \tag{14.1}$$

For brevity, we write this as

$$\tilde{\mathbf{y}}_0 = D_0 \mathbf{y}_0.$$

Had we looked for solutions in a neighbourhood of $x = 1$ the same method would have produced solutions that are power series in $1 - x$ possibly multiplied by a power of $1 - x$, and indeed we would have found that a basis of solutions in this neighbourhood is given by

$F(\alpha, \beta, \alpha + \beta - \gamma + 1, 1 - x)$ and $(1 - x)^{\gamma-\alpha-\beta} F(\gamma - \alpha, \gamma - \beta, \gamma - \alpha - \beta + 1, 1 - x)$.

We shall call these solutions y_{11} and y_{12}.

The radius of convergence of all the four power series we have written down is 1, and this means that in a neighbourhood of $x = \frac{1}{2}$, say, we have four solutions, and therefore two are linearly expressible in terms of the other two. Let us say that

$$y_{01} = a_{11} y_{11} + a_{12} y_{12}$$

$$y_{02} = a_{21} y_{11} + a_{22} y_{12},$$

where the coefficients a_{jk} are constants.

We can write this information in matrix form as

$$\begin{pmatrix} y_{01} \\ y_{02} \end{pmatrix} = \begin{pmatrix} a_{11} & a_{12} \\ a_{21} & a_{22} \end{pmatrix} \begin{pmatrix} y_{11} \\ y_{12} \end{pmatrix}. \tag{14.2}$$

For brevity, we write this as

$$\mathbf{y}_0 = A \mathbf{y}_1.$$

We now propose to look at what happens to the solutions y_{01} and y_{02} as x is taken on a small circle around the point $x = 1$. We can do this by watching what happens as y_{11} and y_{12} undergo the same journey. We know what they do individually, by analogy with Eq. 14.1:

$$\begin{pmatrix} \tilde{y}_{11} \\ \tilde{y}_{12} \end{pmatrix} = \begin{pmatrix} 1 & 0 \\ 0 & e^{2\pi i(\gamma-\alpha-\beta)} \end{pmatrix} \begin{pmatrix} y_{11} \\ y_{12} \end{pmatrix}. \tag{14.3}$$

Again, for brevity, we write this as

$$\tilde{\mathbf{y}}_1 = D_1 \mathbf{y}_1.$$

So conducting x around the point $x = 1$ returns y_{01} and y_{02} as

$$\begin{pmatrix} 1 & 0 \\ 0 & e^{2\pi i(\gamma-\alpha-\beta)} \end{pmatrix} \begin{pmatrix} a_{11} & a_{12} \\ a_{21} & a_{22} \end{pmatrix} \begin{pmatrix} y_{01} \\ y_{02} \end{pmatrix}, \tag{14.4}$$

which is

$$D_1 A \mathbf{y}_0.$$

But this expresses them in terms of y_{11} and y_{12}. To express the answer in terms of y_{01} and y_{02}, we must use the inverse of the transformation in Eq. (14.2). This inverse is given by the matrix

$$A^{-1} = \frac{1}{a_{11}a_{22} - a_{12}a_{21}} \begin{pmatrix} a_{22} & -a_{12} \\ -a_{21} & a_{11} \end{pmatrix}$$

and so the final result is that

$$\tilde{\mathbf{y}}_0 = A^{-1} D_1 A \mathbf{y}_0.$$

Written out in full, this says

$$\begin{pmatrix} \tilde{y}_{01} \\ \tilde{y}_{02} \end{pmatrix} = \frac{1}{a_{11} a_{22} - a_{12} a_{21}} \begin{pmatrix} a_{22} & -a_{12} \\ -a_{21} & a_{11} \end{pmatrix} \begin{pmatrix} 1 & 0 \\ 0 & e^{2\pi i (\gamma - \alpha - \beta)} \end{pmatrix} \begin{pmatrix} a_{11} & a_{12} \\ a_{21} & a_{22} \end{pmatrix} \begin{pmatrix} y_{01} \\ y_{02} \end{pmatrix}.$$
$$(14.5)$$

To conclude that y_{01} and y_{02} are branches of the same function, it is enough to show that the matrix

$$\begin{pmatrix} a_{22} & -a_{12} \\ -a_{21} & a_{11} \end{pmatrix} \begin{pmatrix} 1 & 0 \\ 0 & e^{2\pi i (\gamma - \alpha - \beta)} \end{pmatrix} \begin{pmatrix} a_{11} & a_{12} \\ a_{21} & a_{22} \end{pmatrix}$$

is not diagonal, and a routine calculation shows that it will be diagonal if and only if

$$e^{2\pi i (\gamma - \alpha - \beta)} = 1,$$

that is, if and only if $\gamma - \alpha - \beta$ is an integer.

A very similar calculation involving a small circuit around $x = \infty$ can be carried out. In this case, the basis of solutions valid near $x = \infty$ that we choose is

$$y_{\infty 1} = x^{-\alpha} F(\alpha, \alpha - \gamma + 1, \alpha - \beta + 1, 1/x), \quad y_{\infty 2} = x^{-\beta} F(\beta, \beta - \gamma + 1, \beta - \alpha + 1, 1/x).$$

Under analytic continuation around $x = \infty$ these solutions return as

$$\tilde{y}_{\infty 1} = e^{-2\pi i \alpha} y_{\infty 1} \text{ and } \tilde{y}_{\infty 2} = e^{-2\pi i \beta} y_{\infty 1},$$

which we can write as

$$\begin{pmatrix} \tilde{y}_{\infty 1} \\ \tilde{y}_{\infty 2} \end{pmatrix} = \begin{pmatrix} e^{-2\pi i \alpha} & 0 \\ 0 & e^{-2\pi i \beta} \end{pmatrix} \begin{pmatrix} y_{01} \\ y_{02} \end{pmatrix}. \qquad (14.6)$$

So if we repeat the above calculation but extending a basis of solutions valid near $x = \infty$ analytically around the point $z = 1$ we find that the corresponding matrix is diagonal, and the two solutions are not branches of the same function, if and only if $\alpha = \beta$.

Likewise, if we extend a basis of solutions near $x = 1$ we find that the members of the basis are branches of the same function unless $\gamma = 1$.

We conclude that, these special cases aside, the members of a basis of solutions to the hypergeometric equation turn out, under analytic continuation, to be branches of the same function.

14.1.2 The Riemann Sphere

One of Riemann's simple but productive innovations was to work with the plane of complex numbers augmented by a point at infinity. This gave him a sphere, and he regarded the connection between the plane and the sphere as being given by stereographic projection. This is a map from the sphere to a tangent plane at a point S (which we shall call the South Pole) that is defined as follows. Let N (the North Pole) be the point on the sphere diametrically opposite to S, and let P be any point on the sphere other than N. The map from the sphere to the plane maps P to P', the point where the line NP meets the plane (Fig. 14.1).

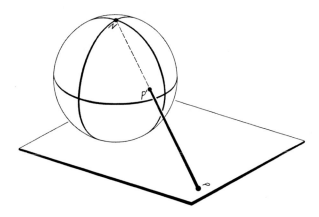

Fig. 14.1 Stereographic projection of the Riemann sphere onto the plane, *Anschauliche Geometrie*, 2nd edition, Springer, 1996

14.1.2.1 Result 2

We shall now see that analytically continuing a solution in a loop that winds once around all three branch points returns the original solution.

Suppose that we have a function with exactly three branch points, the points $z = a, b, c$ or, if you prefer, the points $z = 0, 1, \infty$. This means that as the function is continued analytically around the point $z = a$ it returns multiplied by a factor $e^{2\pi i \alpha}$, as it is continued analytically around the point $z = b$ it returns multiplied by a factor $e^{2\pi i \beta}$, and as it is continued analytically around the point $z = c$ it returns multiplied by a factor $e^{2\pi i \gamma}$.

Suppose it is continued analytically around the point $z = a$ and then around the point $z = b$. It now returns multiplied by a factor $e^{2\pi i \alpha} e^{2\pi i \beta} = e^{2\pi i (\alpha + \beta)}$. By the deformation principle, you can think of this path as consisting of a loop starting at a point P (other than one of the branch points) that goes around the point $z = a$, returns to P, and then goes around the point $z = b$ and returns to P, or as a loop that goes both a and b.

If it is continued analytically around the point $z = a$, then around the point $z = b$, and then around the point $z = c$ it returns multiplied by a factor

$$e^{2\pi i \alpha} e^{2\pi i \beta} e^{2\pi i \gamma} = e^{2\pi i (\alpha + \beta + \gamma)}.$$

But now something more can be said. For simplicity, suppose that we take a, b, and c to be points on the Riemann sphere.

The following argument is clearest if we suppose that the three points are near the North Pole, and the path around them lies entirely south of them. Now we can imagine that the path is gradually deformed by moving it south until it lies arbitrarily close to the South Pole. It is now clear that continuing the function along this path returns it unchanged, and by the deformation principle stated above this means that the function returned unchanged along its original path. Therefore, $\alpha + \beta + \gamma$ must be an integer.

14.2   Riemann's P-Functions

Riemann began his [235] by specifying geometrically the functions he intended to study. Any such function P is to satisfy the following three properties: [1]

1. It has three distinct branch points at a, b, and c, but each branch is finite at all other points[2]
2. A linear relation with constant coefficients exists between any three branches P', P'', P''' of the function $c' P' + c'' P'' + c''' P'''$.
3. There are constants α and α', called the exponents, associated with the branch point a, such that P can be written as a linear combination of two branches P^{α} and $P^{\alpha'}$ near a, $(z - a)^{-\alpha} P^{\alpha}$ and $(z - a)^{-\alpha'} P^{\alpha'}$ are single-valued, and neither zero nor infinite at a. Similar conditions hold at b and c with constants β, β' and γ, γ', respectively.

To eliminate troublesome special cases, Riemann further assumed that none of α, α', β, β', γ or γ' are integers, and that the sum $\alpha + \alpha' + \beta + \beta' + \gamma + \gamma' = 1$. He denoted such a function of z

$$P \begin{pmatrix} a & b & c \\ \alpha & \beta & \gamma \ z \\ \alpha' & \beta' & \gamma' \end{pmatrix} \text{ or something like } P \begin{pmatrix} \alpha & \beta & \gamma \\ \alpha' & \beta' & \gamma' \end{pmatrix} (z)$$

when $(a, b, c) = (0, \infty, 1)$.

---

[1] Riemann's notation comes from his reading of Gauss's still unpublished paper, which Riemann explained in his note (1857b) he had read in Gauss's *Nachlass*. Gauss had died in 1855.

[2] This corrects a rare slip in the English translation of Riemann's work; Riemann said, in somewhat obscure words, that the P-function is locally single-valued except at the point a, b, c.

In terms of the singular points at a, b, and c the first and third conditions say, for example, that $P^{(\alpha)}$ is branched like $(x - a)^\alpha$. The second condition says that there are at most two linearly independent determinations of the function under analytic continuation of the various separate branches. One of the results Riemann established is that, as with differential equations, this information specifies a *P*-function up to a constant multiple.

The analogy between *P*-functions and hypergeometric functions becomes clear if we take $a = 0$, $b = 1$, and $c = \infty$. There are two linearly independent solutions of the hypergeometric equation at each singular point; they are branched according to certain expressions in α, β, and γ; and any three solutions are linearly dependent. Riemann showed that information of this kind about the solutions determines the differential equation completely, which goes some way to explain the great significance of the equation, as well as its special character.

It will be enough to record the results that he achieved. A more detailed look at how Riemann arrived at them is given in the next section of this chapter.

Result 3: If P_1 is another *P*-function branched at the same three points a, b, c and with the same six exponents $\alpha, \alpha', \beta, \beta', \gamma, \gamma'$ then P_1 is a constant multiple of P.

Result 4 If P and P_1 are two *P*-functions with the same branch points a, b, c and exponents the corresponding pairs of which differ by integers, then P_1 is obtained from P by multiplying by powers of $z - a$, $z - b$, and $z - c$, the precise powers being determined by the differences in the exponents.

Result 5 A *P*-function satisfies a differential equation and when $a, b, c = 0, 1, \infty$ and the exponents of the *P*-function are suitably chosen the differential equation is precisely the hypergeometric equation.

The conclusion is that, just liken algebraic function, a *P*-function is specified up to a constant multiple by the information about its branching.

The most immediate and important difference between Riemann's approach and that of his predecessors is the relative lack of computation. As he remarked, his new method "allows results that were formerly obtained in part only after somewhat troublesome calculations to be derived almost immediately from the definition".[3] Rather than starting from a hypergeometric series he began with a *P*-function having ∞^2 branches and three branch points. To be sure, any two linearly independent branches have expansions as hypergeometric series, and any three branches are linearly dependent, but the argument employed by Riemann inverted that of Gauss and Kummer. His starting point was the set of solutions, functions which are shown to satisfy a certain type of equation. Their starting point was the equation from which a range of solutions are derived. Riemann showed that a very small amount of information, the six exponents at the branch points, entirely characterises the equation and defines the behaviour of the solutions. The hypergeometric equation is special in this respect,

---

[3] See Riemann ([235], 3).

as Fuchs [110] was able to explain, and consequently the task of generalising the theory to cope with other differential equations was to be quite difficult.

14.3   Riemann's Arguments

The analytic continuation of a P-function is determined by what happens at the branch points, because any closed path can be written as a product of loops around $a, b,$ and c. When two linearly independent branches P' and P'', say, are continued analytically in a loop around the branch point a in the positive (anti-clockwise) direction they return as two other branches, $\tilde{P}'$ and $\tilde{P}''$, say. But then, by the second defining property of a P-function,

$$\tilde{P}' = a_1 P' + a_2 P''$$

$$\tilde{P}'' = a_3 P' + a_4 P''$$

for some constants $a_1, a_2, a_3, a_4,$ so the matrix $A = \begin{pmatrix} a_1 & a_2 \\ a_3 & a_4 \end{pmatrix}$ describes what happens at the point a. This is very like what was described in Result 1.

Let B and C be the matrices which describe the behaviour of P' and P'' under analytic continuation around b and c, respectively. Then, as Riemann noted (in line with Result 2), a circuit of a and b can be regarded as a circuit of c in the opposite direction, so

$$CBA = \begin{pmatrix} 1 & 0 \\ 0 & 1 \end{pmatrix}.$$

So, as Riemann remarked "the coefficients of A, B and C completely determine the periodicity of the function".

Now, for definiteness, Riemann supposed $a=0, b = \infty, c = 1,$ and chose branches $P^{(\alpha)}, P^{(\alpha')}, P^{(\beta)},$ etc. as in (3). A circuit around a in the positive direction returns $P^{(\alpha)}$ as $e^{2\pi i \alpha} P^{(\alpha)}$ and $P^{(\alpha')}$ as $e^{2\pi i \alpha'} P^{(\alpha')},$ so

$$A = \begin{pmatrix} e^{2\pi i \alpha} & 0 \\ 0 & e^{2\pi i \alpha'} \end{pmatrix}.$$

To express the effect on $P^{(\alpha)}$ and $P^{(\alpha')}$ of a circuit around $b = \infty,$ he replaced them by their expressions in terms of $P^{(\beta)}$ and $P^{(\beta')},$ conducted the new expressions around $\infty,$ and then changed them back into $P^{(\alpha)}$ and $P^{(\alpha')}$ by writing

$$\begin{pmatrix} P^{(\alpha)} \\ P^{(\alpha')} \end{pmatrix} = B' \begin{pmatrix} P^{(\beta)} \\ P^{(\beta')} \end{pmatrix}.$$

Then

$$B = B' \begin{pmatrix} e^{2\pi i \alpha} & 0 \\ 0 & e^{2\pi i \alpha'} \end{pmatrix} B'^{-1},$$

with a similar expression for what happens near c.

Since $CBA = I$, it follows on taking determinants that

$$\det(A)\det(B)\det(C) = 1 = e^{2\pi i(\alpha + \alpha' + \beta + \beta' + \gamma + \gamma')}, \tag{14.7}$$

which is why Riemann had assumed $\alpha + \alpha' + \beta + \beta' + \gamma + \gamma' = 1$.

Equation (14.7) is an equation in 2×2 matrices, so it yields four equations for the eight entries in the two matrices B' and C' in terms of the six parameters of the P-function $\alpha, \alpha', \dots, \gamma'$.

Riemann wrote the equations out explicitly, and showed that four of the eight entries determine the other four. Indeed, he did better: in July 1856 he wrote down how to express the entries in the matrices B' and C' in terms of the six coefficients $\alpha, \alpha', \beta, \beta', \gamma, \gamma'$, but in the published paper he merely gave the various ratios $\frac{\alpha_\gamma}{\alpha'_\gamma}$, etc.

These were enough to prove the next result, Result 3:

Result 3

If P_1 is another P-function branched at the same three points a, b, c and with the same six exponents $\alpha, \alpha', \beta, \beta', \gamma, \gamma'$ then P_1 is a constant multiple of P.

This makes sense because, on the one hand, the branches of P and P_1 near $z = a$ behave in the same way, and, on the other hand, the analytic continuation of the branches of P and P_1 around the other singular points is given by the same matrices, so they should behave in exactly the same way everywhere.

More precisely, Riemann first showed that the ratio P_1^α / P^α is constant in a neighbourhood of $z = a$ and then because the exponents are the same the analytic continuation of the branch P_1^α is the same as that of P and so P_1 is a constant multiple of P. This is where he made tacit use of Liouville's theorem (see Sect. E.4).

As Riemann then remarked, a very similar argument deals with two P-functions with the same branch points a, b, c and exponents the corresponding pairs of which differ by integers.[4] Now, although the analytic continuation is the same for the two functions, the quotient P_1^α / P^α is not constant in a neighbourhood of $z = a$, and instead $(z-a)^\delta P_1^\alpha / P^\alpha$ is constant in that neighbourhood for a suitable integer power δ. This gave him Result 4:

Result 4

If P and P_1 are two P-functions with the same branch points a, b, c and exponents the corresponding pairs of which differ by integers, then P_1 is obtained from P by multiplying by powers of $z - a$, $z - b$, and $z - c$, the precise powers being determined by the differences in the exponents.

---

[4] These are Riemann's versions of Gauss's contiguous functions.

Only now did Riemann satisfy himself that P-functions exist. He did this in Sect. 7, where he deduced the next result:

Result 5

A P-function satisfies a differential equation and when $a, b, c = 0, 1, \infty$ and the exponents of the P-function are suitably chosen the differential equationis precisely the hypergeometric equation.

More precisely, $P = y$, $P_1 = \dfrac{dy}{dx}$, and $P_2 = \dfrac{d^2y}{dx^2}$ are three such P-functions and, as Riemann showed, they satisfy a linear relationship with coefficients that are certain rational functions in x. Explicitly, in the case $\gamma = 0$, he found that P satisfies the hypergeometric equation in this form[5]:

$$(1 - z)\frac{d^2y}{d \log z^2} - (A + Bz)\frac{dy}{d \log z} + (A' - B'z)y = 0.$$

Accordingly, Riemann connected his P-functions with the functions $F(\alpha, \beta, \gamma, z)$ of Gauss:

$$F(\alpha, \beta, \gamma, z) = const\, P^{(\alpha)} \begin{pmatrix} 0 & a & 0 \\ 1 - c & b & c - a - b \end{pmatrix} (z)$$

So Riemann had shown that the branching data of the P-function determined its monodromy relations, i e. the group generated by the matrices A, B, C.[6] Furthermore, Riemann had established that the hypergeometric equation is the only second-order linear equation whose solutions satisfy the geometric conditions of his three postulates.

Riemann concluded by illuminating the relationship between P and F, its hypergeometric series representation. Since α and α' may be interchanged

$$P \begin{pmatrix} \alpha & \beta & \gamma \\ \alpha' & \beta & \gamma \end{pmatrix} (z) = P \begin{pmatrix} \alpha' & \beta & \gamma \\ \alpha & \beta & \gamma \end{pmatrix} (z)$$

---

[5]The logarithmic derivative $\frac{d}{d \log x}$ satisfies $\frac{d}{d \log x} = x\frac{d}{dx}$ and $\frac{d^2}{d \log x^2} = x\frac{d}{dx} + x^2\frac{d^2}{dx^2}$, as you can check by setting $u = \log x$ and using the fact that $\frac{df}{du} = \frac{df}{dx}\frac{dx}{du}$.

[6]Monodromy matrices were introduced by Hermite in response to Puiseux [229], a paper that had carefully examined the effect of analytic continuation on a branch of an algebraic function around one of its branch points with a view to elucidating the integration an algebraic function over a closed path containing a branch point; Cauchy reported on this work in Cauchy [38]. Riemann did not cite this work; as was the custom of the time, Riemann seldom gave references—except, in his case, to Gauss and Dirichlet—but he was well read all the same, and was undoubtedly one of those mathematicians who absorbed the work of others and then rederived it in his own way. The term monodromy group was first used by Jordan in his *Traité* ([154], 278) and its subsequent popularity derives from its successful use by Jordan and Klein in the 1870s.

there are eight P-functions for each hypergeometric series in z, say

$$P\begin{pmatrix} \alpha & \beta & \gamma \\ \alpha' & \beta & \gamma \end{pmatrix}(z) = z^{\alpha}(1-z)^{\gamma}F(\beta+\alpha+\gamma, \beta'+\alpha+\gamma, \alpha-\alpha'+1, z).$$

There are six choices of variable, so 48 representations of a function as a P-function.

14.4 Exercises

1. Look back at the table of Kummer's 24 solutions to the hypergeometric equation and determine their domain of convergences as functions of a complex variable.
2. Find values of α, β, and γ so that these solutions are only two- or three-, or five-valued functions.

Questions

1. The properties of Riemann's P-functions are obviously derived from the properties of solutions of the hypergeometric equation. In what sense are they the essential properties?

Chapter 15
Schwarz and the Complex Hypergeometric Equation

15.1 Introduction

A second-order linear ordinary differential equation has a basis of two solutions, and it was to turn out that their quotient has interesting properties. Indeed, the set-theoretic inverse function has still more interesting properties that bring out a strong family resemblance between the regular or Platonic solids, plane lattices, and tessellations of the non-Euclidean disc.

15.2 Quotients of Solutions

First, we write the hypergeometric equation, or indeed an arbitrary second-order linear ordinary differential equation, in the form

$$f'' + pf' + qf = 0, \tag{15.1}$$

where p and q are functions of z.

To investigate the behaviour of two linearly independent solutions u and v of Éq. (15.1), it turned out to be convenient to introduce their quotient $w = u/v$. Then

$$w' = \frac{u'v - uv'}{v^2}.$$

It is important to observe that

$$w' = 0 \Leftrightarrow u'v - uv' = 0 \Leftrightarrow \frac{u'}{u} = \frac{v'}{v} \Leftrightarrow u = kv,$$

© The Author(s), under exclusive license to Springer Nature Switzerland AG 2021
J. Gray, *Change and Variations*, Springer Undergraduate Mathematics Series,
https://doi.org/10.1007/978-3-030-70575-6_15

for some constant k. Because u and v are a basis of solutions, we infer that w' never vanishes and so w is locally one-to-one away, that is, from any singularities of u or v.

Now we restrict our attention to the hypergeometric equation, for which

$$p = \frac{\gamma - (\alpha + \beta + 1)z}{z(1 - z)}, \quad q = -\frac{\alpha\beta}{z(1 - z)}.$$

We are going to follow Schwarz and make some assumptions about α, β, and γ as the argument proceeds. We start by assuming that they are all real. This means that the solutions of the hypergeometric equation take real values when z is restricted to real values.

In 1872, Schwarz investigated the question of when the solutions to this differential equation are algebraic functions of z. That is to say, they are functions $w(z)$ such that there is polynomial equation $F(z, w) = 0$. (For example, in real variables, the function $f(x, y) = x^2 + y^2 - 1$ is an algebraic function of x and y.) This means that for every value of z there is a finite number of values of w. We have seen that the quotient can be made to take real values, so the images of the segments $(-\infty, 0)$, $(0, 1)$, and $(1, \infty)$ can be real.

Schwarz knew from Gauss's work and Riemann's that the solutions of the hypergeometric equation have three singular points, at $0, 1, \infty$, and that the solutions are of the form $x^a f_0(z)$ near $x = 0$, $(1 - x)^a f_1(1 - x)$ near $x = 1$, and $(1/x)^a f_\infty(1/x)$ near $x = \infty$, where the exponent a is one of the numbers in Kummer's table and the fs are analytic functions that do not vanish at the point in question.

When a function is algebraic there are only finitely many values of w for each value of z. The quotient of two solutions of the hypergeometric equation, like the individual solutions, is usually many valued, because it is of the form z^a times a holomorphic function. This immediately implies that the exponents a must be rational numbers, and this imposes conditions on the coefficients α, β, γ of the hypergeometric equation. Schwarz found that on setting

$$(1 - \gamma)^2 = \lambda^2, \ (\alpha - \beta)^2 = \mu^2, \ (\gamma - \alpha - \beta)^2 = \nu^2,$$

where λ, μ, and ν are real and positive, the image of the upper half-plane under a quotient of two solutions of the hypergeometric equation has angles of $\lambda\pi$, $\mu\pi$, and $\nu\pi$. After imposing some further restrictions on α, β, and γ, he could even insist that λ, μ, and ν be reciprocals of integers. We shall now follow him part of the way.

Schwarz observed that two linearly independent solutions of a hypergeometric equation that are each algebraic will have a quotient that is algebraic, because it can only have finitely many values at each point z.

The quotient is singular at the points $0, 1, \infty$ where the individual solutions are singular. Otherwise, because α, β, and γ are real, the hypergeometric equation has solutions that are real-valued on each of the segments $(-\infty, 0)$, $(0, 1)$, and $(1, \infty)$ of the real axis. The same is, therefore, true of the quotient: it can take real values on the real axis. But, because the quotient is many-valued, these segments have other images.

Before we find what they are, let us look at the effect of the quotient of a neighbourhood of the singular points. The upper half line can be considered as a triangle with vertices at $0, 1, \infty$, where the angles at each vertex are π. We can now see that the quotient w maps these angles to angles $\lambda\pi, \mu\pi, \nu\pi$.

Now we need to check some things:

- that each segment is mapped monotonically onto its image,
- that no two of the sides cross, and
- that the upper half-plane is mapped onto the interior of the triangle.

These properties hold because we are dealing with a second-order ordinary differential equation, and so we know that w' can only vanish at the singular points, and elsewhere is one to one. In particular, it is monotonic on the segments.

To find the images of the $(-\infty, 0)$, $(0, 1)$, and $(1, \infty)$, in general, we let a basis of solutions consist of two functions, say $u(z)$ and $v(z)$, and look at the quotient $w(z) = u(z)/v(z)$. Suppose the complex variable z is taken on a loop and returns to its starting point. The functions $u(z)$ and $v(z)$ return as new solutions, and therefore can be expressed as linear combinations of $a(z)$ and $v(z)$, say functions, say $au(z) + bv(z)$ and $cu(z) + dv(z)$, where a, b, c, and d are constants determined by the loop. So the quotient returns as

$$\frac{aw(z) + b}{cw(z) + d}.$$

So w has been subject to a Möbius transformation.

Now we use the fact that a Möbius transformation sends a straight line or a circle to a straight line or a circle (see Appendix F). So if the images of one of these segments can be a straight line segment, and any other image is the transform of that one by a Möbius transformation, then the other images are either straight lines or circles.

Consider now the segments $(-\infty, 0)$ and $(0, 1)$. They meet at an angle of π at $x = 0$, and locally the quotient is a complex function multiplied by a power of z, say z^λ. So these two segments are mapped to straight lines meeting at an angle of $\pi\lambda$. What happens at the other two meeting points is this: the argument about the angle is exactly the same, but the segment joining them is the images of the real axis not by w but some Möbius transformation of w, so it is a straight line or circular arc. As a result, the image of the whole upper plane is a straight-sided or circular-arc triangle with angles of $\pi\lambda$, $\pi\mu$, and $\pi\nu$.

Analytic continuation moves this region around by Möbius transformations, and the result is a net of circular-arc triangles, which are the images of the upper and lower half-planes by w as z is conducted around the plane, avoiding the branch points at $z = 0, 1, \infty$.

It is one thing to know that the image of the upper (or lower) half-plane by the quotient $w(z)$ is a circular-arc triangle, and another to know what happens as the variable z is led on one loop after another. We get a succession of images of the

Fig. 15.1 In every triangle
the angles are either $\pi/2$ or
$\pi/3$ or $\pi/6$, and cumulatively
they cover the plane

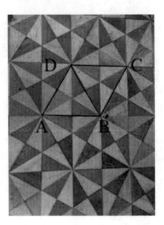

upper and lower half-planes and so a succession of circular-arc triangles. How do
they fit together? Do they fit together like pieces of a jigsaw puzzle, or do they overlap
any old how?

The crucial consideration is the angles at the vertices. If one of the angles is,
say $\pi/6$, then it is reasonable to suspect that 12 will fit together at the vertex, each
obtained from the one before by a turn through $\pi/6$, and this is indeed what happens.
If a vertical angle is π/n for some integer n then $2n$ fit together at the vertex, each
obtained from the one before by a turn through π/n. So if all three vertical angles
are of the form π/n for integers n then we expect that successive images form a web
of triangles. But if any vertical angle is not of that form then complicated overlaps
will occur, which is why Schwarz (and Poincaré after him) excluded those cases.

Suppose, for example, that the quotient we are looking at maps the upper half-
plane to a triangle with angles $\pi/2$, $\pi/3$, and $\pi/6$. These angles sum to π, so all three
sides of the image will be straight. Suppose for definiteness that

- the point $z = 0$ is mapped to the point A where the angle is $\pi/2$.
- the point $z = 1$ is mapped to the point B where the angle is $\pi/3$.
- the point $z = \infty$ is mapped to the point C where the angle is $\pi/6$.

Suppose that z goes on a path starting from $z = i$ and exiting the upper half-plane
between $z = 0$ and $z = 1$. Then the image of the lower half-plane will be a triangle
ABC' congruent to triangle ABC and attached along the edge AB.

Let z re-enter the upper half-plane between $z = 1$ and $z = \infty$. The new image of
the upper half-plane will be another triangle $BC'A'$ congruent to ABC and attached
to the previous one along the edge BC'.

The process continues in the fashion shown in Fig. 15.1.

Each light triangle is the image of the upper half-plane, and each dark triangle the
image of the lower half-plane.

The figure may equally well be regarded as a web of equilateral triangles (each
containing three light and three dark triangles, such as ABD and BCD), correspond-

ing to a different hypergeometric equation in which the angles at the vertices are all $\pi/3$.

It can also be seen as a web of parallelograms (made by joining two equilateral triangles together, such as $ABCD$). Now we have a figure with four vertices, so we are no longer dealing with the hypergeometric equation but with a different ordinary differential equation.

In each of these cases, the point $z = i$ has an image in the first, third, fifth, ..., triangles, and the point $z = -i$ has an image in the second, fourth, sixth, ..., triangles. We get some sort of a map from the complex plane to the web of triangles we have constructed, but it is not easy to say what is the image of $z = i$ because it seems to have infinitely many images. Suppose for a moment that they have been marked $P_1, P_3, P_5, \ldots$, each of them in the same position in the appropriate triangle.

What is much clearer is the set-theoretic inverse map, from the web of triangles or parallelograms back to the complex plane. This is reminiscent of the arcsin function. There are infinitely many angles whose sine is $\frac{1}{2}$, but the sine function treats them alike:

$$\sin(\pi/6) = \sin(\pi/6 + 2k\pi) = \frac{1}{2} = \sin(5\pi/6 + 2k\pi) \ldots$$

for any integer k.

For definiteness, let us return to the example where the upper and lower half-planes are each mapped to triangles with angles $\pi/2$, $\pi/3$, and $\pi/6$. In each light triangle is a point P_{2j+1}. The function that is the set-theoretic inverse of φ in this case maps them all to the point i. Similarly, φ has mapped any given point ζ in the upper half-plane to a point in each light triangle, and the inverse function maps all those points back to ζ.

In the parallelogram case, the simplest thing that can happen is that every point ζ in the upper half-plane has two images in each parallelogram.[1] The inverse function is now one that takes every value twice (this is again reminiscent of the sine function). But what is more interesting is that the points where it takes a given value form into two families. In each family, the points form a lattice; they are separated by integer multiples of the lengths AB and AD. If we represent AB by the complex number ω_1 and AD by the complex number ω_2, then it follows that the inverse function—let us call it F—satisfies these equations:

$$F(z) = F(z + \omega_1) = F(z + \omega_2)$$

for every z. By analogy with the sine function, the function F is said to be doubly periodic. It is also an elliptic function.[2]

---

[1] This is a consequence of the Cauchy integral theorem in complex function theory.

[2] The efforts to identify doubly periodic and elliptic functions involved many mathematicians from Gauss to Riemann and Weierstrass. The connection between elliptic integrals—integrals of the form $\int_{z_0}^{z} \frac{dt}{\sqrt{f(t)}}$ where $f(t)$ is a quartic in t—and elliptic functions was one of the great discoveries of Abel and Jacobi in the 1820s. See, for example, Botttazzini and Gray [22] and more briefly Gray [127].

Fig. 15.2 In every triangle
the angles are all $2\pi/5$, so the
triangles are all congruent, in
Nash ([203], 101), copyright
Elsevier (2014)

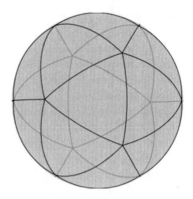

Now, Schwarz's problem was to arrange for there to be only finitely many image triangles, so the condition he was led to discover was that $\lambda + \mu + \nu > 1$. The only triangles that meet this condition are the so-called digons (triangles with angles $\pi/2, \pi/2, \pi/n$ that fit together like $2n$ segments of an orange bisected at the equator) and the triangles that fit together to form the faces of one of the regular solids (so the triangles are a decomposition of a regular solid into congruent triangles).

If we set the digons aside, the only examples of congruent triangles with an angle sum greater than π and angles of the form $2\pi/n$ for some integer n—the condition that n of them meet at each vertex—are these:

- angles of $2\pi/3$, $2\pi/3$, and $2\pi/3$.
- angles of $\pi/2$, $\pi/2$, and $\pi/2$.
- angles of $2\pi/5$, $2\pi/5$, and $2\pi/5$.

These form the faces of the regular tetrahedron, octahedron, and icosahedron (see Fig. 15.2), respectively, as they appear on the sphere.

Schwarz also noticed that when $\lambda + \mu + \nu = 1$ the triangles are Euclidean and fit together to cover the plane. And he gave one example when $\lambda + \mu + \nu < 1$: $\lambda = \frac{1}{5}$, $\mu = \frac{1}{4}$, $\nu = \frac{1}{2}$ (Fig. 15.3), but he missed its true significance, and was to become very cross with Poincaré when he pointed it out. Poincaré's entirely reasonable opinion, expressed to a mutual acquaintance, was that missing this discovery was Schwarz's fault, and there was nothing that could be done about it.

15.3   Exercises

1. What nets of congruent triangles can you draw on the surface of a sphere? What are the angles at each vertex?
2. What nets of congruent triangles can you draw on the plane? What are the angles at each vertex?

Fig. 15.3 The triangles with angles $\frac{\pi}{5}, \frac{\pi}{4}, \frac{\pi}{2}$ cover a disc, Schwarz *Gesammelte Mathematische Abhandlungen*, vol. 2, 240

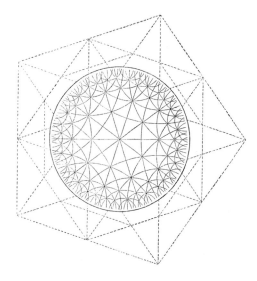

3. What do you make of Schwarz's net in Fig. 15.3? Look on the web for other figures of this kind.

Questions

1. We have seen that the map $\zeta(z) = z^{1/2}$, which maps the complex z-plane to the complex ζ-plane, is not too hard to understand, but even this simple case can be confusing. The set-theoretic inverse map $z(\zeta) = \zeta^2$ is a much simpler $2:1$ map. If you accept that a quotient of solutions to the hypergeometric equation maps the upper half-plane to a triangle, what does its set-theoretic inverse do?

2. What properties does the set-theoretic inverse map have when considered as a map of the whole net of triangles?

Chapter 16
Complex Ordinary Differential Equations: Poincaré

16.1 Introduction

Poincaré (Fig. 16.1) took up the hypergeometric equation in order to enter a prize competition, but it led him to his discovery of what he called Fuchsian functions and the role of non-Euclidean geometry in complex analysis, and it made his name as a mathematician of the first rank.[1]

16.2 Poincaré and Linear Ordinary Differential Equations

Henri Poincaré was born at Nancy on 29 April 1854, the son of a professor of medicine at the university there. Apparently, he had a happy childhood, and his mother, a very active and intelligent women, consistently encouraged him intellectually. His brilliance at mathematics became apparent in the final years at school and he entered the École Polytechnique at the top of his class, despite a poor performance in drawing. He had a lifelong capacity to immerse himself completely in abstract thought; it was said of him that he thought all the time. Although he was extremely prolific, he seldom bothered to resort to pen and paper, he disliked taking notes and gave the impression of taking ideas in directly, and having a perfect memory for details of all kinds. When asked to solve a problem he could reply, it was said, with the swiftness of an arrow.

In 1875, he graduated from the École Polytechnique, only second because of another low mark in drawing, and proceeded to the École des Mines. When he graduated from there he became a mining inspector, and had to write a report on a mining disaster in which 22 people were killed. In 1878, he presented his doctoral thesis to the faculty of Paris on the subject of partial differential equations. Darboux,

---

[1]For more detail, see Gray [124], upon which this account is based.

© The Author(s), under exclusive license to Springer Nature Switzerland AG 2021

J. Gray, *Change and Variations*, Springer Undergraduate Mathematics Series,
https://doi.org/10.1007/978-3-030-70575-6_16

Fig. 16.1 Jules Henri
Poincaré (1854–1912)

one of the examiners and an early supporter of Poincaré, said that the thesis contained
enough ideas for several good theses, although some points in it still needed to be
corrected or made precise; Poincaré never did this. By now he had decided on a
career as a mathematician, and by December 1879 he was in charge of the analysis
course at the Faculty of Sciences at Caen.

That year the prize competition of the Académie des Sciences called for a contribution to the theory of differential equations, and on March 22 Poincaré submitted
his essay 'Mémoire sur les courbes définies par une équation différentielle'. In it he
considered first-order non-linear differential equations

$$\frac{dx}{X(x, y)} = \frac{dy}{Y(x, y)},$$

where X and Y are real polynomial functions of real variables x and y, and investigated the global properties of their solutions. But he withdrew the essay on 14 June
1880, and the examiners never reported on it.[2]

Instead, he turned his attention to some work by the German mathematician
Lazarus Fuchs, and submitted an essay on a topic connected with it on 28 May
1880. Fuchs was a former student of Kummer's who had then taken up Weierstrass's
complex function theory, and was particularly interested in linear differential equations in the complex domain. In his major papers (1865) and (1866), he had shown
how to generalise Riemann's insights into the hypergeometric equation to linear
ordinary differential equations of any order, and he had successfully characterised
those equations all of whose solutions are holomorphic everywhere except for a finite
number of points where the coefficients of the differential equation become infinite.

In 1880, Fuchs had returned to the topic with a new question, and as part of this
work he had considered a second-order linear differential equation with a basis of

---

[2]It did however lead to a series of papers initiating the subject of flows on surfaces, see, e.g. Poincaré
[213, 214], and the great memoir on celestial mechanics [215].

solutions $f_1(z)$ and $f_2(z)$ and investigated what happens to their quotient $\zeta(z) = \frac{f_1(z)}{f_2(z)}$ under analytic continuation. It is clear that when z is taken on a loop that encloses a singular point, the quotient returns in the form

$$\frac{a_{11}f_1(z) + a_{12}f_2(z)}{a_{21}f_1(z) + a_{22}f_2(z)},$$

and so ζ is a many-valued function of z.

Poincaré read Fuchs's work, but was not persuaded by it. He entered into a correspondence with Fuchs that is too technical to describe here.[3] But it is clear that while Fuchs was deeply immersed in Weierstrass's complex function theory, with its insistence on power series methods, which are essentially local in scope, Poincaré had immediately picked up on the global nature of the solutions to differential equations.

Poincaré looked at the simplest cases, among them the hypergeometric equation, which has three singular points. Suppose one of the singular points is at the origin, and a basis of solutions is given by the functions $z^{\rho_1}f_1(z)$ and $z^{\rho_2}f_2(z)$, where f_1 and f_2 are holomorphic and non-zero in a neighbourhood of the origin, then their quotient is $z^{\rho_1 - \rho_2}f_1(z)/f_2(z)$. The factor $f_1(z)/f_2(z)$ is holomorphic and non-zero near the origin, so the behaviour of the quotient near the origin is governed by the exponent difference, $\rho_1 - \rho_2$.

Poincaré wrote to Fuchs to say that if the exponent differences were ρ_1, ρ_2, and ρ_3 at infinity, then either $\rho_1 + \rho_2 + \rho_3 > 1$, in which case z is rational in ζ, or $\rho_1 + \rho_2 + \rho_3 = 1$, in which case z was doubly periodic. Even in this case there were difficulties, and Poincaré supplied an example to show that Fuchs's theorem was still wrong. But, if a requirement that Fuchs had imposed on the differential equations he was considering was dropped, then the case $\rho_1 + \rho_2 + \rho_3 < 1$ could be included, which, Poincaré remarked, gives a "much greater class of equations than you have studied, but to which your conclusions apply. Unhappily my objection requires a more profound study, in that I can only treat two singular points". However, z is still single-valued, and

> These functions I call Fuchsian, they solve differential equations with two singular points whenever ρ_1, ρ_2, and ρ_3 are commensurable with each other. Fuchsian functions are very like elliptic functions, they are defined in a certain circle and are meromorphic inside it.

On the other hand, he concluded, he knew nothing about what happened when there were more than two singular points.

When Poincaré wrote again to him again on the 30 July his own researches on the new functions he led him to discover that they

> present the greatest analogy with elliptic functions, and can be represented as the quotient of two infinite series in infinitely many ways. Amongst those series are those which are entire series [which] converge in a certain circle and do not exist outside it, as thus does the Fuchsian function itself.

---

[3]See Gray [124, 125].

and he went on to explain how the new functions were the solutions of an extensive class of differential equations.

Before we turn to discuss what Fuchsian functions are, note that they only illuminate the study of a differential equation if they can be defined independently of the equation. This Poincaré did by introducing Fuchsian and theta-Fuchsian series. Moreover, by calling these functions Fuchsian and not Schwarzian Poincaré was showing that he had not read Schwarz's paper. He was to be much criticised by Klein for this, but he refused to back down, and he only ended the quarrel by naming a related but new class of functions Kleinian, even though Klein had not had much to say about them.

In the essay Poincaré submitted to the competition, he considered when the quotient $\zeta = \zeta(z)$ of two independent solutions of the differential equation $\frac{d^2y}{dz^2} + Qy = 0$ defines, by inversion, a meromorphic function z of ζ—note that y, z, and ζ are all complex variables. He showed that Fuchs's conditions were not necessary and sufficient. Rather, for z to be meromorphic on some domain it was necessary and sufficient that the exponent differences at each singular point, including infinity, differ by an aliquot part of unity (i.e. $\rho_1 - \rho_2 = \frac{1}{n}$, for some positive integer n). If the domain is to be the whole complex sphere then this condition is still necessary, but it is no longer sufficient. He found that there were too many special cases for Fuchs's methods to work easily, and so he proposed to take a new approach, beginning with Fuchs's example of a differential equation in which there are two finite singular points a_1 and a_2, where the exponent differences are 1/3 and 1/6, and the exponent difference at ∞ is 1/2.

In this case, he found the change in z was of the form

$$z \mapsto z', \quad \frac{z' - \alpha}{z' - \beta} = e^{2\pi/3}\left(\frac{z - \alpha}{z - \beta}\right)$$

under analytic continuation around a_1, and

$$z \mapsto z', \quad \frac{z' - \gamma}{z' - \delta} = e^{\pi/3}\left(\frac{z - \gamma}{z - \delta}\right)$$

under analytic continuation around a_2, and $\frac{1}{z} \mapsto -\frac{1}{z}$ around ∞. Note that the first two of these maps keep the points α and β fixed and are otherwise like rotations.[4]

Accordingly, z is a meromorphic single-valued function of ζ mapping a parallelogram composed of eight equilateral triangles onto the complex sphere, and $\zeta = \infty$ is its only singular point, so z is an elliptic function. The differential equation, Poincaré showed, has in fact an algebraic solution

$$y_1 = (x - a_1)^{1/3}(x - a_2)^{3/2}$$

and a non-algebraic solution y_2 such that

---

[4]For information about coaxial circles, see Appendix F.

Fig. 16.2 A quadrilateral in
the unit disc (Poincaré,
Oeuvres 1, 365)

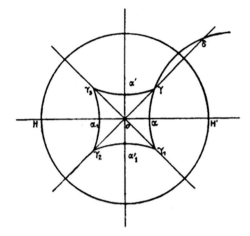

$$\frac{y_2}{y_1} = \int (x - a_1)^{-2/3}(x - a_2)^{-5/6}dx.$$

This result agrees with Fuchs's theory.

However, it might be that the domain of ζ could not be the whole ζ sphere. Poincaré showed that this could happen even when the differential equation had only two finite singular points. For example, if the exponent differences were $\frac{1}{4}$, $\frac{1}{2}$, and $\frac{1}{6}$ at ∞, then as long as z crosses no cuts ζ stays within a quadrilateral $\alpha_0\alpha'\gamma$ (see his Fig. 2, p. 86, given as Fig. 16.2).

Furthermore, however z is conducted about in its plane, ζ cannot escape the circle HH'. Poincaré described the quadrilateral as "mixtiligne", the circular-arc sides meet the circle HH' at right angles. This geometric picture is quite general, curvilinear polygons are obtained with non-re-entrant angles and circular-arc sides orthogonal to the boundary circle. Thus, the domain of x is $|\zeta| < OH$, and Poincaré then investigated whether z is meromorphic. This reduces to showing that, as ζ is continued analytically, the polygons do not overlap. This does not occur if the angles satisfy conditions derived from Fuchs's theory, unless the overlap is in the form of an annular region.

However, if the angles are not re-entrant, this cannot happen, and so z is meromorphic. Poincaré's proof of this is of incidental interest. He projected the circle HH' stereographically onto the southern hemisphere of a sphere, and then projected the image orthogonally back onto its original plane. The circular arcs orthogonal to HH' become straight lines, which renders the theorem trivial. This result virtually concluded Poincaré 's essay. As he said in his letter to Fuchs, his understanding was limited essentially to the case of two finite singular points.

Poincaré also wrote three anonymous supplements to the essay bearing the motto "Non inultus premor" that were received by the Académie on 28 June, 6 September, and 20 December.[5] We shall look at them shortly for the surprising connection they make between the theory of linear differential equations and non-Euclidean geometry.

In due course, Poincaré's essay was awarded second prize.[6] Hermite, one of the judges, said of the essay that[7]:

> the author successively treated two entirely different questions, of which he made a profound study with a talent by which the commission was greatly struck. The second ...concerns the beautiful and important researches of M. Fuchs, ... The results ...presented some lacunas in certain cases that the author has recognized and drawn attention to in thus completing an extremely interesting analytic theory. This theory has suggested to him the origin of transcendents, including in particular elliptic functions, and which has permitted him to obtain the solutions to linear equations of the second order in some very general cases. A fertile path is there that the author has not entirely gone down, but which manifests an inventive and profound spirit. The commission can only urge him to follow up his researches in drawing to the attention of the Academy the excellent talent of which they give proof.

16.3   Poincaré's Breakthrough and Non-Euclidean Geometry

Poincaré himself has left us one of the most justly celebrated accounts of the process of mathematical discovery, which concerns exactly his route to the theory of Fuchsian functions.[8] Poincaré gave this account in a lecture he gave to the Société de Psychologie in Paris 1908, and it was later published as the third essay in his volume *Science et Méthode* [221], with the title "L'invention mathématique" (Mathematical discovery).

Poincaré began by doubting that Fuchsian functions could exist, but shortly came to the opposite view. He tells us in the lecture ([221], p. 50) that:

> For two weeks I tried to prove that no function could exist analogous to those I have since called the Fuchsian functions: I was then totally ignorant. Every day I sat down at my desk and spent an hour or two there: I tried a great number of combinations and never arrived at any result. One evening I took a cup of coffee, contrary to my habit; I could not get to sleep, the ideas surged up in a crowd, I felt them bump against one another, until two of them hooked onto one another, as one might say, to form a stable combination. In the morning I had established the existence of a class of Fuchsian functions those which are derived from the hypergeometric series. I had only to write up the results, which just took me a few hours.

---

[5]This is the motto of Poincaré's home town of Nancy; it means "No-one touches me with impunity". The supplements are to be found in the Poincaré dossier in the Académie des Sciences, but for whatever reason Nörlund did not publish them when he published the essay in *Acta Mathematica*, and they were not included in Poincaré 's *Oeuvres*. The supplements confirm and greatly amplify what Poincaré said in the lecture 28 years later, and have since been published as Poincaré [225].

[6]It was ranked behind one by Halphen, and was not published until Nörlund edited it for *Acta Mathematica* Vol. 39, 1923, 58–93, in *Oeuvres*, I, 578–613.

[7]Quoted in Poincaré, *Oeuvres*, II, 73.

[8]For more detail, see Poincaré [225].

Then he had to find an independent description of the new functions, so that it is a meaningful remark that they solve certain differential equations. He went on ([221], 51):

> I then wanted to represent the functions as a quotient of two series; this idea was perfectly conscious and deliberate; the analogy with elliptic functions guided me. I asked myself what must be the property of these series, if they exist, and came without difficulty to construct the series that I called theta-fuchsian.

Next, he said that ([221], 51–52)

> At that moment I left Caen where I then lived, to take part in a geological expedition organized by the École des Mines. The circumstances of the journey made me forget my mathematical work; arrived at Coutances we boarded an omnibus for I don't know what journey. At the moment when I put my foot on the step the idea came to me, without anything in my previous thoughts having prepared me for it; that the transformations I had made use of to define the Fuchsian functions were identical with those of non-Euclidian geometry. I did not verify this, I did not have the time for it, since scarcely had I sat down in the bus than I resumed the conversation already begun, but I was entirely certain at once. On returning to Caen I verified the result at leisure to salve my conscience.

The supplements go beyond the essay precisely in their use of non-Euclidean geometry. In the first supplement, Poincaré began by reviewing the tessellation of the disc by "mixtiligne" quadrilaterals obtained by successively operating on one, which he called Q, by transformations M and N. He observed (p. 9) that these transformations form a group, and remarked:

> There are close connections with the above considerations and the non-Euclidean geometry of Lobachevskii. In fact, what is a geometry? It is the study of a group of operations formed by the displacements one can apply to a figure without deforming it. In Euclidean geometry the group reduces to rotations and translations. In the pseudogeometry of Lobachevskii it is more complicated. Indeed, the group of operations formed by means of M and N is isomorphic ('isomorphe') to a group contained in the pseudogeometric group. To study the group formed by means of M and N is therefore to do the geometry of Lobachevskii. Pseudogeometry will consequently provide us with a convenient language for expressing what we will have to say about this group. (Poincaré 's emphasis.)

Poincaré's realisation on boarding the bus at Coutances can be described very simply. He realised that the straightened version of the "mixtiligne" figures described at the end of his Prize essay was identical with the figures in Beltrami's description of non-Euclidean geometry [5, 6]; that, therefore, the original figures were conformally accurate representations of non-Euclidean figures; and finally that this meant the transformations formed from M and N were non-Euclidean isometries. Beltrami's detailed discussion of the non-Euclidean differential geometry of the disc enabled Poincaré to give a new meaning to his previously analytical transformations. Consequently, on p. 20, he remarked that:

> The Fuchsian functions are to the geometry of Lobachevskii what the doubly periodic functions are to that of Euclid.

Poincaré remained stuck on the case of the hypergeometric equation at least until his fourth letter, 30 July. Liberation came from an unexpected source, arithmetic ([221], 52, 53):

I then undertook to study some arithmetical questions without any great result appearing and without expecting that this could have the least connection with my previous researches. Disgusted with my lack of success, I went to spend some days at the sea-side and thought of quite different things. One day, walking along the cliff, the idea came to me, always with the same characteristics of brevity, suddenness, and immediate certainly, that the arithmetical transformations of ternary indefinite quadratic forms were identical with those of non-Euclidean geometry.

Once back at Caen I reflected on this result and drew consequences from it; the example of quadratic forms showed me that there were Fuchsian groups other than those which correspond to the hypergeometric series; I saw that I could apply them to the theory of theta-Fuchsian series, and that, as a consequence, there were Fuchsian functions other than those, which derived from the hypergeometric series, the only ones I knew at that time. I naturally proposed to construct all these functions; I laid siege systematically and carried off one after another all the works begun; there was one however, which still held out and as the chase became involved it took pride of place. But all my efforts only served to make me know the difficulty better, which was already something. All this work was quite conscious.

The second supplement is given over to a more rigorous description of non-Euclidean geometry, and to tessellations of the disc by polygons with angles π/m for integers m. When the polygon is a triangle, he also discussed more carefully the ways of constructing Fuchsian functions in this case and was led to conjecture a result which he said he was not yet in any state to prove—the Riemann mapping theorem! Then, on p. 17, he abruptly stated the connection with the theory of quadratic forms, a subject upon which Hermite was an expert.

He let T be a matrix ("substitution") with integer coefficients which preserved an indefinite ternary quadratic form Φ and S be a substitution sending $\zeta^2 + \eta^2 - \xi^2$ to Φ. Then STS^{-1} preserves $\zeta^2 + \eta^2 - \xi^2$ and sends (ζ, η, ξ) to $(\zeta', \eta', \xi)'$, say.

The quantities

$$ z = \frac{\xi + \sqrt{-1}\eta}{\zeta}, z' = \frac{\xi' + \sqrt{-1}\eta'}{\zeta'} $$

are related by transformation $z' = zK$ of the non-Euclidean plane provided $\zeta^2 + \eta^2 - \xi^2 < 0$. He did not prove that a sheet of the hyperboloid of two sheets provides a model of non-Euclidean geometry—which is easy enough to establish—and remarked only that (p. 19): "All the points zK are the vertices of a polygonal net obtained by decomposing the pseudogeometric plane into polygons pseudogeometrically equal to each other".

The third supplement, of only 12 pages, was received on 20 December. Its main result is the extension of the method of polygonal decomposition to include cases where the angles are zero, and the roots of the indicial equation differ by integers. The notable example is Legendre's equation. Poincaré 's method is to push the polygons outwards until one or more vertices are "at infinity", i.e. are on the boundary of the disc, and the corresponding angles vanish. Since the polygons hitherto studied had angles which were only rational multiples of π, Poincaré 's argument relies heavily on its geometrical plausibility.

The unpublished work makes abundantly evident the astounding clarity of Poincaré 's mind, coupled to an almost equally dramatic ignorance of contemporary

Fig. 16.3 A tessellation of
the non-Euclidean disc by
triangles. (Klein,
*Gesammelte mathematische
Abhandlungen*, vol. III,
p. 126)

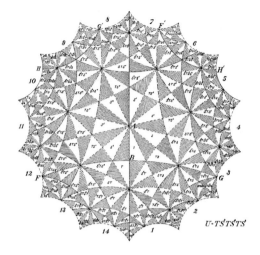

mathematics. There is no mention of the work of Schwarz on the hypergeometric
equation, nor is there any mention of the work of Dedekind or Klein, and even Hermite's work on modular functions, which he must have known, seems to have been
forgotten. In fact, these omissions are not mere oversights; Poincaré genuinely did
not then know the German work.

It marks two things: Poincaré's arrival on the mathematical scene, and the recognition of non-Euclidean geometry as an important tool in mathematics and not merely
as an unexpected feature of geometry.

16.4   Non-Euclidean Geometry

For us to be able to say that all the triangles in Fig. 16.3 are congruent, we have to
define a sense of distance and a group of distance-preserving transformations such
that corresponding sides have the same lengths and corresponding angles are equal.
The convenient thing about this kind of picture of non-Euclidean geometry is that it
represents angles correctly. This requires us to use angle-preserving transformations
of the disc to establish congruences, so we invoke Möbius transformations. But what
about lengths?

To investigate a formula for distance that is invariant under such transformations,
we begin by looking at maps of the above kind that also map the real axis to itself.
They are Möbius transformations of the form

$$z \mapsto \frac{z + r}{rz + 1} = \mu(z), \quad r \in \mathbb{R}.$$

We want a formula for the distance between two points on the real line with the property that the distance from 0 to a is the same as the distance from $\mu(0) = r$ to $\mu(a) = \frac{r+a}{ra+1}$. If we write d for distance, then we want d to be such that

$$d(0, a) = d(\mu(0), \mu(a)) = d\left(r, \frac{r+a}{ra+1}\right).$$

We notice that if $r = \tanh \rho$ and $a = \tanh \alpha$ then the formula we want to say that

$$d(0, \tanh \alpha) = d(\tanh \rho, \tanh(\rho + \alpha)).$$

This will be the case if we define d by the formula

$$d(z_1, z_2) = \tanh^{-1} \left| \tanh \left(\frac{z_2 - z_1}{1 - z_1 z_2} \right) \right|,$$

for then $d(0, \tanh \alpha) = \alpha$ and

$$d(\tanh \rho, \tanh(\rho + \alpha)) = \tanh^{-1} \tanh(\rho + \alpha - \alpha) = \tanh^{-1} \tanh \alpha = \alpha,$$

as required.

If we represent Euclidean distances from the centre by r and non-Euclidean distances from the centre by ρ, then the above formula says that

$$\rho = \tanh^{-1}(r), \quad \text{or } r = \tanh \rho,$$

so

$$dr = \text{sech}^2 \rho \, d\rho = (1 - \tanh^2 \rho) d\rho$$

and therefore

$$d\rho = \frac{dr}{1 - r^2}.$$

This makes good sense. As a point moves outwards its distance from the centre tends to 1 and so $1 - r^2$ tends to zero. Therefore, equal non-Euclidean steps of $d\rho$ are represented by steadily smaller Euclidean steps of dr.

It can now be proved that in non-Euclidean geometry:

- geodesics appear as arcs of circles perpendicular to the boundary circle;
- circles appear as circles;
- the angle sum of a triangle is less than π;
- the area of a triangle with angles α, β, γ is proportional to $\pi - (\alpha + \beta + \gamma)$;
- there are many parallels to a given line through a point not on that line.

Theorems like these gave the new geometry its name, and occasioned much debate in the nineteenth century about its logical coherence and its physical applicability.

With the profusion of new geometries in the twentieth century it gradually lost its position in the panoply of geometry, and is now usually known by the name Klein gave it: hyperbolic geometry.

16.4.1 Summary

Whether the web of triangles is on the sphere, on the plane, or on the non-Euclidean disc, the triangles in each case are mutually congruent. There is a small number of distinct spherical triangles, a small number of distinct Euclidean ones, and an infinite number of distinct non-Euclidean ones that can be used. In each case, the web can be mapped to itself by a group of isometries, and there are functions F on the sphere, plane, or non-Euclidean disc with the property that if g is a member of the appropriate group of isometries then $F(gz) = F(z)$.

This generalises the periodic behaviour of the trigonometric functions. For example, $\sin(x + 2k\pi) = \sin(x)$. In this case, the sine function is defined on the real line and the group of integers acts as follows: $k \in \mathbb{Z}$ sends $x \in \mathbb{R}$ to $x + 2k\pi$.

This was not exactly how it was seen before Poincaré. Schwarz's discovery of the connection between the hypergeometric equation and the regular solids was new. The case of elliptic functions was well known, but the double periodicity of these functions was not seen as related to an action of the group $\mathbb{Z} \oplus \mathbb{Z}$ on the complex plane. Poincaré's introduction of non-Euclidean geometry was wholly new and surprising.

16.5 Exercises

1. Show that a Möbius transformation that maps the unit disc to itself and fixes the points $z = -1, 0$, and 1 is the identity map.
2. Call the arc of a circle that lies inside the unit disc and meets the unit circle at right angles (and also any diameter of the unit circle) a d-line. Show that a Möbius transformation that maps the unit disc to itself maps d-lines to d-lines.
3. Show that any Möbius transformation that maps the unit disc to itself cannot map a segment of a d-line to a proper subset of itself.

Questions

1. Explain why the above exercises show that it is possible to speak of the length of a segment of a d-line. Why did Poincaré regard the presence of a group of Möbius transformations that map the unit disc to itself as almost synonymous with the existence of a (non-Euclidean) geometry in the disc?
2. Find what you can about what is called the Kleinian view of geometry.

Chapter 17
More General Partial Differential Equations

17.1 Introduction

As clarity grew about the existence of solutions for linear ordinary differential equations, and the existence of solution methods for them advanced, it became clear to mathematicians that the corresponding story for partial differential equations was woefully underdeveloped. The first to prove any kind of general existence theorem for partial differential equations was Cauchy, who in 1819 successfully treated the general first-order partial differential equation. His work on initial conditions established the framework that later became known as the Cauchy problem. Ampère also did important work on the subject at the same time in his [2]. Cauchy then returned to the topic in a paper of 1842 and gave an argument to show that a partial differential equation of any order defined by one or more analytic equations has an analytic solution in a neighbourhood of a suitably chosen initial curve (or, if the equation has more than two independent variables, a hypersurface). This is the origin of the idea of the Cauchy hypersurface condition for hyperbolic partial differential equations, but Cauchy's account left much for later mathematicians to do.

Cauchy's ideas were rediscovered and extended by Sonya Kovalevskaya, who also documented unexpected issues with initial conditions, and how it can fail, and in the 1870s Darboux further improved the analysis.

Meanwhile, in the 1860s, Riemann had given a clear example of how the method of characteristics can show that solutions will cease to exist, and applied this in the study of the propagation of sound to show how shock waves can form. His paper, which is discussed in Chap. 22, also contains innovative ideas about the use of Green's function methods in the study of hyperbolic partial differential equations.

© The Author(s), under exclusive license to Springer Nature Switzerland AG 2021   191
J. Gray, *Change and Variations*, Springer Undergraduate Mathematics Series,
https://doi.org/10.1007/978-3-030-70575-6_17

17.2   Cauchy's Method in 1819

Cauchy was a student at the École Polytechnique from 1805 to 1807, where he was taught analysis by Lacroix, whose *Traité Élementaire de Calcul Différentiel et Intégral* was required reading.[1] The account it gives of partial differential equations was by then fairly standard, and owed much to Monge's approach—as one might expect at the École Polytechnique. The equation

$$Pp + Qq = R,$$

for a function $z(x, y)$, where P, Q, and R are functions of x, y, and z, $p = \frac{\partial z}{\partial x}$ and $q = \frac{\partial z}{\partial y}$ was solved by eliminating p from the equation and the identity $dz = p\,dx + q\,dy$ to obtain

$$Pdz - Rdx = q(Pdy - Qdx).$$

Lacroix then distinguished the simpler case, where the differential $Pdz - Rdx$ only involves x and z and the differential $Pdy - Qdx$ only involves x and y, from the general case. The simple case is solved by the method of integrating factors, but in the general case the method fails, although ad hoc changes of variable can sometimes help, as Lacroix proceeded to show. Monge's geometric approach was relegated to a footnote in Sect. 348, where it was described as "very ingenious".

Monge's geometric method was redescribed by Cauchy, who explained it at length for equations in two variables and then showed how to overcome the problems of extending it to any number of variables. You can read Cauchy's paper [34] in Sect. 31.1; it makes an instructive comparison with that of Monge.[2] It is likely, given Cauchy's growing appetite for mathematics, that he read Monge's account.

Neither Monge nor Cauchy specified what conditions on the function defining the partial differential equation are necessary for their proofs to work, but it is likely that Monge assumed that everything is analytic in something like the sense that every function admits a power series expansion, and that Cauchy assumed that functions were no more differentiable than necessary. That would put his paper of 1819 on a par with his paper a couple of years later on ordinary differential equations and with his introduction of epsilon-delta analysis at the École Polytechnique in 1821. That said, as was typical in Cauchy's work, he let conditions on the function f emerge in the course of the proof. In fact, although his [34] is an existence proof Cauchy never used the term "exist" and never stipulated what hypotheses on f he used, namely, that f be continuously differentiable.

One of the assessment questions on this part of the course is to give an account of Cauchy's proof in his paper [34] that first-order partial differential equations have solutions; there is a translation of the paper in Chap. 31. It would therefore be inappropriate to give a detailed explanation of it here, but we can note that it opens

---

[1]The École Polytechnique had just been reorganised by Napoleon as a military school; Cauchy entered third in the ranking of the 125 entrants.

[2]The paper is reprinted in his *Oeuvres* series 2, volume 2, 1958.

with a clever change of variables argument that greatly simplifies the equation to be solved, then there is an investigation of a necessary condition, then a quick, analytical derivation of the equations that Monge had exhibited a decade before, and then an investigation of the initial conditions.

17.3 Cauchy and the General Partial Differential Equation

We must also be sketchy in our account of Cauchy's papers of 1842 on the existence of solutions to the general partial differential equation for two reasons: it is a much more difficult paper, and what Cauchy provided is itself little more than an outline at crucial points.

He had already stated his aims in the paper (Cauchy [36]) (I take this translation from Cooke ([47], 25):

> In the theory of equations mathematicians have properly considered fundamental the question whether every equation has a root. Similarly in the integral calculus one of the most important questions, a fundamental question, is obviously whether every ordinary or partial differential equation can be integrated. But – and this ought to surprise us at first sight – despite the numerous works of mathematicians on the integral calculus, this question, important though it be, is nowhere solved in full generality. To be sure the existence of general integrals of ordinary differential equations, which contain only one independent variable, is now established by two different methods which I have given, the first in my lectures at the *Ecole Polytechnique*, the second in a lithographed memoir of 1835. .... In addition, the existence of general integrals of partial differential equations is established in certain cases where one is able to integrate these equations, for example when the equation reduces either to a single equation of first order or to linear equations in which the coefficients of the unknowns and of their derivatives remain constant. But does an arbitrary system of ordinary or partial differential equations always admit a corresponding system of general integrals? Such is the problem which seemed to me worthy of the attention of mathematicians. The present solution is based on considerations which I shall explain briefly.
>
> For a long time mathematicians, supposing without proof that every ordinary or partial differential equation admits a general integral, have considered Taylor's formula as the means of developing this integral in a series of increasing integer powers of an increment i given to an independent variable t, which can be considered as representing time. Further, using a theorem which I proved in 1831 relating to the development of functions, one can be sure that in the case where the series so obtained is convergent, the sum of the series satisfies, as an integral, the ordinary or partial differential equation, at least for real or complex values of the increment i whose moduli do not exceed a fixed bound. Moreover the same remark applies to the sums of the series obtained when, assuming the existence of general integrals of a system of ordinary or partial differential equations, one sets about developing them in Taylor series. But in all cases it remains to be proved that the series so obtained is convergent, at least for i of sufficiently small modulus. Now this end can be achieved using a fundamental theorem which not only determines a bound beneath which the modulus of i may vary arbitrarily without causing the series obtained to diverge, but also determines a bound on the error caused by terminating each series after a certain number of terms. The proof of this theorem is based, as will subsequently be seen, on the principles of the new calculus which I have called "calcul des limites" and on a device of analysis which can be given many useful applications.

Cauchy then went on to develop his "calcul des limites", which we call the method of majorants, for determining if a series converges by comparison with another series that has larger terms but does converge.

An application of these ideas to the theory of partial differential equations came in his [37]. He began by claiming that any partial differential equation can be reduced to a system of first-order (and what we would call) quasi-linear partial differential equations by introducing more unknown functions, and then said that therefore it was enough to show how to solve a system of such first-order partial differential equations.

In his [37] gave a careful argument to show that a single such equation can be solved if the equation has analytic coefficients and certain analytic initial data is given, and in his [37] he dealt with a system of such equations.

In his [37], Cauchy set himself the task of showing that a partial differential equation of the form

$$u_t = a_1 u_{x_1} + a_2 u_{x_2} + \cdots + a_n u_{x_n} + v \tag{17.1}$$

for an unknown function u has a solution, where $a_1, a_2, \ldots, a_n$ and v are analytic functions of the independent variables $x_1, x_2, \ldots x_n$, and t, and the value of u is prescribed in a neighbourhood of a point in which $t = \tau$, a constant. Thus, the initial data is the value of $u(x_1, x_2, \ldots, x_n, \tau)$ and its partial derivatives with respect to the other variables $x_1, x_2, \ldots, x_n$, namely, $u_{x_j}(x_1, x_2, \ldots, x_n, \tau)$, $j = 1 \ldots n$.

He now investigated the consequences of assuming the partial differential equation has a solution that is a power series in powers of $t - \tau$ when t equals some value τ. This means that near $t = \tau$ the solution u is a function w of the form

$$w = 1 + I_1(t - \tau) + I_2(t - \tau)^2 + \cdots . \tag{17.2}$$

The coefficient I_n is given by

$$I_n = \frac{D_t^n}{n!},$$

where D_t^n is the nth derivative of w with respect to t evaluated at the point $t = \tau$. His task was now to show that this series converges for suitably small absolute values of $t - \tau$.

Cauchy interpreted the partial differential equation as saying that

$$D_t = a_1 D_{x_1} + a_2 D_{x_2} + \cdots + a_n D_{x_n} + v$$

and so the coefficients I_n are expressions in various D_{x_j} acting on various a_k. His question now was how to estimate them and obtain the convergence result that he wanted.

He let the variables vary by small amounts and considered the maximum effect their variation has on the coefficients $a_1, a_2, \ldots, a_n$ and v. He then observed that this effect is produced by a particularly simple partial differential equation and so a study of this partial differential equation could be used to study Eq. (17.1). This is the equation of the same form as Eq. (17.1) in which

$$a_j(x_1, x_2, \ldots, x_n, t, w) = \alpha_j(x_1 x_2 \ldots x_n t w)^{-1}, \quad j = 1, \ldots, n$$

in which $\alpha_1, \alpha_2, \ldots \alpha_n$ are constants.

This is an equation of the kind that his paper of 1819 applies to, and it can be solved by passing to a system of ordinary differential equations. These equations have a solution for some non-zero values of the variables. This solution dominates the conjectured power series solution (17.2), and so it converges and, as Cauchy checked, it defines a solution of the original partial differential equation (17.1) in a neighbourhood of the given system of initial values.

As Cooke ([47], 27) remarked, Cauchy did not discuss the uniqueness of the solution, when it exists. He seems to have assumed that the solution is determined by the initial values. But these papers form what is regarded as Cauchy's contribution to the Cauchy–Kovalevskaya theorem, so historical generosity seems to have been at work.

17.4 Kovalevskaya's Theorem and Her Counter-Example

In 1875, the Russian mathematician Sonya Kovalevskaya (Fig. 17.1) published a paper in the Berlin *Journal für die reine und angewandte Mathematik* [165] that conveyed the results she had written the year before as a private student of Weierstrass's in Berlin, and would undoubtedly have led to the award of a Ph.D. at the University of Berlin had women been eligible to study for degrees there at all. But they were not, and so Weierstrass persuaded one of his former students, Lazarus Fuchs, by then a professor at Göttingen, to see that she was awarded a Ph.D. there.[3]

In this paper, Kovalevskaya gave a new proof of Cauchy's theorem on the existence of solutions to a first-order quasi-linear partial differential equation in the analytic case. She says that she had learned this result from Weierstrass's lectures, and it seems that neither of them knew of Cauchy's much earlier proof of the same result. Her ignorance is understandable, although she does mention work by Briot and Bouquet [23], who do cite Cauchy, but it is striking that Weierstrass could claim that he did not.

She also indicated how the theorem might be extended to systems of partial differential equations, and therefore to partial differential equations defined by a polynomial equation in n variables and the partial derivatives of the unknown function. She showed that there is always a convergent power series in the variables about a point $(a_1, a_2, \ldots, a_n)$ that satisfies the equation at that point, and that if such a function satisfies the partial differential equation then the coefficients of its Taylor series expansion can be determined from the partial differential equation. This theorem,

---

[3] German universities were admirably relaxed about where people had studied for their degree.

Fig. 17.1 Sonya
Kovalevskaya (1850–1891).
Acta Mathematica, Table
générale des tomes 1–3,
1882–1912, p. 153

which was independent of Cauchy's, is her contribution to the Cauchy–Kovalevskaya theorem.

She then went on to surprise her supervisor with a novel, and indeed disturbing, observation about initial conditions for partial differential equations. Her example ([165], 22) was all the more disturbing because it concerned the one-dimensional heat equation, which one might have thought was well understood. The equation is

$$u_t = u_{xx},$$

and everything could be expected to be well behaved if $u(x_0, t)$ and $u_t(x_0, t)$ are given as analytic functions of t. Kovalevskaya took as initial conditions the requirement that $u(x, 0) = (x - 1)^{-1}$, and observed that the partial differential equation $u_t = u_{xx}$ is formally solved by the infinite series

$$\sum_{j=0}^{\infty} \left(\frac{t^j}{j!}\right) \cdot \left(\frac{d^{2j}}{dx^{2j}} f(x)\right),$$

which reduces to the function $f(x) = (x - 1)^{-1}$ when $t = 0$. However, the power series solution diverges for all $t \neq 0$.

It might be objected that the boundary condition involves a function that becomes infinite when $x = 1$. Could this be the reason that the solutions diverge? It is not, for, as Cooke ([47], 33) observes, one can conduct a similar analysis when $f(x) = (1 + x^2)^{-1}$. In this case, the formal solution to the partial differential equation is

$$u(x, t) = \sum_{m,n=0}^{\infty} (-1)^{m+n} \frac{(2m + 2n)!}{(2m)!n!} x^{2m} t^n.$$

When $t = 0$ this reduces to the series

$$u(x, 0) = \sum_{m=0}^{\infty} (-1)^m x^{2m},$$

which is indeed the power series for $(1 + x^2)^{-1}$, a series that is never infinite (provided x is real, which the partial differential equation surely requires). However, when $x = 0$ the series reduces to

$$u(0, t) = \sum_{n=0}^{\infty} (-1)^n \frac{(2n)!}{n!} t^n,$$

which diverges for all $t > 0$.

Cooke goes on to quote a solution in the form of a Fourier integral:

$$u(x, t) = \int_0^{\infty} e^{-y-y^2 t} \cos(xy) dy,$$

which is analytic only if $t > 0$.

That year, 1875, the French were also caught ill-informed. The young Gaston Darboux published a four-page paper in the *Comptes Rendus* For January 1875 in which he rederived Cauchy's result from 1835 on the solution of ordinary differential equations and extended it to partial differential equations, noting that Briot and Bouquet had given a new proof and explored the consequences of the theorem. He also noted that one still lacked a perfectly general theory of equations of this kind, and promised a subsequent paper in which he would explain the theory of characteristics, which he attributed to Monge.

Darboux's argument was not that different from Cauchy's: first show that there is a formal power series that satisfies the equation and the given initial conditions; second show that for a certain range of the variables this series converges.

Within the month the Italian mathematician Angelo Genocchi had written in with "some observations". He admired Darboux's talent, but he noted that Cauchy had written about the problem for systems of partial differential equations in a series of papers in the *Comptes Rendus* for 1842, so Cauchy deserved the credit for the first proof. Then in 1873 the French mathematician Puiseux had made some important remarks about implicit functions that an Italian mathematician called Félix Chio had amplified, also in the *Comptes Rendus*. Genocchi added that Cauchy also deserved credit for the theory of higher dimensional spaces "about which there is so much noise at present", and that there was also the delicate point discussed by German mathematicians of what they called "convergence in equal degree" (and we call uniform convergence today).

Genocchi's note was published, and the perpetual secretary of the Académie des Sciences, Joseph Bertrand, took the opportunity to press for the prompt publication of the *Oeuvres* of Cauchy. The opportunity was also given to Darboux to further develop his method.

All this burst of activity came as a surprise to Weierstrass, as Cooke relates. It seems that Weierstrass had not renewed his subscription to the *Comptes Rendus* on time, and so only got the relevant copies of it some time after they had appeared in early 1875. He quickly wrote to Kovalevskaya to tell her what was going on Cooke ([47], 35):

> So you see, my dear, that this question is one which is awaiting an answer, and I am very glad that my student was able to anticipate her rivals in time and at least not fall behind them in working out the problem.

> Darboux mentions several exceptional cases which are of special interest; I am inclined to think that he also has encountered the difficulties (as in the equation $\frac{\partial \varphi}{\partial t} = \frac{\partial^2 \varphi}{\partial x^2}$) which gave you so much trouble at first and which you later overcame so successfully.

He also sent a copy of her dissertation to Hermite, Darboux's mentor, and it so impressed both men that they became staunch advocates for her later on. This surely contributed to the high opinion that Poincaré was to have of her work. In the course of his prize-winning paper on celestial mechanics that made his name internationally, he wrote (Poincaré [215], Sect. 3): "Mme Kovalevski has considerably simplified Cauchy's demonstration and has given the theorem its definitive form".

It might seem that this means only that Kovalevskaya improved on Cauchy's method of proof, and indicated that it can break down. In fact, twentieth-century mathematicians found that her theorem pointed to a number of subtle developments of which one is worth mentioning here precisely because it was not sufficiently appreciated at the times: boundary conditions matter greatly when solving partial differential equations. We shall return to this point in later chapters.

17.5   Exercises

The central points of this lecture are to establish that Cauchy opened up the study of general partial differential equations (of order greater than one) but only in the analytic case, and that by a more careful analysis Kovalevskaya was able to show that his method of finding a transversal hypersurface did not always work.

It seems to me that these points would be obscured by working through mathematical examples, so none are provided, except for this one:

1. Find examples of power series with a zero radius of convergence.

Questions

1. What does the reception of Cauchy's ideas about partial differential equations tell us about how mathematical ideas circulated in the mid-nineteenth century?

Chapter 18
Green's Functions and Dirichlet's Principle

18.1 Introduction

Green introduced the functions that have come to bear his name in an attempt to solve problems in potential theory. Here we shall see how he used them, and how Dirichlet and Riemann used them to study Laplace's equation. In this chapter, more than is often the case when dealing with mathematics as it was discovered, the results are imprecise and quite some effort from later mathematicians was needed to make them rigorous.

18.2 Green's Theorems and Green's Functions

Green introduced these functions in his famous *Essay* [128], where he claimed that given the value of a function $\bar{V}$ on a closed surface σ there is a unique continuous extension to a function V defined on the interior which satisfies the Laplace equation and has no singular values inside the surface. This is, of course, the Dirichlet problem (before Dirichlet).

He defended this claim as follows (I have somewhat modified his language). He supposed that there is a function U that is a harmonic except at the point P, where it becomes infinite like $1/r$ near the origin and is zero on the surface σ. Then it followed from a general theorem of his that

$$V(P) = \frac{1}{4\pi} \int_\sigma V \frac{dU}{d\mathbf{n}},$$

where the expression $\frac{dU}{d\mathbf{n}}$ denotes the normal derivative of the function U. As he remarked, this shows that the value of V at P is known when its values are known on the surface.

© The Author(s), under exclusive license to Springer Nature Switzerland AG 2021
J. Gray, *Change and Variations*, Springer Undergraduate Mathematics Series,
https://doi.org/10.1007/978-3-030-70575-6_18

Note, however, that Green's function U depends on a parameter that locates the singular point. If you can determine a Green's function for a given region that has its singular point at an arbitrary point P then this formula does indeed define a harmonic function in the region—but it is Green's function that varies, not the values of a *single* Green's function.

But why does such a function exist? Green answered this question this way ([116], 32):

> To convince ourselves that there does exist such a function as we have supposed U to be; conceive the surface to be a perfect conductor put in communication with the earth, and a unit of positive electricity to be concentrated in the point P, then the total potential function arising from P and from the electricity it will induce upon the surface, will be the required value of U. For, in consequence of the communication established between the conducting surface and the earth, the total potential function at this surface must be constant, and equal to that of the earth itself i. e. to zero (seeing that in this state they form but one conducting body). Taking, therefore, this total potential function for U, we have evidently $0 = \bar{U}, 0 = \nabla(U)$, and $U = 1/r$ for those parts infinitely near to P. As moreover, this function has no other singular points within the surface, it evidently possesses all the properties assigned to U in the preceding proof.

This argument is an appeal to physics pure and simple. Nonetheless, the introduction of a function of a particular kind that solves what was to become called the Dirichlet problem was to become a dominant idea in work on potential theory. Such functions are today called Green's functions. Their use derives from what has become known as Green's identity.

Let us introduce the notation ∇U of a function $U = U(x, y, z)$ to mean

$$\nabla U = (U_x, U_y, U_z),$$

and

$$\nabla^2 U = U_{xx} + U_{yy} + U_{zz},$$

the Laplacian of U.[1]

Green began by considering $\nabla U.\nabla V$. It is a sum of three terms, and integrating each by parts gave him

$$\int_{vol} \nabla U.\nabla V = \int_{surf} V(\mathbf{n}.\nabla U) - \int_{vol} V\nabla^2 U,$$

where integrals $\int_{vol}$ are taken over a region D in $\mathbb{R}^3$ and integrals $\int_{surf}$ are taken over the surface C of the region D, and $\mathbf{n}$ is the outward unit normal vector at a point on C.

Likewise,

$$\int_{vol} \nabla V.\nabla U = \int_{surf} U(\mathbf{n}.\nabla V) - \int_{vol} U\nabla^2 V,$$

---

[1] Green wrote everything out in full.

but switching U and V does not change the integral on the left-hand side, so

$$\int_{surf} V(\mathbf{n}.\nabla U) - \int_{vol} V\nabla^2 U = \int_{surf} U(\mathbf{n}.\nabla V) - \int_{vol} U\nabla^2 V,$$

and this rearranges to give *Green's identity*:

$$\int_{vol} U\nabla^2 V - \int_{vol} V\nabla^2 U = \int_{surf} U(\mathbf{n}.\nabla V) - \int_{surf} V(\mathbf{n}.\nabla U).$$

The Poisson problem asks for a function V with these properties

- $\nabla^2 V = F$ in D and
- $V = f$ on C

for given functions F and f. It reduces to the Dirichlet problem when $F = 0$.

Green's method transforms the Poisson problem into another that might be easier to solve. He looked for a function U such that

- $\nabla^2 U = 0$ except at one point P in D, where it is infinite like $1/r$ (and r is the radial distance from (x, y, z) to P) and
- $U = 0$ on C.

We plug these into Green's identity and get

$$\int_{vol} V\nabla^2 U - \int_{vol} U\nabla^2 V = \int_{surf} V(\mathbf{n}.\nabla U) - \int_{surf} U(\mathbf{n}.\nabla V).$$

We take these integrals in turn.

- $\int_{vol} V\nabla^2 U = u(P)$
- $\int_{vol} u\nabla^2 V = \int_{vol} U F$
- $\int_{surf} V(\mathbf{n}.\nabla U) = \int_{surf} f(\mathbf{n}.\nabla U)$
- $\int_{surf} v(\mathbf{n}.\nabla V) = 0$.

So we deduce that

$$V(P) = \int_{vol} U F + \int_{surf} f(\mathbf{n}.\nabla U).$$

This expresses the function V in terms of the given data f and the function U, which is known as a *Green's function*. The hope is that U can be found because it depends only on the shape of D. For simple shapes, Green's function can often be found explicitly, thus yielding a specific solution of the Dirichlet problem. Likewise, theorems that establish the existence of a Green's function for a large class of domains (such as Harnack's theorem, see Sect. 19.3) also solves the Dirichlet problem for those domains, and that can be easier to do.

18.3   Dirichlet Principle and Problem

The young William Thomson seems to have been the first to state something like this principle. In a paper in Liouville's Journal for 1847, he claimed that a harmonic function on a given region bounded by a surface could be found for which the normal derivative at every point on the surface took a given value F.

He claimed that among the functions for which $\int_{surf} V F dS = A$, where A is an arbitrary constant, there will be one for which the integral

$$\int_{vol} \left(\left(\frac{\partial V}{\partial x} \right)^2 + \left(\frac{\partial V}{\partial y} \right)^2 + \left(\frac{\partial V}{\partial z} \right)^2 \right) dx dy dz$$

takes a minimum value, and for this function V

$$\frac{\partial^2 V}{\partial x^2} + \frac{\partial^2 V}{\partial y^2} + \frac{\partial^2 V}{\partial z^2} = 0,$$

and the normal derivative of V is a constant multiple of the given value F. The only proof he gave was that the result followed by the calculus of variations, which in fact only connects the double and triple integrals but does not guarantee the existence of a minimum.

The Dirichlet problem specifies a simply connected domain T with a boundary ∂T, and a continuous function defined on ∂T. It then asks for a harmonic function defined on the interior of T that continuously extends the function defined in ∂T. It is then easy to show that the function is unique—if it exists. Dirichlet had suggested in lectures in 1856–1857 at Göttingen that a solution would always exist because of a general argument that became known as the "Dirichlet principle".[2]

The Dirichlet principle, as given by Grube, states[3]:

> For every bounded connected domain T there are clearly infinitely many functions u continuous together with their first-order derivatives, for x, y, z which reduce to a given value on this surface. Among these functions there will be at least one which reduces the following integral $U = \int_{vol} \left((\frac{\partial u}{\partial x})^2 + (\frac{\partial u}{\partial y})^2 + (\frac{\partial u}{\partial z})^2 \right)$ extended over the domain T, to a minimum; it is evident that this integral has a minimum since it cannot become negative. We can now show the following:
>
> 1. Every such function u which minimizes U, satisfies the differential equation $\frac{\partial^2 u}{\partial x^2} + \frac{\partial^2 u}{\partial y^2} + \frac{\partial^2 u}{\partial z^2} = 0$ everywhere in the domain T. This already makes it clear that there always exists a function u having the desired property, namely that function for which U becomes a minimum.
> 2. Every function u which satisfies the [above, JJG] differential equation within the domain T, minimizes the integral U.

---

[2]The lectures were published posthumously in an edition by Grube in 1876.

[3]From Dirichlet ([64], 127–128), quoted in Bottazzini, *The Higher Calculus*, 300.

3. The integral U can have only one minimum. It follows from 2 and 3 that there is only one function u with the desired property.

As is often remarked, the problem with these claims is that they assume that there *is* a function that minimises the integral simply on the grounds that the integral can never be negative. But that is to confuse the existence of a lower bound with the existence of a function for which the integral attains its lower bound. (Compare the behaviour of the function $f(x) = 1/x$ on the positive real axis: it is bounded below by zero, but there is no value of x for which $f(x) = 0$.)

So Thomson had shown that the Dirichlet principle is a claim in the calculus of variations, and indeed the Euler–Lagrange equation for the integral U leads to the Laplace equation. But that does not vindicate the principle—it merely locates it in a family of plausible but unproved claims. We shall now look briefly at the first attempts to prove it, and then at other attempts on the Dirichlet problem.

We also have Dedekind's statement of the Dirichlet principle, which is quoted in a paper by Weierstrass [271].[4] There Dedekind wrote that

> Given any finite surface, one can always, and in only one way, endow it with mass so that the potential at any point of the surface has an arbitrary, continuously varying, value.

This is not very clear, but Dedekind followed with a mathematical interpretation:

> As a proof, we offer the following theorem.
>
> Given any finite connected space t, there is always one and only one function w that, together with its first derivatives, is everywhere continuous in t, and on the boundary of t takes arbitrarily prescribed, continuously varying values, and satisfies the equation
>
> $$\frac{\partial^2 f}{\partial x^2} + \frac{\partial^2 f}{\partial y^2} + \frac{\partial^2 f}{\partial z^2} = 0$$
>
> everywhere in t.

Dedekind then noted that an exactly analogous situation holds in the subject of heat diffusion, where it is also intuitively evident. Then he went on

> We prove the theorem by drawing on pure mathematical evidence. It is in fact reasonable that among all functions u that, together with their first derivatives, are everywhere continuous in t, and take arbitrarily prescribed values on the boundary of t, there must be one (or more) that give the integral taken over the entire space t its least value.

Dedekind first proved that such a minimiser has the required property, and then that it is unique.

18.4 Riemann on Green's Theorem

Riemann lectured on this material in the summer semester of 1861 at Göttingen. His lectures were, one could say, his own reworking of what Dirichlet had done, and they

---

[4]Dedekind would have learned this material, as Riemann did, from Dirichlet's lectures.

were published posthumously in an edition by his former student Karl Hattendorff in 1875, under the title *Schwere, Elektricität und Magnetismus* (Gravity, Electricity and Magnetism).

In Sect. 21, Riemann showed how to construct Green's functions to solve the Dirichlet problem. First, he derived Green's theorem following Green's own argument, to which he referred. Then he assumed that there is a function $U_1(x, y, z)$ that satisfies Laplace's equation inside a domain T bounded by a closed surface S and that takes the value $-1/r$ on S, where r is the distance from a point (x, y, z) to a fixed but arbitrary point $P' = (x', y', z')$ in T. He deferred proof of the existence of the function U_1 until Sect. 34. Then the function

$$U = U_1 + \frac{1}{r}$$

satisfies the Laplace equation in T except at the point P', where it becomes infinite like $1/r$, and it takes the value zero on S.

He now surrounded the point P' with a small sphere of radius c that lies entirely in T. He labelled the interior of this sphere T_2, and the complement of this region in T he called T_1.

Inside T_1 the functions U and V satisfy the conditions of Green's theorem, and Riemann considered the limit as $c \to 0$. The integral

$$\int_T V \nabla^2(U) dx dy dz$$

is zero over all of T and can be ignored. It remained to consider

$$-\int_T U \nabla^2(V) dx dy dz.$$

The space T being closed and bounded the integral taken over T_1 is bounded for any value of c. Inside T_2 the volume element may be taken to be $r^2 \sin\theta dr d\theta d\varphi$, and because U only becomes infinite like $1/r$ the contribution of the integral over T_2 to the whole integral remains finite as $c \to 0$.

The integral over S, the outer boundary of T_1, is

$$\int_S V \frac{\partial U}{\partial n} d\sigma,$$

where $\frac{\partial U}{\partial n}$ is the normal derivative of U. The integral over the inner boundary, the small sphere of radius c, reduces to $-4\pi V(P')$ as $c \to 0$. So Riemann deduced that

$$4\pi V(P') = -\int_T U \nabla^2(V) dx dy dz + \int_S V \frac{\partial U}{\partial n} d\sigma.$$

Riemann next looked at what happens when the function V becomes discontinuous in various ways, corresponding in potential theory to the presence of mass on a surface, on a line, or concentrated at a point.

He also showed that the potential function with given boundary values is unique, and showed how to find it explicitly for regions of various shapes. He showed that if $Q' = (u', v', w')$ is another point in T and $U_{P'}$ and $U_{Q'}$ are Green's functions that become infinite at the points P' and Q', respectively, then

$$U_{P'}(Q') = U_{Q'}(P'),$$

and remarked that this meant that U was a symmetric function of (x', y', z') and (u', v', w').

18.5 Riemann on the Dirichlet Principle

Riemann discussed the existence of a potential function in Sect. 34. He noted that Green had established it by an appeal to physics, but this left a gap that Gauss had filled. Gauss had argued in his ([117], Sects. 31–34) that given any closed surface S in three-space and a continuous function U on the surface, there is a mass distribution M on the surface (which may even be negative) such that the potential function V of this mass distribution differs from the given function U by a constant, and indeed that the mass distribution can be chosen so that $V = U$. In other words, there is a distribution of mass such that the function $V + \frac{1}{r}$ vanishes everywhere on S.

But, said Riemann, this proof was too close to potential theory, and a purely analytic proof was needed. That, he said, had been provided by Dirichlet in his lectures, and he proceeded to describe it. The claim is that any single-valued, finite, continuous function v on a closed surface S in three-space can be extended in a unique way into the interior T so as to remain single-valued, finite, and continuous and satisfy the Laplace equation

$$\nabla^2 v = 0.$$

To prove this theorem, he considered the integral over the inside of S

$$\Omega(u) = \int_T \nabla u . \nabla u \, dx \, dy \, dz,$$

where u agrees with v on S and is continuously differentiable inside S. Evidently there are infinitely many such functions, and Riemann denoted one by u_1 and any other by $u = u_1 + hs$, where h is an arbitrary constant and s is a function of x, y, z that vanishes on S and has the same properties as u inside S.

Then, said Riemann, the integral for $\Omega(u)$ depends on the function u but is always positive and finite. Therefore, he went on, there is a particular function v for which the integral $\Omega(v)$ takes its least value. This value cannot be zero, which is a constant, for then the values on S must all be the same.

Riemann now considered the function $u = v + hs$ and deduced that

$$\Omega(v + hs) = \Omega(v) + hI + h^2\Omega(s),$$

where

$$I = \int_T \nabla(v).\nabla(s)dxdydz.$$

This forces $I = 0$, for otherwise when h is very small and I is negative one could have

$$\Omega(v + hs) < \Omega(v),$$

contradicting the minimality of $\Omega(v)$.

From this Riemann deduced that $I = 0$ is the necessary, and also sufficient, condition for $\Omega(v)$ to take its minimum value. Integration by parts then shows that this condition is the same as

$$\int_T s\nabla^2(v) = 0,$$

and because s is arbitrary therefore that

$$\nabla^2(v) = 0.$$

Riemann now deduced that the function v is continuously differentiable, and that it is unique.

He then showed that there is a unique Green's function U that also solves the Dirichlet problem. He considered the function

$$U = U_1 + \frac{1}{r},$$

where r denotes the distance of the point (x, y, z) from the point $P' = (x', y', z')$ inside S where the function U is infinite, and U_1 satisfies Laplace's equation away from P' and agrees with $-\frac{1}{r}$ on the boundary of the domain. The function $1/r$ satisfies Laplace's equation and so the above function is a Green's function that vanishes on S and is harmonic everywhere except at the point (x', y', z') in ST. The uniqueness of a harmonic function with given boundary conditions had been proved by Dirichlet himself.

18.6 Exercises

1. Green's functions are not easy to compute; find instructive examples on the web.

Questions

1. A chicken and egg question: Which is mathematically more fundamental, the existence of a Green's function for a given domain or a solution of the Dirichlet problem for that domain? Or are they equivalent?
2. Are there examples of problems in physics that lead to an unsolvable Dirichlet problem, or is the problem one for mathematicians only?

Chapter 19
Attempts on Laplace's Equation

19.1 Introduction

The work of Green, Gauss, and Riemann showed very clearly that the study of real functions of two and three variables was a rich domain that would be essential in the study of physics (gravitation and electro-magnetism), and which (in two dimensions) was a powerful tool in the emerging subject of complex function theory. For physicists such as Thomson, Helmholtz, and Maxwell, nature provided the existence and uniqueness theorems upon which the theory rested, but for mathematicians, and especially those a step or more away from theoretical physics, those theorems looked increasingly insecure.

19.2 Weierstrass, Prym, and Schwarz

Weierstrass was not persuaded by the Dirichlet principle, even for planar regions. In a paper he read to the Royal Academy of Sciences in Berlin in 1870, but which was not published until the second volume of his collected works in 1895, he agreed that if the Dirichlet integral exists and attains its minimum then the minimising function is harmonic and unique. But when he turned to the existence question, he offered what he called a simple example to show the inadmissibility of Dirichlet's reasoning. He observed that

$$J = \int_{-1}^{1} \left(x \frac{d\varphi}{dx} \right)^2 dx,$$

where $\varphi(-1) = a \neq b = \varphi(1)$ is always positive and can take any non-zero value however small, but cannot take the value zero unless $\frac{d\varphi}{dx}$ vanishes on the interval $[-1, 1]$, which is ruled out by the boundary conditions.

Weierstrass's example was the function

© The Author(s), under exclusive license to Springer Nature Switzerland AG 2021
J. Gray, *Change and Variations*, Springer Undergraduate Mathematics Series,
https://doi.org/10.1007/978-3-030-70575-6_19

Fig. 19.1 The graph of
arctan(x/0.01)/(arctan(1/0.01)),
$-1 \leq x \leq 1$

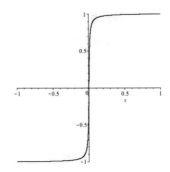

$$\varphi(x) = \frac{a+b}{2} + \frac{b-a}{2} \frac{\arctan(x/\varepsilon)}{\arctan(1/\varepsilon)},$$

where ε is arbitrarily small and positive.[1] The graph of this function with $\varepsilon = 0.01, a = -1, b = 1$, in Fig. 19.1 suggests that this function is likely to do the trick.

$$\varphi'(x) = \frac{b-a}{2\arctan(1/\varepsilon)} \frac{\varepsilon}{x^2 + \varepsilon^2},$$

so the integral is

$$\left(\frac{b-a}{2\arctan(1/\varepsilon)}\right)^2 \int_{-1}^{1} \left(\frac{x\varepsilon}{x^2 + \varepsilon^2}\right)^2 dx.$$

The integrand is always less than $\left(\frac{x}{\varepsilon}\right)^2$ and $\left(\frac{\varepsilon}{x}\right)^2$, and takes its maximum value of $\frac{1}{4}$ at $x = \pm\varepsilon$.

Weierstrass now argued that the integrand is positive and less than $\varepsilon/(x^2 + \varepsilon^2)$, so the integral is less than

$$\frac{\varepsilon}{2} \frac{(b-a)^2}{\arctan(1/\varepsilon)}.$$

So the integral can be arbitrarily small for a suitable function $\varphi(x)$ which has a continuous first derivative, but can never be zero. So, he concluded,

The Dirichlet principle leads in this case to a false result.

It is curious that Weierstrass did not use the simpler straight line version, given by

$$y = \begin{cases} 1 & \text{if } 1/n \leq x \leq 1 \\ nx & \text{if } -1/n \leq x \leq 1/n \\ -1 & \text{if } -1 \leq x \leq -1/n. \end{cases}$$

In this case,

---

[1]It helps to recall that $\frac{d}{dx}\arctan(x/\varepsilon) = \varepsilon/(\varepsilon^2 + x^2)$.

$$J = \int_{-1}^{1} \left(x \frac{d\varphi}{dx} \right)^2 dx = \int_{-1/n}^{1/n} n^2 x^2 dx = \frac{2}{3n},$$

which becomes arbitrarily small as $n \to \infty$. It is unlikely he did not think of it. More likely, he rejected it because the theory of admissible curves that were not smooth was much under discussion at the time and he did not want to weaken his critique of the Dirichlet principle with extraneous considerations of that kind.

Weierstrass's argument destroys the belief that any integral that is bounded below attains its bounds. But it leaves open the possibility that the Dirichlet integral may do so, and that the Dirichlet principle leads to a solution of the Dirichlet problem. However, a former student of Riemann's, Friedrich Prym, showed in his [228] that Riemann's use of the Dirichlet principle can fail, even when Dirichlet's problem can be solved.

Prym did so by exploiting an idea of Riemann's hitherto ignored in studies of the problem, namely, that a continuous function may oscillate wildly. Such functions may depart from their Fourier series representations, unlike the ones studied by Dirichlet that had only finitely many maxima and minima.

Prym considered (in Sect. 8) a disc in the plane of radius $R < \frac{1}{2}$ centred on the origin, and took polar coordinates ρ and τ centred on the point $(-R, 0)$, so ρ takes every value from 0 to $2R < 1$ and τ every value from $-\pi$ to π. On this disc he defined the function u as the real part of the complex function

$$u + iv = i\sqrt{-\ln(R + x + iy)}$$

that was taken to satisfy $-\ln(R + x + iy) = -\ln \rho - i\tau$. He now showed that this function is defined and continuous at the origin of the polar coordinates.

The Dirichlet problem is solved, because the functions u and v are everywhere defined and single valued, even on the boundary of the disc, and the function u is harmonic because it is the real part of a complex function. However, as Prym then showed, Dirichlet's integral $L(u, 0)$ is infinite, because the function u oscillates infinitely often in any neighbourhood of the point $\rho = 0$.

Prym's contribution left open the question of whether the Dirichlet problem could be solved. The leading figure here was Hermann Amandus Schwarz (Fig. 19.2), and in a series of papers around 1870 he was able to solve the problem in a fair degree of generality.[2]

In his paper [244], Schwarz first solved the problem when the domain T is the unit disc. He considered an arbitrary function f on the unit circle that is finite, continuous, and real-valued everywhere (so it corresponds to a periodic function defined on $\mathbb{R}$). He then wrote down the function $u(r, \phi)$ that solves the Dirichlet problem. It is defined by the following equations: $u(1, \phi) = f(\phi)$ and

---

[2] Another way forward was an iterative process described by Carl Neumann, whose work will not be discussed here.

Fig. 19.2 Hermann
Amandus Schwarz
(1843–1921)

$$u(r, \phi) = \frac{1}{2\pi} \int_0^{2\pi} f(\psi) \frac{1 - r^2}{1 - 2r\cos(\psi - \phi) + r^2} d\psi, \quad 0 \le r < 1$$

A careful analysis shows that it is finite and continuous in every closed unit disc
$0 \le r < 1$, and converges to a function that is finite and continuous on the whole
closed unit disc. Once this is established, straightforward differentiation shows that
the function $u(r, \phi)$ satisfies the differential equation $\Delta u = 0$ in the interior of the
disc.

It follows that the Dirichlet problem is solved for any domain conformally equivalent to the unit disc, but Riemann's claim that this was true of any simply connected
domain was far from understood or accepted (or, it should be said, precise). Indeed,
one of Schwarz's earliest papers [242] had been to establish the equivalence precisely
for a disc and a square, thus giving a new twist to the famous problem of squaring the
circle.[3] (This is an early example of what was to become the Schwarz–Christoffel
theorem at work.) So Schwarz next presented a method of extending the solution
from domains where the problem was solved to domains formed by overlapping
such domains in the plane. This gave him a large class of domains for which the
Dirichlet problem had a solution.

---

[3]For his proof that a square can be mapped analytically onto a circle, see a translation of his paper
below, in Sect. 31.3.

19.2.1 Schwarz's Alternating Method

Schwarz in his [243] considered two domains, T_1 and T_2 that overlap in such a way that $T^* = T_1 \cap T_2$, the region common to both of them, is two-dimensional and the boundaries of the regions cross with distinct tangents.

He supposed that the Dirichlet problem on T_1 can be solved for any prescribed (finite, continuous, real-valued) function g_1 on the boundary of T_1, and likewise that it can be solved on T_2 for any function g_2 on the boundary of T_2. He then showed that there is a a solution to the Dirichlet problem for the domain $T_1 \cup T_2$ with any prescribed function u on the boundary $L_0 \cup L_3$. This function will be bounded below, and so without loss of generality Schwarz assumed that $u > 0$.

He divided the boundary of T_1 into two parts, one outside T_2 that he called L_0, and one inside T_2 that he called L_2. Similarly, he called L_3 the boundary of T_2 that lies outside T_1 and L_1 the part of the boundary of T_2 that lies inside T_1. The region $T_1 \cup T_2 \setminus T^*$ he called T. (For a picture, see Fig. 31.1.)

Schwarz's idea was that an air pump (his term) could be imagined that pumped air from the region T^* alternately into the regions $T_1 \setminus T^*$ and $T_2 \setminus T^*$, through the membranes L_1 and L_2. In mathematical terms, he supposed that the Dirichlet problem is first solved on T_1 with the boundary values $u_1 = u$ on L_0 and a constant, k on L_2 (k will later be chosen to be the minimum value of the function u on the boundary $L_0 \cup L_3$). The solution is a harmonic function u_1, say, inside T_1. He then chose the solution of the Dirichlet problem for the region T_2 for a function which took the same values as the function u on L_3 and the values of u_1 on L_1. He then treated the region T_1 as he had just treated T_2 to obtain a harmonic function u_3, then turned to T_2 again to obtain a harmonic function u_4, and so on.

Why does this help? The maximum and minimum values of a harmonic function are taken on the boundary of its domain, and the maximum and minimum values of the function that is u on $L_0 \cup L_3$ are, say, g and k, respectively, and define $G := g - k$, and we note that on L_2 the maximum value of the function $u_2 - u_1$ is less than $g - k$.

Next, the function $u_3 - u_1$ solves the Dirichlet problem where the boundary function is 0 in L_0 and $u_2 - k$ on L_2. So $u_3 - u_1$ is never negative inside T_1, its maximum value is less than G, and its maximum value on L_1 is less than Gq_1, where $q_1 < 1$. A simple scaling argument shows that q_1 depends linearly on the maximum value of the boundary function on L_2.

Similarly, the maximum absolute value of $u_4 - u_2$ on L_1 is less than Gq_1 and on L_2 is less than Gq_1q_2, where $q_2 < 1$.

Continuing in this way, Schwarz obtained two sequences of functions, $\{u_{2i-1}\}$ defined on T_1 and $\{u_{2i}\}$ defined on T_2, with the properties that along L_1, $u_{2i-1} = u_{2i}$ and along L_2, $u_{2i+1} = u_{2i}$. He then defined two new functions

$$u' = u_1 + (u_3 - u_1) + (u_5 - u_3) + \ldots + (u_{2i+1} - u_{2i-1}) + \ldots ,$$

$$u'' = u_2 + (u_4 - u_2) + (u_6 - u_4) + \ldots + (u_{2i+2} - u_{2i}) + \ldots .$$

These series converge unconditionally because successive terms diminish faster than a geometric progression with ratio $q_1 q_2$. (The geometric progression arises from the linearity observation made above.) The function u' is harmonic inside T_1, and the function u'' is harmonic inside T_2, and they agree on the entire boundary of T^*, which is L_1 and L_2. Therefore, the functions u' and u'' agree on T^*, and so they define a harmonic function on the whole of $T_1 \cup T_2$ that has the prescribed boundary values of the function u on the boundary $L_0 \cup L_3$. So the Dirichlet problem is solved for the union of the two regions, which is what Schwarz set out to do.

The alternating method provides a solution to the Dirichlet problem for a large class of regions, including all plane domains with polygonal boundaries with finitely many sides. In general, the boundary can be made up of piecewise analytic arcs crossing transversally, but it is not clear what can happen in the limit, so the case of arbitrary boundaries, even rectifiable ones, was left unresolved. That said, Constantin Carathéodory praised Schwarz in Schwarz's *Festschrift* volume ([30], 20) for separating out the interior part and the boundary part of the Riemann mapping theorem, and indeed Poincaré's method of sweeping out (1890b) similarly made certain simplifying assumptions about the boundary but left the extent of the method unresolved.

On the other hand, as Archibald ([1], 83) points out, Schwarz's paper required familiarity with Weierstrassian methods to understand, and such knowledge was only available to those with access to copies of Weierstrass's lectures and notes taken by the few students capable of doing so. Archibald, quoting ([65], 154), records the astute opinion of Gösta Mittag-Leffler, who was a significant figure in spreading the Weierstrassian model of analysis:

> The Germans themselves are not in general sufficiently familiar with Monsieur Weierstrass's ideas to be able to grasp without difficulty an exposition made strictly on the classical model that the great geometer has given. Take, for example, Monsieur Fuchs ...he regards [Weierstrass's] methods as thoroughly superior to the method of Riemann. And yet he always writes in the manner of Riemann. All this evil derives from the fact that M. Weierstrass has not published his courses. It is true that the Weierstrassian method is taught in several German universities, but everyone is not yet a pupil of Weierstrass or a pupil of one of his pupils.

19.3 Harnack

The study of the Dirichlet problem for general two-dimensional domains was much advanced by Axel Harnack in his book [138].[4] He began by reviewing the theories of Schwarz and Neumann, and observed that these authors had not fully studied the nature of the boundary before admitting, however, that he had not been able to extend their methods. Therefore, he had adopted a different approach using Green's functions. He established existence theorems for functions with prescribed singularities, derived the general theorems in Riemann's paper on Abelian functions, and showed how his ideas led to a proof of the Riemann mapping theorem.

---

[4]Harnack restricted his attention to this case because of the availability of conformal mappings.

Harnack's book was well received, and subsequent mathematicians often made use of what became known as Harnack's theorem ([138], 67). Harnack considered a sequence of harmonic functions u_n that are defined on a surface F and restrict to continuous functions U_n on the boundary of F. He furthermore supposed that for every arbitrarily small δ and for every point s of the boundary there is a finite domain partly bounded by a piece of the boundary of F containing s and that contains interior points of F and is such that the values each function u_n takes on this domain (including its boundary) vary by less than δ. As he remarked, it is a necessary condition that the functions U_n be continuous. Harnack first established the lemma that if the sum $U := \Sigma U_n$ converges uniformly, then the sum $u := \Sigma u_n$ converges at every interior point of the surface F to a harmonic function. From this he deduced Harnack's theorem:

Theorem 19.3 *If a sequence of harmonic functions u_n all have the same sign (say, positive) and the sum $u := \Sigma u_n$ converges at an interior point of the surface F, then it converges at every interior point of the surface F to a harmonic function.*

Alternatively, if a sequence of harmonic functions u_n tend from below to the values of a function u, then u is a harmonic function.

Informally, the sequence of harmonic functions either converges to a harmonic function or it fails to converge at all.

Harnack's book contains a number of advances in what later became point-set topology. He defined a domain to be connected if any two points in it can be joined by a finite polygonal arc that can be covered by overlapping discs all lying in the interior of the domain. Later authors refined this idea and separated the idea of connectedness (here, path-connectedness) from that of a domain (a region for which every point has a disc-like neighbourhood lying entirely in the region). Harnack also defined boundary points to be those points every neighbourhood of which contains some points belonging to the domain and some that do not. He then claimed that a simply connected domain has a continuous boundary, although the boundary may have corners and cusps, be nowhere differentiable and may, implicitly, need not even be rectifiable.[5] A boundary might also have arbitrary number of incisions, lines drawn inwards from boundary points, which would be traversed twice by any circuit of the boundary (Fig. 19.3).

It was with this idea of a domain and its boundary in place that Harnack then proved that there is a Green's function for every bounded region with an arbitrary boundary. He argued in three stages. First, he accepted Neumann's approach establishing the existence of a unique harmonic function that agrees with a given continuous function on the boundary for polygonal regions with no re-entrant angles. Then he showed that if the given function on the boundary is always finite but has isolated jump discontinuities then there is still a harmonic function agreeing with the given one at points on the boundary where the given function is continuous. Finally, he used an approximative argument, which he attributed ultimately to Schwarz, to deal with the

---

[5]Riemann in his [234] had defined a domain as simply connected if any curve in it joining two boundary points divides the domain into two pieces.

Fig. 19.3 A circuit of the
disc with the incision AB
traversed AB twice—once
going up and once coming
down

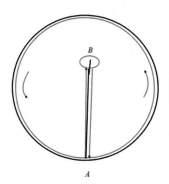

general simply connected domain. He considered that the domain can be steadily
approximated by polygonal regions. He used his theorem from p. 67 to establish
that the sequence of harmonic functions on these domains converged to a harmonic
function on the given domain. Arbitrary bounded domains were then patched together
out of simply connected pieces. To prove the Riemann mapping theorem Harnack
used his Green's function approach to establish the existence of a suitable harmonic
function and consequently of a complex function mapping the given bounded domain
onto a circle. He also showed in this way that non-simply connected domains can
be mapped onto a domain bounded by several circles (a problem Riemann had also
investigated in work that was still unpublished).

In the event, it was to turn out that the Dirichlet problem can be solved for a
very large class of boundaries of a two-dimensional, disc-shaped region, but that the
problem in three dimensions can only be solved for a restricted class of boundaries
(without spikes, for example). For that reason, and because of the strong connection
to complex function theory, I have kept the story that follows to two dimensions, and
even then the full history of potential theory in the period is too rich to describe here.
For a look at some of the major issues that were raised, and how some of them were
solved, see Appendix D.

19.4 Exercises

1. Find some domains homeomorphic to a disc whose boundaries are not homeomorphic to circles.

Questions

1. Try to follow through the first few stages of Schwarz's alternating method when
   the initial values on L_0 are 1 and the initial values on L_3 are 3, noting the values
   assigned at each stage to L_1 and L_2. Does it strike you as likely that the method
   will lead to good approximations to the sought-for harmonic function?

Chapter 20
Applied Wave Equations

20.1 Introduction

The wave equation is one of the most useful in physics. Here we look at the dramatic story of the trans-Atlantic cable and the later introduction of the telegraphist's equation.

The best resource for the history and context of the trans-Atlantic cable up to the present day is surely the *History of the Atlantic Cable & Undersea Communication* at atlantic-cable.com. See, among other things, Bern Dibner's *The Atlantic Cable* (1959). When I gave this course in 2017 it seemed to me that everything was here except an explanation of the mathematics. Since then I am pleased to say that Liam Morris, a student on the course that year, has posted an account of the mathematics.

20.2 The Trans-Atlantic Cable

The wave equation was at the mathematical heart of a dramatic nineteenth-century story: the struggle to connect Britain and America by a trans-Atlantic cable.

The first working electric telegraphs were produced in the 1830s, and soon a network of cables crossed Europe and spread throughout the eastern seaboard of the United States. Information could now be sent reliably, and very much faster than by a man on a string of horses, and typically it came transcribed letter by letter into a stream of short and long pulses—such as the dots and dashes of Morse Code.

However, it was not so easy to connect Britain to Continental Europe. It was discovered that placing a cable under water increased its capacitance (the ability of a body to store electric charge). The increased capacitance caused the signal to spread out, so that the gap between one item and the next had to be increased, causing the time for a message to be transmitted to increase. The English Channel is not very wide, at its narrowest it is only some 22 miles. It was much riskier to run a cable

© The Author(s), under exclusive license to Springer Nature Switzerland AG 2021
J. Gray, *Change and Variations*, Springer Undergraduate Mathematics Series,
https://doi.org/10.1007/978-3-030-70575-6_20

across the Atlantic, but it would surely be extremely valuable and therefore attempts had to be made.

To understand the problem, it is necessary to consider what is called the *telegraphist's equation*. This equation describes the current u at any point x in a straight wire at any time t during the transmission of electric signals down the wire. It is

$$KL\frac{\partial^2 u}{\partial t^2} + (KR + LS)\frac{\partial u}{\partial t} + RSu = \frac{\partial^2 u}{\partial x^2}, \tag{20.1}$$

where the constants that appear in the equation involve the capacitance K, the self-inductance L, the resistance R, and the leakage S of the wire. (Self-inductance is the induction of a voltage in a current-carrying wire as the current changes.)

It was written down for the first time by the German physicist Gustav Kirchhoff in 1857, and profoundly studied by the brilliant but eccentric English physicist Oliver Heaviside in 1876, but prior to them William Thomson had cleverly exploited a simpler equation to design a cable that would work across the Atlantic.[1]

The equation Thomson derived for the flow of electricity down a wire was the heat equation. Although he did this on physical grounds, it was nonetheless the case that this equation and Fourier's study of it were at the core of his thinking. This was not Eq. (20.1) later derived by Kirchhoff, but it can be obtained from it when the inductance L is negligible by comparison with the resistance R, so the constant KL may be taken to be zero, and the equation becomes the one-dimensional heat equation,

$$K\frac{\partial u}{\partial t} + Su = \frac{1}{R}\frac{\partial^2 u}{\partial x^2}.$$

Thomson, however, was not aware of self-inductance so his methods take no account of it.

It is worth noting that mathematicians often seek to understand a complicated equation such as this one by simplifying it. For example, in the case at hand, if one assumes there is no resistance ($R = 0$) and no leakage ($S = 0$) then the equation reduces to the wave equation:

$$KL\frac{\partial^2 u}{\partial t^2} = \frac{\partial^2 u}{\partial x^2}.$$

Thomson's simpler equation implies that an instantaneous pulse sent down a wire of length x lasts for a time T proportional to x^2 seconds (see below), and so two separate pulses must be transmitted T seconds apart in order to be received as distinct signals at the far end. But in 1855 Thomson's advice was ignored.

Attempts to lay the cable were dogged by failure. The first cable, laid in 1857, snapped after 338 miles. A second cable, laid in 1858, succeeded, and a 99-word message was sent from Queen Victoria to President Buchanan to mark the event,

---

[1]See Thompson [256], Kirchhoff [155], Heaviside [140], and Rayleigh's *Theory of Sound*, Vol. 1 p. 466.

Fig. 20.1 William
Thomson, later Lord Kelvin
(1824–1907), *Memorials of
the Old College of Glasgow*,
Glasgow, 1871

but this only revealed a greater failure: the message took $16\frac{1}{2}$ hours to transmit. Misguided attempts to improve performance made things worse, and after a month the cable had to be abandoned. As the mathematician Thomas Körner put it, "2500 tons of cable and £350, 000 of capital lay useless on the ocean floor".[2]

Thomson (Fig. 20.1) had opposed the original design, and was now placed in a position to insist on his insights being implemented. Now, "Half a million pounds was being staked on the correctness of the solution to a partial differential equation".[3]

As before, the first attempt at laying the cable, which was much heavier than the earlier one, failed when the cable broke, and another one had to be laid. But they were able to recover the ends of the broken cable and reconnect it, and on 8 September 1866 America and Europe were joined by two cables that, moreover, worked as planned. Signals could be transmitted at roughly eight words a minute, a decisive improvement on the earlier, and by then defunct, cable. Thomson was knighted and further rewarded with considerable amount of money, some of which he used to buy an ocean-going yacht—he was a keen sailor.

Thomson's telegraphist's equation is the heat equation. There is only space here to describe his solution, not to prove it.[4]

The problem is that of the distribution of heat in a semi-infinite one-dimensional rod $x \geq 0$, which may not make much sense as a problem about heat but makes good sense if we think of the rod as a wire.

Define the functions $f_-(w) = \exp\left(-\frac{(x-w)^2}{4Kt}\right)$ and $f_+(w) = \exp\left(-\frac{(x+w)^2}{4Kt}\right)$. Then the function

$$\theta(x, t) = \frac{\theta_0}{2\sqrt{\pi K t}} \int_0^\infty (f_-(w) - f_+(w))dw$$

is a solution of the heat equation

---

[2] See Körner ([164], 334).

[3] See Körner ([164], 336).

[4] For the mathematical details, see Körner ([164], Chap. 62).

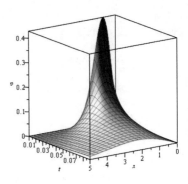

$$\frac{\partial \theta}{\partial t}(x, t) = K \frac{\partial^2 \theta}{\partial x^2}$$

for which $\theta(x, t) \to \theta_0$ as $t \to 0$ for all $x > 0$ and $\theta(0, t) = 0$ for all $t > 0$.

If instead it is required that $\theta(0, t) \to f(t)$ as $x \to 0+$ for all $t > 0$, then the solution becomes

$$\theta(x, t) = \frac{1}{2\sqrt{\pi K}} \int_0^\infty f(s) \frac{x}{(t - s)^{3/2}} \exp\left(-\frac{x^2}{4K(t - s)}\right) ds.$$

So if the question becomes what is the result of briskly heating one end—by a function $f(t)$ that is zero outside of a small interval (say $f(t) = 0$, $t > a$)—then the above expression is the answer.

In these circumstances, the solution is well approximated by

$$\theta(x, t) = \int_0^a f(s) P(x, t) ds,$$

where

$$P(x, t) = \frac{x}{2\pi^{1/2}(Kt)^{3/2}} \exp\left(-\frac{x^2}{Kt}\right).$$

So if the input is an initial, short blip—a pulse concentrated in a very short interval—the output at a point x at time t is given by the graph in Fig. 20.2. Slices for constant t show the shape of the pulse at that time; slices for constant x show what happens at that point as time goes by.

If $t = t_0$, a constant then $P(x, t_0)$ is a function of x, and

$$\frac{\partial}{\partial x} P(x, t_0) = \frac{1}{2\pi^{1/2}(Kt_0)^{3/2}} \exp\left(-\frac{x^2}{Kt_0}\right)\left(1 - \frac{2x^2}{Kt_0}\right).$$

The maximum value occurs at

$$x = \sqrt{\frac{K t_0}{2}} \quad \text{and is} \quad \frac{\sqrt{2}}{4\pi^{1/2}} \frac{e^{-1/2}}{K t_0}.$$

So when t_0 is small the maximum is quite large and occurs for a small value of x, but when t_0 is large the maximum is quite small and occurs for a large value of x. This confirms what the graph suggests, that the signal travels down the wire getting weaker as it goes.

If $x = x_0$ is a constant then $P(x_0, t)$ is a function of t, and

$$\frac{\partial}{\partial t} P(x_0, t) = \frac{-3}{4\pi^{1/2}(Kt)^{5/2}t} x_0 \exp\left(-\frac{x_0^2}{Kt}\right)\left(Kt - 2\frac{x_0^2}{3}\right).$$

The maximum value occurs at

$$t = \frac{2}{3K} x_0^2 \quad \text{and is} \quad \left(\frac{3e^{-3/2}}{2^{5/2}\pi^{1/2}}\right)\frac{1}{x_0^2}.$$

So when x_0 is small the maximum is reached very quickly and is quite large, but when x_0 is large the maximum is quite small and occurs for a large value of t. This again confirms what the graph suggests, that the signal travels down the wire getting weaker as it goes.

We see that the output is a pulse that rises very quickly from zero to a maximum and then steadily and slowly declines to zero. As a result, the solution of the heat equation is immediately non-zero everywhere, rises to a maximum that is inversely proportional to x^2 for a time proportional to x^2, and then declines.

We can also see that when x_0 is large, the pulse broadens as it travels by an amount proportional to the square of the length of the distance it has travelled. For in $\frac{\partial}{\partial t} P(x_0, t)$ the first term in t is $t^{-7/2}$ and so very nearly zero, and in particular almost a constant, and the exponential term is likewise very nearly 1 and therefore also essentially a constant. Therefore, the variation in $P(x_0, t)$ around its maximum is determined by the final term in t, which is quadratic in x_0.

Moreover, all these values depend on the value of K, so that is where the physics comes in: finding the materials that give the best value for the shape of the pulse. In particular, K should be as small as possible to keep the pulse sharp.

20.3 Poincaré's Solution

The full telegraphist's equation was solved for the first time by Poincaré in 1893.[5] In his short paper [217], he took the telegraphist's equation in the form

---

[5]Poincaré had been fascinated by telegraphy as a boy and was eager to explain how it worked to anyone, especially family members, how it worked. He continued this interest throughout his life, writing on wireless telegraphy too.

$$A\frac{\partial^2 V}{\partial t^2} + 2B\frac{\partial V}{\partial t} = C\frac{\partial^2 V}{\partial x^2}.$$

He then supposed that the physical units were so chosen that this equation becomes

$$\frac{\partial^2 V}{\partial t^2} + \frac{\partial V}{\partial t} = \frac{\partial^2 V}{\partial x^2},$$

and the velocity of the signal is that of light (and is 1 in these units). He then set $V = Ue^{-t}$ and reduced the equation (as Fourier had done in his study of heat) to

$$\frac{\partial^2 U}{\partial t^2} = \frac{\partial^2 U}{\partial x^2} + U.$$

In so doing, he assumed that $B^2 - 4AC$ is non-zero.

He now looked for a solution corresponding to the initial conditions $U = f$ and $\frac{\partial U}{\partial t} = f_1$ where f and f_1 vanish outside the interval $a \le x \le b$ and are polynomials in between. He obtained the solutions by the method of Fourier transforms, which lies outside this course, and was able to show that the solutions are a combination of two basic types, one in which the initial conditions are $f = 0$, $f_1 \ne 0$ and the other in which $f \ne 0$, $f_1 = 0$.

In the first case, the solution is a certain Bessel function $\Lambda(x, t)$ for $-t < x < t$ and zero otherwise.[6] This is an interval of length $2t$.

In the second case, the solution is a more complicated expression involving the same Bessel function that is non-zero outside the interval $a - t, b + t$, which is also an interval of length $2t$.

This means that what goes in as a pulse of width $b - a$ comes out as an interval of increasing width. However, Poincaré was able to show that if $b - a$ is very small then the solutions are

$$U(x, t) = \frac{1}{2}f(x - t), \quad a + t < x < b + t,$$

$$U(x, t) = \frac{1}{2}f(x + t), \quad a - t < x < b - t,$$

and $U(x, t) = 0$ otherwise. More precisely, the other terms in the solution depend on $b - a$ and are negligible if $b - a$ is very small. The same is true of the solution to the original equation: $V(x, t) = U(x, t)e^{-t}$.

However, if $b - a$ has a finite size, then the solutions will take the form of a pulse with a head and then a tail of length proportional to t and therefore to x, the length of the wire. As he put it, if a pulse of some simple kind is transmitted between times $t = a$ and $t = b$

---

[6]Bessel functions arise in the oscillations of a hanging chain, and are standard fare in applied mathematics.

one sees first of all that the *head* of the perturbation will travel with a certain speed, in such a way that in front of this head the perturbation is zero, contrary to what happens in Fourier's theory of heat and in agreement with the laws of propagation of light or of plane sound waves deduced from the equation of the vibrating string. But there is an important difference with this latter case, because the perturbation, as it propagates, leaves behind a non-zero residue …. If $b - a$ is small …the residue is negligible in front of the principal perturbation, but this is not the case if the perturbation lasts for a long time and if $b - a$ is finite. The residue can then disturb the observations,

Therefore, when an attempt is made to transmit a periodic wave down the wire, the velocity and wavelength depend on the frequency, the waves undergo dispersion, and the head of the disturbance moves with a finite speed. This is also the case with the transmission of light but not of heat, and the head, once it has passed, leaves behind a disturbance which never vanishes, unlike what happens with the wave equation.

Poincaré was apparently unaware of a remarkable discovery that Heaviside had made in 1887, when he showed that the values of the physical constants can be so adjusted that the rate of dispersion is zero. This can be done both mathematically and physically, it merely requires that the leakage be non-zero. Far from being an inconvenience, this condition is necessary for the production of distortionless telephony. The signal becomes fainter over distances, but this can be corrected by fitting amplifiers. Long-distance telegraphy had dealt with distortion by accepting a low transmission rate, so as to separate the pulses. Telephony required much higher frequencies; with some leakage and a deliberately high self-inductance it became distortionless. Long-distance communication was reborn—although the money for the first successful patents went to the American electrical engineer Michael Pupin in 1901, and not to Heaviside.[7]

It is intriguing to see that Poincaré also failed to mention Heaviside's ingenious discoveries in his lecture course in 1894, *Cours sur les oscillations électriques*. There he surveyed a considerable amount of mostly French experimental work, with a view to deciding between the old theory of electro-magnetism (due to Kirchhoff) and the modern theories of Maxwell and Hertz. The reason may have been a misplaced interest in the general case. Poincaré's analysis of the telegraphist's equation depended on the condition $B^2 - 4AC \neq 0$ or $KR \neq LS$, but equality in these cases is exactly the condition upon which Heaviside's insight depends. So the experimental work was given a theoretical twist and technological implications were not mentioned.

20.3.1 Conclusion

The principal three partial differential equations that we have considered, the heat equation, Laplace's equation, and the wave equation, became known as the differential equations of mathematical physics. It is a striking fact that between them they describe so many of the advances in applied mathematics made in the nineteenth century and into the twentieth, and it is fortunate that in many cases they can

---

[7]See Yavetz [276].

be solved when appropriate boundary or initial conditions are specified, for, as the telegraphist's equation indicates, rigorous solution methods for general partial differential equations are hard to find. But it is remarkable how much of the modern world was made possible by the study of the calculus of functions of several variables.

20.4   Exercises

Questions

1. The telegraphist's equation can also be seen as a variant of the wave equation (technically, in the language of a later chapter, it and the wave equation are both hyperbolic partial differential equations). What does it mean that a good—indeed, financially successful—understanding of it can be obtained by treating it as the heat equation?

Chapter 21
Revision

21.1 Revision and Assessment 2

This chapter is given over to revision and discussion of the second assignment, see H.3.

I also recommended that students read some of Sergiu Klainerman's essay from 2000: "PDE as a unified subject". Of course, it is sometimes obscure at this stage. Many of the themes that have driven research into partial differential equations in the twentieth century have not been broached in this course or, very likely, in any undergraduate course. But the first 14 pages, omitting pages 5 and 6, are surprisingly intelligible, and in any case they are part of an answer to the traditional request from better students to be told something about what research mathematicians do. Perhaps more to the point, these 14 pages are a modern reflection on the themes that occupy the final part of this book, and will be worth students thinking about them when writing their final essay.

The essay is available on the web at
https://web.math.princeton.edu/~seri/homepage/papers/telaviv.pdf

© The Author(s), under exclusive license to Springer Nature Switzerland AG 2021   225
J. Gray, *Change and Variations*, Springer Undergraduate Mathematics Series,
https://doi.org/10.1007/978-3-030-70575-6_21

Chapter 22
Riemann's Shockwave Paper

22.1 Introduction

Riemann was very interested in mathematical physics. He published four papers on various aspects of it in his lifetime, and four more were published after his death. Of the papers he published, the one on the formation of shockwaves [236] has at least two major claims to fame. It is the first paper to explore the phenomenon, and it made a contribution to the theory of hyperbolic partial differential equations that is still in use today.

At the start of this paper, Riemann remarked that just as the study of linear partial differential equations had been most fruitful when special physical problems were investigated rather than general ones, so too the study of non-linear problems was likely to benefit from studying physical problems and taking all factors into account.

22.2 Riemann's Paper

Riemann's [236]—as its title indicates—is about plane waves of finite amplitude. In the papers by d'Alembert, Euler, and others, and several later authors, only waves of infinitesimal amplitude had been considered. Poisson had published a long and difficult paper on waves of finite amplitude in 1807, and more recently the leading German physicist, Hermann von Helmholtz had published two more papers on experimental investigations of the subject. In one of them, he was the first person to explain the phenomenon of overtones. Then the subject had passed to British applied mathematicians, as Riemann noted.

What is most interesting about Helmholtz's paper on overtones was his discovery that while the superposition of sound waves in the air is linear when the oscillations are infinitesimal, they are not linear for waves of finite amplitude. Instead, overtones arise when the squared amplitude of the waves exerts a force comparable to that

© The Author(s), under exclusive license to Springer Nature Switzerland AG 2021
J. Gray, *Change and Variations*, Springer Undergraduate Mathematics Series,
https://doi.org/10.1007/978-3-030-70575-6_22

causing the oscillations, and so the partial differential equation describing the motion is necessarily non-linear.

But dealing with such waves was the least of Riemann's novelties. The paper is famous for two things:

1. dealing with a non-linear second-order partial differential equation with discontinuous solutions, and showing that the zone of influence is wedge shaped. The equation is hyperbolic, and its characteristics represent the path in space-time of the signals.
2. his methods, which proved of lasting significance in investigating hyperbolic partial differential equations.

Riemann considered a compressible gas in which motion takes place along the x-axis. At time t and position x, the density is ρ, the pressure p, and the velocity is u. The relation between pressure and density is given by a function

$$p = \varphi(\rho),$$

where all that is known about $\varphi(\rho)$ is that its derivative is always positive: $\varphi'(\rho) > 0$. This says, reasonably enough, that pressure increases with density.

He obtained these differential equations for ρ and u (Sect. 1, p. 147):

$$\rho_t = -\frac{\partial}{\partial x}(\rho u),$$

$$\rho(u_t + u u_x) = -\varphi'(\rho)\rho_x.$$

In terms of $\lambda = \log \rho$, the first of these equations can be written as

$$\lambda_t + u\lambda_x = -u_x$$

and the second as

$$u_t + u u_x = -\varphi'(\rho)\lambda_x.$$

To simplify these equations, he defined

$$f(\rho) = \int \varphi'(\rho)d\lambda$$

and

$$r = \frac{1}{2}(f(\rho) + u), \text{ and } s = \frac{1}{2}(f(\rho) - u).$$

The new variables r and s will be shown to be the coordinate variables that simplify the partial differential equation. They are also a pair of characteristics, and it may be helpful to look at the much simpler topic of Burgers' equation (see Appendix B.2) before proceeding.

For brevity, let us also write

$$\alpha = u + \sqrt{\varphi'(\rho)}, \ \beta = u - \sqrt{\varphi'(\rho)}.$$

Riemann now deduced in a few lines that

$$r_t = -\alpha r_x$$

$$s_t = -\beta s_x.$$

From the formula $dr = r_x dx + r_t dt$, he deduced that

$$dr = r_x(dx - \alpha dt)$$

and so r is constant along the curve defined by

$$\frac{dx}{dt} = \alpha$$

and so the point with a constant value of r moves forward with velocity α in the direction of increasing x. Similarly, points with constant s move backwards with velocity $-\beta$ in the direction of decreasing x.
As he put it

a particular value of r, or of $f(\rho) + u$, moves towards larger values of x with velocity $\sqrt{\varphi'(\rho)} + u$, while a particular value of s, or of $f(\rho) - u$, moves towards smaller values of x with velocity $\sqrt{\varphi'(\rho)} - u$.

A definite value of r will gradually meet with each value of s lying ahead of r, and the velocity of its progress will depend at a given moment on the value of s with which it meets.

A further calculation, the details of which I omit, led Riemann to observe that the differential

$$(x - \alpha t)dr + (x - \beta t)ds$$

is exact, and if it is set equal to dw then w satisfies the partial differential equation

$$w_{rs} = m(w_r + w_s), \tag{22.1}$$

where m is a function of $r + s$. In fact, on setting $f(\rho) = r + s = \sigma$,

$$m = -\frac{1}{2}\frac{d\log\frac{d\rho}{d\sigma}}{d\sigma}.$$

However, if standard hypotheses about gases are admitted (Poisson's and Boyle's law) then, Riemann showed, it is possible to reduce to the situation where $m = -\frac{1}{2a}$, where a is a constant of proportionality in Boyle's law. This formulation depends on

r and s not being constants, and Riemann also looked quickly at the situation where either r or s is constant (if r is constant then w is a function of s alone, and if s is constant then w is a function of r alone).

The change of coordinates from x and t to r and s depends on r and s not being constant, and so, as Riemann noted, his method gives no information about any region in which r or s is constant. Moreover, the coordinate change is only valid where the Jacobian is finite and non-zero, and this Jacobian is

$$2\sqrt{\varphi'(\rho)}r_x s_x.$$

The cases $r_x = 0$ and $s_x = 0$ have been discussed. What is much more interesting is Riemann's argument about ρ as a function of x. From its definition it follows that the graph of ρ as a function of x varies in time, and the higher values of ρ increase faster than the lesser values. So if the graph is an increasing function it evolves as time goes by into a less steep function. But if the graph is that of a decreasing function, it can evolve into the graph of a multi-valued function of x, which is absurd. For this to happen, Riemann showed, it is enough that $r_x = \infty$.

Riemann had now arrived at the linear second-order partial differential equation (22.1). He now began to single out a number of important features of its solutions, but first, in Sect. 4, he made some general observations of lasting significance.

We treat first of all the case where the initial disturbance of equilibrium is restricted to a finite region defined by the inequalities $a < x < b$. Thus outside this interval, u and ρ, and consequently r and s, are constant. The values of these quantities for $x < a$ are denoted with suffix 1; for $x > b$ suffix 2. The region in which r is variable gradually moves forward according to Section 1, its lower bound having velocity $\sqrt{\varphi'(\rho_1)} + u_1$, while the upper bound of the region, in which s is variable, moves backward with velocity $\sqrt{\varphi'(\rho_2)} - u_2$. After a time interval

$$\frac{b - a}{\sqrt{\varphi'(\rho_1)} + \sqrt{\varphi'(\rho_2)} + u_1 - u_2},$$

the two regions separate, and between them a gap forms in which $s = s_2$ and $r = r_1$, and consequently the gas particles are again in equilibrium. Thus from the initially disturbed location, two waves issue in opposite directions. In the forward wave, $s = s_2$; accordingly, to a particular value ρ of the density is associated the velocity $u = f(\rho) - 2s_2$, and both values [i.e. of density and velocity, JJG] move forward with constant velocity

$$\sqrt{\varphi'(\rho)} + u = \sqrt{\varphi'(\rho)} + f(\rho) - 2s_2.$$

In the wave moving backward, on the other hand, the velocity $-f(\rho) + 2r_1$ is associated to the density ρ, and these two values move backward with velocity $\sqrt{\varphi'(\rho_1)} + f(\rho) - 2r_1$. The rate of propagation is greater for greater densities, because both $f(\rho)$ and $\sqrt{\varphi'(\rho)}$ increase with ρ.

If we think of ρ as the ordinate of a curve for the abscissa x, then each point of this curve moves forward parallel to the x-axis with constant velocity. Indeed the greater the ordinate, the greater the velocity will be. It is easy to see that, according to this law, points with greater ordinates would finally overtake preceding points, with smaller ordinates, so that to a given value of r would correspond more than one value of ρ. Since this cannot occur in physical reality, a condition must enter that renders the law invalid. In fact, the derivation of the differential equation is based on the assumption that u and ρ are continuous functions of r having finite derivatives. However, this assumption ceases to hold as soon as the density curve is perpendicular to the x–axis at some point. From this moment on, a discontinuity appears in this curve, so that a greater value of ρ immediately succeeds a smaller value. This case will be discussed in the next section.

The compression waves, that is, the parts of the wave in which the density increases in the direction of propagation, become ever narrower with their forward progress and finally become compression shocks. However, the width of the expansion waves grows in proportion to elapsed time.

We may easily show, at least under the assumption of Poisson's (or Boyle's) law, that in the case when the initial disturbance of equilibrium is not confined to a finite region, compression shocks must also form in the course of the motion, excluding quite special cases. The velocity with which value of r moves forward is

$$\frac{k+1}{2}r + \frac{k-3}{2}s$$

under this hypothesis. Thus larger values will, on average, move with greater velocity. A larger value r' must eventually overtake a preceding smaller value r'', unless the value of s corresponding to r'' is, on average, smaller by

$$(r' - r'')\frac{1+k}{3-k}$$

than the value of s simultaneously corresponding to r'. In this case, s becomes negatively infinite for positive infinite r, and thus for $x = +\infty$, the velocity u is $+\infty$ (or instead, the density, according to Boyle's law, becomes infinitely small). Thus excluding special cases, it must always transpire that a value of r, larger by a finite amount, follows immediately after a smaller value. Consequently, since $\frac{\partial r}{\partial x}$ becomes infinite, the differential equations lose their validity, and forward-moving compression shocks must occur.

In the next sections, Riemann showed how the compression shocks propagate, he showed that the values of u and ρ on either side of the shock are linked. Riemann, however, failed to ensure conservation of energy; the relevant relations were provided in Rankine [232], Rayleigh [233], and Hugoniot [147, 148]. But Riemann did notice that the shocks must be supersonic with respect to the state in front of them and subsonic behind.[1]

The analysis naturally depends on the initial conditions, and he showed that when u and ρ each have two different constant values on $x < 0$ and $x > 0$ two waves emerge from the point of discontinuity and each could be either a compression or a rarefaction wave. He analysed all four cases.

In the final sections of the paper, Riemann did not show how to solve the partial differential equation (22.1) but how to transplant the method of Green's functions from the elliptic to the hyperbolic case. His method of defining and using the adjoint of the given partial differential equation in order to solve the equation has since become standard.

His method was extremely ingenious. He wished to solve a partial differential equation by a function that, with its first derivatives, takes given values on a given (non-characteristic) curve. To do this, he introduced a new partial differential equation

---

[1]Rankine cited four previous authors: Poisson, Stokes, Airy, and Earnshaw, but not Riemann; Riemann cited only Helmholtz; Rayleigh and Hugoniot did not mention Riemann. The first person to follow Riemann was E.B. Christoffel in his [41].

(technically, this is the adjoint of the original partial differential equation) where he was free to specify its boundary conditions, provided they meet certain constraints. He then showed that a solution to the equation he wished to solve can be found if a solution of the new equation can be found. But that equation can be solved—although it is very similar to the first one—because the boundary conditions can be suitably chosen. Therefore, the original equation can be solved.

22.3   Darboux on Riemann's Approach to the Shockwave Equation

To complete the story, we now follow Darboux's use of Riemann's method to solve Euler's equation (4.6); I shall consider only the case $\beta \neq \beta'$.

The adjoint equation to that equation is

$$\frac{\partial^2 u}{\partial x \partial y} + \frac{\beta'}{x - y}\frac{\partial u}{\partial x} - \frac{\beta}{x - y}\frac{\partial u}{\partial y} - \frac{\beta + \beta'}{(x - y)^2} = 0.$$

On setting $u = (x - y)^{\beta + \beta'} v$ the last term disappears, so the adjoint equation can be treated in the form

$$\frac{\partial^2 v}{\partial x \partial y} - \frac{\beta}{x - y}\frac{\partial v}{\partial x} + \frac{\beta'}{x - y}\frac{\partial v}{\partial y} = 0. \tag{22.2}$$

Let us denote a solution of this equation by $Z(\beta', \beta)$, so

$$v = (x - y)^{\beta + \beta'} Z(\beta', \beta).$$

We already know that $x^\lambda F(-\lambda, \beta', 1 - \lambda - \beta, y/x)$ is a solution of Eq. (4.6), so switching β and β' gives $v = x^\lambda F(-\lambda, \beta, 1 - \lambda - \beta', y/x)$ as a solution of Eq. (22.2).

This can be souped up by applying Möbius transformations to x and y into a more general solution

$$v = (y_0 - x)^\lambda (x - x_0)^{-\beta' - \lambda} F(-\lambda, \beta, 1 - \lambda - \beta', \sigma),$$

where $\sigma = \dfrac{(x - x_0)(y - y_0)}{x - y_0)(y - x_0)}$. Therefore,

$$u = (y - x)^{\beta + \beta'}(y_0 - x)^\lambda (x - x_0)^{-\beta' - \lambda} F(-\lambda, \beta, 1 - \lambda - \beta', \sigma)$$

is a solution to the adjoint equation.

Still following Darboux, we now seek a solution u of the adjoint equation that is

$$e^{\int_{y_0}^{y} a\, dy} = \left(\frac{y - y_0}{y_0 - x_0} \right)^{\beta'}$$

when $x = x_0$, and is $\left(\dfrac{y_0 - x}{y_0 - x_0} \right)^{\beta}$ when $y = y_0$. When $x = x_0$ we have $\sigma = 0$, the series for F reduces to 1, and the factor $(x_0 - x)^{-\lambda + \beta'}$ that makes u either zero or infinite unless $\lambda = -\beta'$.

Setting $\lambda = -\beta'$ we find that

$$u = (y_0 - x)^{-\beta'} (y - x)^{\beta + \beta'} (y - x_0)^{-\beta} F(\beta, \beta', 1, \sigma);$$

and indeed if $x = x_0$ then $u = \left(\dfrac{y - y_0}{y_0 - x_0} \right)^{\beta'}$, and if $y = y_0$ then $u = \left(\dfrac{y_0 - x}{y_0 - x_0} \right)^{\beta}$. So u is the solution of the adjoint equation that we seek.

Finally, to solve Eq. (4.6) in the most general form one substitutes this value of u in either Eq. (31.32) or (31.33). For example, from (31.33) one deduces

$$z_{x_0, y_0} = (uz)_{x_1, y_1} + \int_{x_0}^{x_1} u_{x, y_1} f_1(x)\, dx + \int_{y_0}^{y_1} u_{x_1, y} f_2(y)\, dy.$$

Recall that in this notation, f_1 and f_2 are two arbitrary functions that depend on the boundary conditions for z and $\Phi_{\alpha, \beta}$ is what is obtained by replacing x and y by α and β in the function $\Phi(x, y)$.

22.4 Telegraphy

At this point in the second edition of his *Leçons* Darboux showed how to connect these ideas to the study of telegraphy, which we considered in Chap. 20.

He first showed how to deduce a solution to Euler's equation (4.6) in the case when $\beta = \beta'$ from a solution when $\beta \neq \beta'$. The equation becomes

$$\frac{\partial^2 z}{\partial x \partial y} = \frac{\beta(1 - \beta)}{(x - y)^2} z,$$

and the solution he obtained in this fashion is

$$u = (y_0 - x)^{-\beta} (y - x)^{2\beta} (y - x_0)^{-\beta} F\left(\beta, \beta, 1, \frac{(x - x_0)(y - y_0)}{(x - y_0)(y - x_0)} \right).$$

He replaced β by $\beta - x$ in the equation, and let $\beta \to \infty$ when the equation becomes

$$\frac{\partial^2 z}{\partial x \partial y} = z,$$

which he remarked "is a transform of the so-called *telegraphist's equation*".

This allowed him to solve the telegraphist's equation by what he called indirect methods, but he said it was better to apply Riemann's methods.

He began with the telegraphist's equation in the form

$$\frac{\partial^2 U}{\partial x^2} = \alpha \frac{\partial^2 U}{\partial t^2} + 2\beta \frac{dU}{dt},$$

where α and β are positive constants and $U = U(x, t)$ denotes the potential at time t and a distance x. He wrote

$$U = e^{-\beta t/\alpha} u,$$

and chose units in which the speed of light is unity, and so the equation became

$$\frac{\partial^2 u}{\partial x^2} - \frac{\partial^2 u}{\partial t^2} + u = 0. \tag{22.3}$$

The characteristics of this equation are

$$x + t = const., \quad x - t = const.$$

Darboux then continued Sect. 361 as follows:

This equation is its own adjoint. If u and v are two distinct solutions of it, then

$$u \frac{\partial^2 v}{\partial x^2} - v \frac{\partial^2 u}{\partial x^2} - u \frac{\partial^2 v}{\partial t^2} + v \frac{\partial^2 u}{\partial t^2}$$

$$= \frac{\partial}{\partial x} \left(u \frac{\partial v}{\partial x} - v \frac{\partial u}{\partial x} \right) - \frac{\partial}{\partial t} \left(u \frac{\partial v}{\partial t} - v \frac{\partial u}{\partial t} \right).$$

So the integral

$$\int \left(u \frac{\partial v}{\partial x} - v \frac{\partial u}{\partial x} \right) dt + \left(u \frac{\partial v}{\partial t} - v \frac{\partial u}{\partial t} \right) dx$$

vanishes on any contour.

Let us then, as in Sect. 359, form a contour partly composed of characteristics. Let Ox be the x-axis and Ot be the t-axis. The characteristics are represented by lines parallel to the bisectrices of the axes. If A is an arbitrary point of the plane, we draw through this point two characteristics that cut an arbitrary curve in two points β and γ, and we take the above integral around the contour $A\beta\gamma A$.

On $A\beta$ one has $dx = dt$ so one can exchange dx and dt; as a result the corresponding portion of the integral will be

$$\int_A^\beta (u dv - v du).$$

Similarly, the portion of the integral taken over $A\gamma$, on which $dx = -dt$, will be

$$\int_A^\gamma (u dv - v du).$$

If one can find a solution V of the equation that reduces to 1 on the segments $A\beta$ and $A\gamma$, then the sum of the two previous integrals will be

$$2u_A - u_\beta - u_\gamma,$$

and one will have the equation

$$2u_A = u_\beta + u_\gamma - \int_\beta^\gamma \left(u\frac{\partial v}{\partial x} - v\frac{\partial u}{\partial x} \right) dt + \left(u\frac{\partial v}{\partial t} - v\frac{\partial u}{\partial t} \right) dx$$

in which everything is known as soon as one knows the function u and one of its derivatives on the curve K.

It remains to find the solution v that we have supposed to exist. The preceding arguments show us the way, and we are led to look for a solution of Eq. (22.3) that depends only on the variable

$$\theta = \frac{(t - t_0)^2 - (x - x_0)^2}{4}.$$

One finds, for this solution, precisely the function J that satisfies the differential equation

$$J''\theta + J' - J = 0,$$

which is associated to Bessel functions.

Let us apply our general integral to the particular case that is the most important in practice, where our curve K reduces to the x-axis, and let us denote the coordinate of A by x_0. We then have

$$2u_A = u_\beta + u_\gamma - \int_\beta^\gamma \left(u\frac{\partial v}{\partial t} - v\frac{\partial u}{\partial t} \right) dx.$$

Suppose that we are given at the start the potential and its derivative. We then know that at the start

$$u = f(x), \qquad \frac{\partial u}{\partial t} = \varphi(x).$$

If we recall that $x - t$ and $x + t$ remain constant on AB and AC, respectively, we will have

$$2u(x_0, t_0) = f(x_0 - t_0) + f(x_0 + t_0) + \int_{x_0-t_0}^{x_0+t_0} v\varphi(x)dx + \frac{t_0}{2} \int_{x_0-t_0}^{x_0+t_0} f(x)\frac{dv}{d\theta}dx,$$

where v is equal to $J(\theta)$ and θ is $\frac{t_0^2 - (x-x_0)^2}{4}$.

All the details of the propagation can be deduced from this formula.

22.5 Exercises

As with some of my other history courses, there comes a point where the historical significance of some of the conceptual developments under discussion is arguably obscured by strictly mathematical exercises, which would either be too elementary or too hard. This is the case from now on, so only questions will now appear at the end of each chapter.

Questions

1. Look back over the solution to the wave equation in the form $u_{xt} = 0$. The characteristic curves are parallel to the x- and t-axes, and the general solution is $u(x, t) = f(x) + g(t)$. How much information must be given on the x- and t-axes for the solution to be determined in the rectangle defined by the origin and the points $(a, 0)$, $(0, b)$, and (a, b)?

2. Look over the account of Burgers' equation in Sect. B.2 and then again at Riemann's account of the formation of a shockwave.

Chapter 23
The Example of Minimal Surfaces

23.1 Introduction

Few branches of mathematics have the visual charm of the theory of minimal surfaces, which is one of the areas where analysis and differential geometry most profitably intersect. The topic was initiated by Euler and Lagrange but advanced only slowly until the work of Meusnier in the 1780s. The appropriate partial differential equation is difficult because it is non-linear, and it was taken up by Legendre and Monge, but left many secrets that were only to be unlocked in the later nineteenth century, chiefly by Riemann and Weierstrass.[1]

23.2 Euler and Lagrange

Many interesting problems in geometry and analysis arise when a function of some kind is to be minimised. A geodesic on a surface is a curve of shortest length joining two points, and we have already seen in Chap. 7 that problems in the calculus of variations can have attractive solutions. Informally, a minimal surface is a surface of least area spanning a given curve in space, and many elegant surfaces can be obtained by dipping a wire frame into a strong soap solution.[2] More precisely, a minimal surface is a surface with the property that any closed curve drawn on the surface encloses a region of smaller area than any other surface with that curve as boundary.[3] So minimal surfaces are the two-dimensional analogues of geodesics

---

[1] For more detail on all of this material, see the forthcoming book by Gray and Micallef on Jesse Douglas, minimal surfaces, and the first Fields Medal.

[2] As a quick check on Google images will confirm. An hour or two with homemade wire contours in various shapes, such as a trefoil not, or two circles (unlinked and linked) will be highly instructive.

[3] The curve must be a non-self-intersecting and bound a region of the surface.

© The Author(s), under exclusive license to Springer Nature Switzerland AG 2021
J. Gray, *Change and Variations*, Springer Undergraduate Mathematics Series,
https://doi.org/10.1007/978-3-030-70575-6_23

on a surface. Unfortunately, for reasons that have to do with the difficulty of the mathematics and the poor grasp people had of it in the early days, the name is a bit of a misnomer. It emerges from the mathematical formulations of the theory that they should be called extremal surfaces: they may either be of least area, greatest area, or ambiguous in this respect. A precise analogy is with finding a maximum or a minimum of a curve with an equation $y = f(x)$. If you only have the ability to consider solutions of $y' = 0$ you will find the maxima, minima, and horizontal inflection points—the extremal points.

The study of minimal surfaces began in the eighteenth century with Euler, who had the idea that they might be interesting, Lagrange, who gave an analytic version of the theory in the form of a partial differential equation satisfied by a minimal surface given by an equation of the form $z = f(x, y)$, and Meusnier, who gave a geometric condition that a minimal surface must satisfy.

In Sect. V, Sects. 45–47 of his *Methodus inveniendi* (1744) Euler discussed problems involving surfaces of revolution. In Sect. 45, he showed how to find, among all curves passing through two points that enclose a given area with an axis, the curve that on rotation about that axis generates the solid whose surface has the least area. In Sect. 46, he showed how to find, among all curves of a given length and passing through two points, the one that generated the greatest solid on being rotated about an axis.[4] In Sect. 47, he found among the same class of curves the one that generated the solid of either greatest or least area on being rotated about the axis.

In each of these cases, and throughout the book, Euler began with the integral expressing the quantity to be maximised or minimised, looked at the variation of the integral, and deduced a differential equation that characterised the solution. In Sect. 45, the solution curve satisfies the equation

$$dx = \frac{(ny + b)dy}{\sqrt{(1 - n^2)y^2 - 2bny - b^2}}.$$

Here b is a constant of integration and n is a parameter that expresses the effect of the constraints; it is what has come to be called a Lagrange multiplier.

Although Euler could certainly have integrated the above expression, he might well have found the general solution unilluminating, and instead he considered only the special cases where $b = 0$ (b is a constant of integration), $n = 0$, and $n = -1$ (a case we shall ignore).

When $n = 0$

$$dx = \frac{bdy}{\sqrt{y^2 - b^2}},$$

and Euler wrote that "the curve will be a catenary concave to the axis".[5] In this case, the constraint does not enter the problem, and the Euler–Lagrange equation, as we

---

[4] A facsimile of the relevant pages of Problems 45 and 46 will be found in Nitsche ([207], 6).

[5] A catenary, with respect to suitable axes, is given by the equation $y = \cosh x$.

would say, for the problem is for the case of curves of any length. But Euler wrote not a word about that, and dealt only with the other special case, $n = -1$.

In Sect. 47, the solution curve satisfies the equation

$$dx = \frac{cdy}{\sqrt{(b+y)^2 - c^2}},$$

where b is an arbitrary constant determined by the length of the curve and c is a constant of integration, so

$$y = -b + c\cosh(x/c).$$

The corresponding surface is a minimal surface only when $b = 0$, a condition that Euler did not mention (nor did he remark that when $b = 0$ this problem coincides with Problem 45 in the case when $n = 0$). He did, however, conclude that the answer is a catenary, the minimum area deriving from the case when the catenary is concave with respect to the axis, the maximum when it is convex. This surface, in the particular case when it is also a minimal surface, was later called the catenoid by Plateau.

Most books about minimal surfaces credit Euler with being the first to find a surface of revolution that is a minimal surface, and refer to these examples. But it would seem that he did not, in fact, explicitly address the problem of finding the curve that generates the surface of least area on being rotated about an axis, and although its solution appears, it does so only as a special case about which he said nothing.

Lagrange improved on Euler's treatment of the calculus of variations in his essay (1761) and, in particular, in Appendix I of the work he tackled the question of minimal surfaces in this spirit.

He wrote down that the area of a piece of surface given by an equation of the form $z = f(x, y)$ and spanning a fixed boundary was given by $\iint W dx dy$, where $W = \sqrt{1 + p^2 + q^2}$ and, as had become customary, $p = \frac{\partial z}{\partial x}, q = \frac{\partial z}{\partial y}$. He then argued that

$$\delta \iint W dx dy = 0 \Leftrightarrow \iint \delta(W dx dy) = 0 \Leftrightarrow$$

$$\iint \frac{p\delta p + q\delta q}{W} dx dy = 0.$$

This double integral equals

$$\iint \left(\frac{p}{W} \frac{\partial \delta z}{\partial x} + \frac{q}{W} \frac{\partial \delta z}{\partial y} \right) dx dy, \tag{23.1}$$

because, according to the general theory he had developed earlier,

$$\delta p = \delta \left(\frac{\partial z}{\partial x} \right) = \frac{\partial}{\partial x} \delta z$$

with a similar expression involving q. Integrating by parts shows that the first variation of area vanishes when

$$\iint \left(\frac{\partial}{\partial x} \left(\frac{p}{W} \right) + \frac{\partial}{\partial y} \left(\frac{q}{W} \right) dx dy \right) \delta z = 0.$$

Therefore

$$\frac{\partial}{\partial x} \left(\frac{p}{W} \right) + \frac{\partial}{\partial y} \left(\frac{q}{W} \right) = 0, \qquad (23.2)$$

and Lagrange concluded by remarking that, because of (23.2)

$$\frac{p dy - q dx}{W} \qquad (23.3)$$

must be an exact differential.

As he remarked, when $p = 0 = q$ the exact differential condition is satisfied and the surface is a plane, but, as he said, this is a very special case (p. 356) "because the general solution must be such that the boundary of the surface can be determined at will". He was, however, unable to solve any other cases of Eq. (23.2) and therefore he found no other minimal surface, and concluded his account by showing that the sphere solves the problem of finding a surface of least area enclosing a given volume.

So Lagrange had shown that a surface that is the graph of a function $z = f(x, y)$ (the surface is then said to be given in nonparametric or explicit form) and which has the least area among all surfaces with a given perimeter is to be found as a solution of the Euler–Lagrange equation for the area functional:

$$\frac{\partial}{\partial x} \left(\frac{p}{W} \right) + \frac{\partial}{\partial y} \left(\frac{q}{W} \right) = 0. \qquad (23.4)$$

This equation is today called the divergence form of the minimal surface equation (MSE). Lagrange did not write down the corresponding second-order partial differential equation explicitly, most likely because, as we have seen, there was no theory of partial differential equations at the time.[6] It is

$$(1 + q^2) z_{xx} - 2 p q z_{xy} + (1 + p^2) z_{yy} = 0. \qquad (23.5)$$

23.3 Meusnier, Monge, and Legendre

The first mathematician to provide insight into the geometry of minimal surfaces was the 21-year-old Jean Baptiste Marie Charles Meusnier, who was briefly a student of Monge. To understand his contribution, consider, as Euler had done before him,

---

[6]Lagrange did write it in his (1806, 489), albeit in an ad hoc notation.

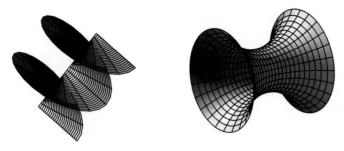

Fig. 23.1 The helicoid (left) and the catenoid (right)

the plane sections of a surface that pass through the normal at a given point, P, say. Suppose that we consistently choose a direction for these normals as P varies (let us call them $\mathbf{n}_P$), as we may at least for small regions of the surface. At each point P, the plane sections containing $\mathbf{n}_P$ cut the surface in curves that pass through P, and each such curve (called a curve of section) has a circle in the plane of section that best approximates it. We say that this circle has a positive radius of curvature if its centre, C, lies on the normal and in the positive direction heading away from P, otherwise negative. The signed magnitude PC, the radius of the circle, is called the radius of curvature of the corresponding curve of section. A saddle-shaped surface will have some curves of section with positive radii of curvature and others with negative radii of curvature. Euler showed that it turns out that for almost all surfaces (except the sphere) and for almost all points on these surfaces, there are precisely two curves of section that have extremal values for the radii of curvature among all curves of section with a common normal. These radii are called the radii of curvature of the surface at the point P (Fig. 23.1).

Meusnier's contribution was to realise that Lagrange's partial differential equation for a minimal surface was the condition for a surface to have the average of its radii of curvature vanish at each point. This average is called the mean curvature.[7] The first two minimal surfaces were in fact discovered by Meusnier, and they are the helicoid and the catenoid.

In his major paper (1784), Monge introduced the principal curves through an arbitrary point on a surface given by an equation of the form $z = f(x, y)$, which he defined (in Sect. 22) as the curves along which consecutive normals intersect, and he observed that, in general, there are two such curves at each point and they are the curves of greatest and least curvature at that point. He went on to give an equation for the radii of curvature R: they satisfy the equation $gR^2 + hkR + k^4 = 0$, where (as he explained more clearly in his (1787, Sect. 3))

$$g = rt - s^2, h = (1 + q^2)r - 2pqs + (1 + p^2)t, \text{ and } k^2 = 1 + p^2 + q^2.$$

---

[7]The term was introduced by Sophie Germain in her (1831).

Here

$$p = \frac{\partial z}{\partial x}, \; q = \frac{\partial z}{\partial y}, \; r = \frac{\partial^2 z}{\partial x^2}, \; s = \frac{\partial^2 z}{\partial x \partial y}, \; t = \frac{\partial^2 z}{\partial y^2}.$$

For a surface of zero mean curvature, the radii of curvature R_1 and R_2 at a point satisfy $\frac{1}{R_1} + \frac{1}{R_2} = 0$, and therefore $R_1 + R_2 = 0$ and so $h = 0$. This is, of course, the equation for a minimal surface in the form

$$(1+q^2)r - 2pqs + (1+p^2)t = 0,$$

or, more explicitly,

$$(1+q^2)\frac{\partial^2 z}{\partial x^2} - 2pq\frac{\partial^2 z}{\partial x \partial y} + (1+p^2)\frac{\partial^2 z}{\partial y^2} = 0.$$

Monge sought to solve this equation in his long paper (1787).[8] In Sect. 23 of his paper (1787), Monge, after noting that Meusnier had shown that a minimal surface was also a surface of zero mean curvature, proceeded to try to solve the partial differential equation. He wrote down the equation for the characteristic curves:

$$(1+q^2)dy^2 + 2pqdxdy + (1+p^2)dx^2 = 0. \tag{23.6}$$

Factorising equation (23.6) then led to the equation for a curve given by

$$dy + \tau dx = 0,$$

along which σ is constant, and another curve, given by

$$dy + \sigma dx = 0,$$

along which τ is constant, where

$$\sigma = \frac{-pq + i\sqrt{1+p^2+q^2}}{1+q^2}, \text{ and } \tau = \frac{-pq - i\sqrt{1+p^2+q^2}}{1+q^2}. \tag{23.7}$$

Each of these led to expressions for x and y in terms of p and q from which Monge obtained expressions for x, y, and z as integrals of functions involving σ and τ. As can be seen, the characteristic curves are complex, not real, and this raises a number of problems. Monge's hope must have been that in the end a real surface could be obtained.

In Sect. XV of the *Applications* Monge repeated the analysis of the principal curves on a surface that he had given in his (1784), and in Sect. XX he turned to the study of surfaces of zero mean curvature. He deduced the equation for a surface

---

[8]The paper was submitted in 1784, but only published in 1787.

of zero mean curvature from his equation for the radii of curvature and noted that this was the equation that Lagrange had shown defined surfaces of least area (but this time made no mention of Meusnier's name—Meusnier had died in the battle of Mainz in 1793).

Once again Monge proceeded to deduce properties of the minimal surface along the characteristic curves. He also presented, as he had before, a second characteristic equation:

$$(1 + q^2)dp^2 - 2pqdpdq + (1 + p^2)dq^2 = 0, \qquad (23.8)$$

which had been mysterious in his paper but had by now been more properly rederived by Legendre. After some algebra, here suppressed, he was able to come up with the result that a solution to the minimal surface equation was of the form

$$x = -\Phi'(\alpha) + \Psi'(\beta) \text{ and } y = -\Phi(\alpha) + \alpha\Phi'(\alpha) + \Psi(\beta) - \beta\Psi'(\beta),$$

where z is found by differentiating these equations to obtain dx and dy and using $dz = pdx + qdy$. The resulting equation can be integrated and the result is that

$$z = \int \Phi''(\alpha)\sqrt{-1 - \alpha^2}d\alpha + \int \Psi''(\beta)\sqrt{-1 - \beta^2}d\beta.$$

For Monge and his contemporaries, the problem with this formula was the apparently ineradicable appearance of imaginary quantities. As we have seen, mathematicians in the eighteenth century had no problem using formal complex methods in real geometrical problems, but the imaginary quantities were required to cancel at the last stage so that the solution could be purely real. Faced with a result where apparently this could not be done Monge did not proceed any further with analysis and his solution did not enable him to find any more examples of minimal surfaces.

The apparently intractable nature of the solution remained, as Poisson was to note in his (1832),

> Monge integrated [the minimal surface equation] in a finite form, but by considerations that did not seem to be admissible and involved him in long discussions with Laplace. Legendre then obtained the same integral, by means of a transformation applicable to a class of second-order equations, which could not leave any doubt as to the exactness of the result. Unfortunately one cannot deduce anything from this integral, which is complicated by imaginary quantities ....

23.4   Riemann and Weierstrass

The topic of minimal surfaces in the nineteenth century is one of the success stories for complex function theory, which itself is major new development in the period. It also illustrates the power of the theory of linear ordinary differential equations, which makes it worthwhile considering it here, and it was a rich field for differential geometers in the decades after Gauss.

Gauss had shown the value of studying a surface embedded in $\mathbb{R}^3$ by looking at the unit normals at each point (and coherently specifying an outward or positive direction). These define a map from the surface to the unit sphere by imagining each normal is moved parallel to itself until it is based at the centre of the unit sphere, and then associating to each point P on the surface the tip of unit normal, which is a point on the sphere. After Gauss's death, this map became called the Gauss map.

The first to study the Gauss map of a minimal surface was Ferdinand Minding in 1850. He showed that it enabled one to transplant the curves of longitude and latitude on the sphere to curves which Gauss had earlier shown were highly convenient in the study of surfaces. Gauss had suggested that a coordinate grid could be imposed on a surface by choosing a curve (usually a geodesic), drawing all the geodesics that meet this geodesic at right angles, and then drawing all the curves orthogonal to those geodesics. If this is done on the sphere starting with the equator, one first obtains the meridians or longitudes, and then the parallels or latitudes. Accordingly, on any surface the curves in the first family, the geodesics, are called meridians, and the curves in the second the parallels. Minding showed that for a minimal surface the inverse of the Gauss map maps meridians to meridians and parallels to parallels. This meant, in particular, that orthogonal curves were mapped to orthogonal curves.

Minding was soon followed by Ossian Bonnet, who considered a different grid on a surface, defined by taking at each point the two curves through that point whose radii of curvature are the extremal values. These curves are called the lines of curvature on the surface. Bonnet showed that the lines of curvature on a minimal surface are mapped by the Gauss map onto curves on the sphere that map by stereographic projection to two families of orthogonal straight lines. It followed, by Gauss's paper of 1825 on conformal mappings, that the Gauss map of a minimal surface was conformal. Bonnet was also able to show in his [20] that the coordinates (or at least the z-coordinate) of a map defining a minimal surface are harmonic.

From conformal maps and harmonic maps to complex analytic maps is in hindsight but a small step, but the examples of the authors just discussed show that it may not have seemed that way in 1860. Indeed, it was taken for the first time only by the two leading complex analysts of their day, Riemann and Weierstrass, independently. Riemann's account was entrusted by him to Hattendorff for editing in April 1866, but apparently dates from 1860 to 1861. The original manuscript consists purely of formulae, and Hattendorff supplied a text; the result was published in 1867. Weierstrass's account was first given in a lecture at the Berlin Academy in 1866. Weierstrass was also the first to give a general account of algebraic minimal surfaces.

Riemann's approach was to define a piece of surface by a map from a patch of $\mathbb{R}^2$ in (p, q)-coordinates into $\mathbb{R}^3$ in the usual (x, y, z)-coordinates, then to map the surface onto the unit sphere by the Gauss map. The area of an infinitesimal piece of the surface in $\mathbb{R}^2$ is related to the corresponding area on the sphere by the Jacobian of the Gauss map. In this way, the area of the entire surface is known as a double integral, and the condition that the surface be a minimal surface is that the first variation of that integral vanishes.

This condition reduced to the statement that a certain differential form was exact, and from that exact differential form Riemann deduced that it was possible to impose isothermal coordinates on the surface in such a way that its Gauss map became complex analytic. It followed that the minimal surface was represented conformally on the sphere and the plane, and Riemann went on to show that mean curvature at each point was zero.

From this insight, Riemann was able to obtain formulae that parametrised the minimal surface:

$$x = Re\left(i \int \left(\frac{du}{d\log\eta}\right)^2 \left(\eta - \frac{1}{\eta}\right) d\log\eta\right) \tag{23.9}$$

$$y = Re\left(\int \left(\frac{du}{d\log\eta}\right)^2 \left(\eta + \frac{1}{\eta}\right) d\log\eta\right) \tag{23.10}$$

and

$$z = Re\left(-2i \int \left(\frac{du}{d\log\eta}\right)^2 d\log\eta\right). \tag{23.11}$$

These equations are equivalent to those known today as the Weierstrass–Enneper equations for the coordinates on a minimal surface, in the form Weierstrass was to give that involves only one function, $u = u(\eta)$ (see Eqs. (23.12), (23.13), and (23.14)). However, Riemann did not pause even to notice that he could now write down infinitely many examples of minimal surfaces in terms of a function $u = u(\eta)$.

Thus, a minimal surface is obtained every time one has an analytic function. The deeper question that then arose was to ask for the minimal surface that spans a given curve in space; this is the so-called Plateau problem.[9]

Riemann tackled this problem by means of a detailed study of the behaviour of the Gauss map, but even he found this task daunting, however, and gave explicit solutions only for simple boundaries: two skew lines in space (the helicoid); two intersecting lines; and a third lying in a plane parallel to the first two, three skew lines (which led to a generalisation of the Riemann P-function); the regular space quadrilateral (later studied by Schwarz), and two circles in two parallel planes.

Weierstrass's approach was different.[10] He started with the expression for the mean curvature, and assumed that the given piece of surface can be defined by a conformal map from a patch U of $\mathbb{R}^2$ with coordinates p and q into $\mathbb{R}^3$ with coordinates x, y, z.

---

[9]It is named for the Belgian physicist Joseph Plateau who, although blind, showed in a series of experiments that the surface of a liquid is in equilibrium if its mean curvature is constant. Weightless films, floating in a different liquid, will therefore have zero mean curvature and locally satisfy Lagrange's equation.

[10]See Weierstrass [267, 268], and the paper [270], which was published for the first time posthumously in his *Mathematische Werke*, vol. 3.

Typically, he gave no references, saying only that it was well known that one could do this; Gauss had indicated in his (1828) that this can always be done. He then showed that x, y, and z must be harmonic functions of p and q, and so the real parts of three complex functions f, g, and h of $u = p + iq$, meromorphic on the whole of U. The conformality condition on the map implies, in terms of f, g, and h, that

$$f'(u)^2 + g'(u)^2 + h'(u)^2 = 0.$$

Weierstrass shrewdly saw that this implies that there are functions $G(u)$ and $H(u)$ such that

$$f'(u) = G^2 - H^2, \quad g'(u) = i(G^2 + H^2), \quad h'(u) = 2GH.$$

He then introduced the complex variable s defined by $s = \dfrac{H(u)}{G(u)}$, and a technical argument, which I omit, finally led Weierstrass to the complex function $G^2(u)\dfrac{du}{ds}$, which he called $S = F(s)$, upon which he based his analysis. He gave explicit power series expressions for the coordinate functions of a minimal surface in terms of S, and also for its Gaussian curvature. He pointed out that the principle of analytic continuation then allowed the coordinate functions to be defined for the entire minimal surface. So he could finally proclaim that to every single-valued analytic function there corresponds a surface with mean curvature everywhere zero.

The parameterisation that does this is known today as the Weierstrass–Enneper equations and it can be given in various forms.[11]

After a little work, the Weierstrass–Enneper representation can be given in the form

$$x = c_1 + \operatorname{Re} \int_{\omega_0}^{\omega} \left(1 - \varpi^2\right) R\left(\varpi\right) d\varpi \tag{23.12}$$

$$y = c_2 + \operatorname{Re} \int_{\omega_0}^{\omega} i\left(1 + \varpi^2\right) R\left(\varpi\right) d\varpi \tag{23.13}$$

$$z = c_3 + \operatorname{Re} \int_{\omega_0}^{\omega} 2\varpi R\left(\varpi\right) d\varpi. \tag{23.14}$$

Here, x, y, and z are the real parts of $f(u)$, $g(u)$, and $h(u)$.

---

[11]See the MIT account, http://ocw.mit.edu/courses/mathematics/18-994-seminar-in-geometry-fall-2004/lecture-notes/chapter18.pdf.

Weierstrass gave no examples; however, for suitable choices of R various well-known minimal surfaces are obtained. For example, setting $R = 1$ leads to Enneper's minimal surface, and $R(\omega) = a/2\omega^2$ to the catenoid.

The alternative expressions in terms of the functions $G(u)$ and $H(u)$ are

$$x = x_0 + \text{Re} \int_{u_0}^{u} \left(G(u)^2 - H(u)^2 \right) du \, , \qquad (23.15)$$

$$y = y_0 + \text{Re} \int_{u_0}^{u} i \left(G(u)^2 + H(u)^2 \right) du \, , \qquad (23.16)$$

$$z = z_0 + \text{Re} \int_{u_0}^{u} 2 \left(G(u)H(u) \right) du \, . \qquad (23.17)$$

The connection between these formulae and Riemann's is given by $\eta = \frac{H(u)}{G(u)}$.

Because Weierstrass's accounts were published and taken up by his students before Riemann's posthumous account appeared, most subsequent authors credited Weierstrass with discovering the intimate connection between minimal surfaces and complex function theory.

As we saw in Chap. 19, Schwarz was the student of Weierstrass who remained most closely associated with the master. This is particularly true of his earliest work, as one would naturally suppose. In 1865, Weierstrass gave a seminar at the University of Berlin on minimal surfaces. Perhaps as part of the seminar, Schwarz took up and solved the problem of finding the minimal surface bounded by a space quadrilateral, unaware that this problem had already been solved by Riemann. Schwarz's mathematical solution was presented to the Berlin Academy by Kummer, accompanied by a Gypsum model that Schwarz, it seems, had made himself, based on experiments with glycerine. On the strictly mathematical side, Schwarz's analysis made considerable use of the theories of elliptic and hyperelliptic functions, because it chose a particular hyperelliptic function for substitution into the Enneper–Weierstrass equations. This gave his surface a natural periodicity: pieces of it could be fitted together to form an annular region with two boundaries, and these pieces could in turn be joined up to form an infinitely extended surface with infinite topological genus. Altogether a remarkable discovery with which to embark on a career in mathematics (Fig. 23.2).

A deep insight into analytic functions underpinned this work. This was the recognition that if a straight line lies in a minimal surface then the surface has that line as a line of symmetry. In other words, and to give the highly plausible physical motivation, if two pieces of minimal surface meet along a common line and have the same normals there then each is the analytic continuation of the other. This is the origin of the Schwarz reflection principle, which Schwarz went on to prove in 1869. Interestingly, Schwarz wrote that he had learned this insight from a conversation with Weierstrass.

Fig. 23.2 A piece of
Schwarz's periodic minimal
surface, S. Schwarz,
*Gesammelte mathematische
Abhandlungen*, vol. I, facing
p. 132

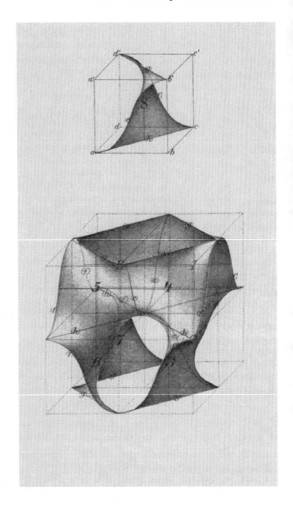

23.5   Simple Solutions of the Plateau Problem

As noted above, the Plateau problem asks for a minimal surface that spans a given
curve in space. In general, given a piece of a minimal surface bounded by a curve
in space, one has no knowledge of the behaviour of the Gauss map on the boundary
curve. But if the boundary curve contains a straight line then along that line the
image of the Gauss map can only be an arc of a circle on the image sphere (although
any part of it may be covered more than once). For this reason, attempts on the
Plateau problem were confined to polygonal curves in space, because the image of
the boundary under the Gauss map is made up of arcs of great circles on the sphere.

The polygon is specified by giving the coordinates of its n vertices. So the problem
is a three-dimensional version of the Schwarz–Christoffel problem, which was under
investigation at the same time.

Accordingly, the solution to the Plateau problem might then be expected to come in two parts. The minimal surface is realised by a map from the upper half line that maps the real axis to the polygonal boundary curve. In the first part, we take arbitrary points $z_1, z_2, \ldots, z_n$ on the real axis and look for a holomorphic map with branch points at these points of the correct order, so that the map will send the upper half-plane onto a polygon with the required angles at each vertex, but not necessarily a polygon of the right size and shape. In the second part, we fix the positions of the points $z_1, z_2, \ldots, z_n$ on the real axis so that the map is onto the given polygon. The side lengths are given by certain integrals, and the task is to use that information to determine the right values of those points on the real axis.

In a note (the "Fortsetzung") added to his first paper on minimal surfaces, Weierstrass approached the problem through the theory of linear differential equations, and tried to define the equation that the functions $G(u)$ and $H(u)$ must satisfy.[12] It is trivial that two functions satisfy a second-order linear differential equation, the important question is what can be said about the coefficients. Weierstrass claimed that the equation satisfied by the functions G and H has rational coefficients that depend on the lengths and directions of the boundary segments. However, "In general", he said,

> the determination of the constants in this differential equation, such as the constants of integration, depend on the solution of transcendental equations; but it is not difficult to find special cases where one can find complete expressions for $G(u)$ and $H(u)$ in terms of known functions.

Weierstrass concluded by promising a full report at a later date, but this was never given. All we have is the "Bestimmung", which is Schwarz's account written in the 1890s of what Weierstrass had in mind when giving that two-page report to the Academy of Sciences in December 1866.

As Schwarz reconstructed it, Weierstrass's approach had been to consider what had to happen at each vertex of the polygon. The polygon has n angles at the corresponding vertices of $\alpha_1, \alpha_2, \ldots, \alpha_n$. Moreover, each angle defines a plane whose orientation is captured by the normal to the angle at the vertex. Schwarz claimed that by working in this way Weierstrass had found some important facts about this differential equation. In particular, he had deduced that at each branch point the equation—regarded as an equation for a function $y(t)$—has a simple pole for the coefficient of $\frac{dy}{dt}$ and a double pole for the y-coefficient. These are the only singular points, and the point at infinity must be what is called an inessential singular point for the differential equation.[13]

This argument fails to determine the required differential equation completely, and as a result it seems to have been abandoned. First, as Weierstrass had indicated, it is necessary to show how the pre-images of the vertices on the boundary are determined, but that problem was not discussed. Second, it says nothing about the important role

---

[12]Lazarus Fuchs was developing the theory of these equations under his influence at the time.

[13]An inessential singular point is one where the solutions of the differential equation remain holomorphic in a neighbourhood of the point even though at least one of the coefficients of the differential equation is singular.

played by the branch point of the Gauss map in the interior of the minimal surface. But these are very difficult problems. Indeed, as late as 1914 Darboux remarked that "Thus far, mathematical analysis has not been able to envisage any general method which would permit us to begin the study of this beautiful question".

23.6 Exercises

Questions

1. Find examples on the web of some of the best-known minimal surfaces: the helicoid, the catenoid, and Enneper's surface.
2. On the (correct) assumption that any closed curve spans a minimal surface, find examples of minimal surfaces that are topologically Möbius bands.

Chapter 24
Partial Differential Equations and Mechanics

24.1 Introduction

Several issues arise here. Does mechanics start with equations of motion, or with principles like the principle of least action? Is the Lagrangian formulation, although completely general, always the best, or can there be others? Is there a worthwhile parallel between mechanics and geometry?

This is a tough chapter, so let me spell out the route through it, and say what is essential. First, I present a clean, modern version of the key mathematical ideas.[1] Given a mechanical problem expressed in terms of a Lagrangian (see Sect. 7.6) in generalised coordinates q_j and $\dot{q}_j$, it can be represented in terms of more symmetric coordinates q_j and p_j and a new function called the Hamiltonian, which is closely related to the Lagrangian (Eq. (24.1)). Important here is what are called Hamilton's equations (Eq. (24.2)).

Hamilton had the brilliant idea of looking at what happens to the time evolution of a Hamiltonian system. Now the upper end point of the Hamiltonian (or Lagrangian) is allowed to be a function of time, and the paths of the qs continually obey their Euler–Lagrange equations. This gave him a function W of the upper end points and the time, and it satisfies a particular first-order partial differential equation (see Eq. (24.3)) that became known as the Hamilton–Jacobi equation after Jacobi saw more deeply into it.

If we suppose that the values $q_1, q_2, \ldots, q_n$ define a point $(q_1, q_2, \ldots, q_n)$ in a space we can call Q, then it turns out as time goes by in a Hamiltonian system these points define a moving hypersurface.

Moreover, solutions of the Hamilton–Jacobi equation satisfy one half of Hamilton's equations, thus making a connection between a first-order partial differential equation and a family of first-order ordinary differential equations. Indeed, the q's as they evolve describe the characteristic curves of the partial differential equation.

---

[1]This one follows www.damtp.cam.ac.uk/user/tong/dynamics.htm, which is David Tong's Cambridge notes.

© The Author(s), under exclusive license to Springer Nature Switzerland AG 2021
J. Gray, *Change and Variations*, Springer Undergraduate Mathematics Series,
https://doi.org/10.1007/978-3-030-70575-6_24

This is what Cauchy appreciated (see Chap. 12), and it opened up, surprisingly, a way of solving some systems of ordinary differential equations by solving a partial differential equation.

This being a history course, after presenting these ideas I document that Hamilton did possess them in his own way, that Jacobi re-interpreted them—and that he had scathing views about much of what had been said about variational principles before him—and that the ordinary differential equation—partial differential equation connection just mentioned was soon appreciated.

24.2   Hamiltonian Dynamics

Let us suppose we have a dynamical problem expressed in terms of a Lagrangian. I shall write $L(q, \dot{q}, t)$ for the Lagrangian, where q stands for an n-tuple in variables $q_1, q_2, \ldots, q_n$ and $\dot{q}$ for the corresponding n-tuple $\dot{q}_1, \dot{q}_2, \ldots, \dot{q}_n$—we can get a long way by thinking of $n = 1$. I shall write Δ rather than δ for the variational symbol, to avoid confusing δ and S, a symbol to be introduced shortly. I shall write t^0 for the initial time and t^1 for the final time.

As we shall see, William Rowan Hamilton defined a function, which he called the principal function, as the time integral of the Lagrangian:

$$S = \int_{t^0}^{t^1} L(q, \dot{q}, t) dt.$$

A standard variational argument shows that

$$\Delta S = \int_{t^0}^{t^1} \Delta L(q, \dot{q}, t) dt$$

$$= \int_{t^0}^{t^1} \left(\frac{\partial L}{\partial q} \Delta q + \frac{\partial L}{\partial \dot{q}} \Delta \dot{q} \right) dt.$$

Note that this is a system of n equations, one for each q_j, $j = 1, \ldots, n$.

We integrate the second term in the integral by parts and obtain

$$\Delta S = \int_{t^0}^{t^1} \left(\frac{\partial L}{\partial q} - \frac{d}{dt} \frac{\partial L}{\partial \dot{q}} \right) \Delta q \, dt + \left[\frac{\partial L}{\partial \dot{q}} \Delta q \right]_{t^0}^{t^1}.$$

If the end points are fixed then the term in square brackets vanishes. So the variation of the integral vanishes if the integrand vanishes for all Δq, and the result is the Euler–Lagrange equations:

$$\frac{\partial L}{\partial q_j} - \frac{d}{dt} \frac{\partial L}{\partial \dot{q}_j}, \quad j = 1, \dots, n.$$

Following Hamilton, we define

$$p_j = \frac{\partial L}{\partial \dot{q}_j},$$

so

$$\frac{dp_j}{dt} = \dot{p}_j = \frac{\partial L}{\partial q_j}.$$

We now define the Hamiltonian

$$H = H(q, p, t) = \sum_j p_j \dot{q}_j - L. \tag{24.1}$$

Therefore

$$\dot{p}_j = \frac{\partial L}{\partial q_j} = -\frac{\partial H}{\partial q_j}, \quad j = 1, \dots, n,$$

and we find

$$\dot{q}_j - \frac{\partial L}{\partial p_j} = \frac{\partial H}{\partial p_j},$$

but L is not a function of the p_j and so $\frac{\partial L}{\partial p_j} = 0$ and so

$$\dot{q}_j = \frac{\partial H}{\partial p_j}.$$

These equations

$$\dot{p}_j = -\frac{\partial H}{\partial q_j} \quad \text{and} \quad \dot{q}_j = \frac{\partial H}{\partial p_j} \tag{24.2}$$

for $j = 1, \dots, n$ are called *Hamilton's equations* or the *canonical equations*. The solution to these equations resolves the dynamical problem at stake.

They have a simple, if perhaps surprising corollary that turns out to be useful. We calculate $\frac{dH}{dt}$:

$$\frac{dH}{dt} = \sum_j \frac{\partial H}{\partial q_j} \frac{dq_j}{dt} + \sum_j \frac{\partial H}{\partial p_j} \frac{dp_j}{dt} + \frac{\partial H}{\partial t}.$$

But by the canonical equations, the RHS equals a sum of terms of the form

$$-\dot{p} \frac{dq_j}{dt} + \dot{q} \frac{dp_j}{dt} + \frac{dH}{dt},$$

so the summations cancel, and it follows that

$$\frac{dH}{dt} = \frac{\partial H}{\partial t}.$$

In other words, if the Hamiltonian H does not involve t explicitly—so $\frac{\partial H}{\partial t} = 0$—then H is constant—$\frac{dH}{dt} = 0$. In conservative systems, the Hamiltonian can be treated as the energy.

Now we turn to a different approach, one that often turns out to have complementary virtues. Hamilton had the ingenious idea of investigating what happened to the path given by the Euler–Lagrange equations as the upper end point varied and as the time taken varied. He may well have done this because he was interested in the passage of light through crystals. So in what follows the $q_j(t)$ that define the path satisfy the Euler–Lagrange equations.

He now regarded his principal function

$$S = \int_0^{t^1} L(q, \dot{q}, t)dt$$

as a function of its upper endpoint $t_1 = T$ and the value of q varies. The initial values of the coordinates q^0 are fixed. We have $\frac{dS}{dt^1} = L$. Suppose that we fix a value of $t^1 = T$. Then we can define a function

$$W = W(q^0, q^1, T) = S(q(T)).$$

The distinction between these two functions is that S is defined on any path but W is a function of the end points and the time the system has been evolving.

A variational argument now gives that

$$\Delta S = \int_0^{t^1} \left(\frac{\partial L}{\partial q} - \frac{d}{dt}\frac{\partial L}{\partial \dot{q}} \right) \Delta q\, dt + \left[\frac{\partial L}{\partial \dot{q}} \Delta q \right]_{t^0}^{t^1},$$

but now we deduce that

$$\Delta S = \frac{\partial L}{\partial \dot{q}},$$

where the term on the right is evaluated at $t = T$. But

$$\Delta S = \frac{\partial W}{\partial q^1}$$

so

$$\frac{\partial W}{\partial q^1} = p^1.$$

Now we allow T to vary, and calculate $\dfrac{\partial W}{\partial T}$. We have

$$\frac{dW}{dT} = \frac{\partial W}{\partial T} + \frac{\partial W}{\partial q^1}\frac{dq^1}{dT} = \frac{\partial W}{\partial T} + p^1\dot{q}^1.$$

We have $\dfrac{dS}{dT} = L$, so, using Eq. (24.1),

$$\frac{\partial W}{\partial T} = -H(q^1, p^1, T).$$

But the upper end points and the time are the only variables involved, so we can drop the labels and write

$$W = W(q, t), \quad \frac{\partial W}{\partial q} = p, \quad \frac{\partial W}{\partial t} = -H(q, p, t).$$

The last equation can be written in this form

$$\frac{\partial W}{\partial t} = -H\left(q, \frac{\partial W}{\partial q}, t\right). \tag{24.3}$$

In this form, it is called the Hamilton–Jacobi equation. It has the property that W does not appear explicitly, so if W is a solution then so is $W + c$, where c is an arbitrary constant.

It remains to show that the solutions of this equation solve half of Hamilton's equations, i.e. that

$$\dot{p} = -\frac{\partial H}{\partial q},$$

on the assumption that we have $\dfrac{\partial H}{\partial p} = \dot{q}$—these are taken as part of the initial conditions. (At each point when $t = 0$ there is a unique curve that obeys the Euler–Lagrange equations and the condition just assumed. As we shall see in Chap. 25, this is closely analogous so the existence of a geodesic on a manifold through a given point in a given direction, and this in turn derives from the fact that the geodesic equation is a second-order equation.)

We write

$$p = \frac{\partial W}{\partial q},$$

so

$$\dot{p} = \frac{d}{dt}\frac{\partial W}{\partial q} = \frac{\partial^2 W}{\partial q \partial q}\dot{q} + \frac{\partial^2 W}{\partial q \partial t}.$$

The second term on the RHS is $-\dfrac{\partial}{\partial q}H(q, p, t)$, and differentiation here of p with respect to q is required because $p = \dfrac{\partial W}{\partial q}$, so we get

$$\dot{p} = \frac{\partial^2 W}{\partial q \partial q}\dot{q} - \frac{\partial H}{\partial q} - \frac{\partial H}{\partial p}\frac{\partial^2 W}{\partial q \partial q}.$$

So we finally obtain

$$\dot{p} = -\frac{\partial H}{\partial q},$$

as required.

So now, if we can solve the Hamilton–Jacobi equation we can solve the canonical equations—and if we can solve the canonical equations we can solve the Hamilton–Jacobi equation.

If we now suppose further that H does not depend explicitly on t then we may write Eq. (24.3) in the form

$$H\left(q, \frac{\partial S}{\partial q}\right) + \frac{\partial S}{\partial t} = 0,$$

for which the general solution will be of the form

$$S(q, \alpha, t) = W(q, \alpha) - \alpha_0 t.$$

With a little more work (but you might think there's been enough) and on the further assumption that H is the total energy, and therefore a constant of the motion, it can be shown that we can therefore think of the surfaces $S = const.$ as wave fronts moving in Q, and their progress analysed in some detail.

We shall investigate the solutions of the Hamilton–Jacobi equation in Sect. 24.4, but we now turn to look at Hamilton's original paper and at Jacobi's comments on it and the theory out of which it grew. His comments are enjoyably fierce.

24.3 Hamilton's and Jacobi's Theories of Dynamics

I first note an annoying sign convention: what we write as U, the potential energy of a system of particles, was denoted $-U$ in the nineteenth century. To minimise confusion, I shall sometimes write $\tilde{U} = -U$ to denote potential energy with the old sign convention. So where we write $L = T - U$ for the Lagrangian, these mathematicians wrote $L = T + \tilde{U}$; in each case, T denotes the kinetic energy of the system.

Hamilton took up the topic of dynamics, the motion of a system of particles under their mutual attraction, in 1834, when he wrote his essay "On a general method in dynamics", and he wrote a "Second essay" on the subject a year later. We shall concentrate on it.[2]

In this essay, he worked from the start in generalised coordinates η_j, and introduced the new coordinates $\varpi_j = \frac{\partial T}{\partial \dot{\eta}_j}$. Here T is expressed with respect to the η_j and $\dot{\eta}_j$; as a function of η_j and ϖ_j the same function is denoted F. Hamilton wrote down the expressions for varying T and varying F and deduced that

$$\frac{\partial (F - \tilde{U})}{\partial \varpi_j} = \frac{\partial F}{\partial \varpi_j} = \dot{\eta}_j, \quad \frac{\partial F}{\partial \eta_j} = -\frac{\partial T}{\partial \eta_j},$$

because the potential function $\tilde{U}$ does not depend on ϖ_j. So Lagrange's equations take the form

$$\frac{d\varpi_j}{dt} = \frac{\partial (\tilde{U} - F)}{\partial \eta_j}.$$

Hamilton set $H = F - \tilde{U}$ and obtained the equations

$$\frac{d\eta_j}{dt} = \frac{\partial H}{\partial \varpi_j}, \quad \frac{d\varpi_j}{dt} = -\frac{\partial H}{\partial \eta_j}.$$

These later became called the canonical equations. This is the first appearance of the canonical equations, although important steps in that direction had been taken by Poisson in a paper Hamilton cited; Hamilton's improvement was the introduction of the ϖ_j.

Hamilton then reintroduced the principal function S as

$$S = \int_0^t \left(T + \tilde{U} \right) dt,$$

which says that the principal function S is the time integral of the Lagrangian. A calculation of the variation of S enabled Hamilton to show that

$$\frac{\partial S}{\partial t} + H = 0$$

(or rather, a set of equations equivalent to that result).

At this point, Fraser and Nakane make several valuable observations. They note ([108], 184) the new derivation did not assume the conservation of mechanical energy,

---

[2] This account closely follows [108] which can profitably be consulted for many interesting insights. They look in some detail at the first essay.

a fact that Hamilton did not notice—he repeatedly stated, indeed, that his method was restricted to cases where that law applied.

Moreover, as it happens, the variational conditions often imposed by Lagrange are the vanishing of the "action" and fixed end points for the curves under consideration. This can have the effect that the only curve considered is the minimiser itself—there are no other candidates! Hamilton allowed the end points to vary and so produced a genuine family of candidate curves.

Finally, they remark that Hamilton also noticed that if the variation of S vanishes then

$$\delta S = \left[\sum \left(\frac{\partial T}{\partial \dot{\eta}_j} \delta \eta_j \right) \right]_0^t - \int_0^t \left(\sum \left(\frac{d}{dt} \frac{\partial T}{\partial \eta_j} - \frac{\partial T}{\partial \eta_j} - \frac{\partial \tilde{U}}{\partial \eta_j'} \right) \delta \eta_j \right) dt.$$

If the end points are fixed then the first sum becomes zero and the other term is simply Lagrange's equations.

24.3.1   Jacobi

Carl Gustav Jacob Jacobi must have read Hamilton's papers in 1836, because he wrote to his brother Moritz about them in September 1836 to say that they had led him to make a deep study of dynamics.[3] Although Jacobi had a genuine interest in mechanics, and had recently made an important breakthrough in the three-body problem, his reworking of Hamilton's ideas was much more that of an analyst.

He established carefully that the principal function S is a function of the variables (x_j, y_j, z_j), the initial positions (a_j, b_j, c_j) and t. Then he wrote down the expression for its variation, and deduced that

$$\frac{\partial S}{\partial t} + \frac{1}{2} \sum_j \left(\left(\frac{\partial S}{\partial x_j} \right)^2 + \left(\frac{\partial S}{\partial y_j} \right)^2 + \left(\frac{\partial S}{\partial z_j} \right)^2 \right) = \tilde{U},$$

an equation he called Hamilton's equation.[4]

A feature of Jacobi's derivation was that it made clear that the forces could depend explicitly on the time. In principle, this extends the study of dynamics to non-conservative systems in which energy is lost, but Jacobi lost interest in this in the 1840s, and it was for his successors to appreciate this advance on the work of Hamilton.

Less clearly, and we cannot discuss this point fully here, Jacobi's method is not wholly variational because he did not discuss the variation of the end points. This

[3] See Koenigsberger ([162], 198), quoted in Hawkins ([139], 205) and Pulte ([230]).
[4] The same equation appears in [135]. Jacobi did not call S the principal function, however.

is not a problem for Jacobi's presentation, because the derivation of "Hamilton's equation" is entirely a piece of partial differential equation theory.

Jacobi had various criticisms of Hamilton's work that we cannot go into here and for which the reader is referred to Fraser and Nakane [108]. It says a lot about the growing place for mathematics in Germany (and, of course, France) and its uncertain place in Great Britain even with a mathematician of Hamilton's talents, that there is much justice in Nakane and Fraser's concluding remarks (p. 220):

> Hamilton was the great creator, and it is unimaginable that Jacobi could have reached the level of remarkable abstract insight that he did without a foundation already in place. Jacobi nevertheless had a better knowledge of contemporary analysis and a better sense for how the new ideas should be developed at an appropriate theoretical level within the calculus of variations, mathematical dynamics and differential equations. He possessed as well a talent for making the new ideas accessible to receptive mathematicians. Although he died some fifteen years earlier than Hamilton, his posthumous *Vorlesungen* would become the most influential work in the history of mathematical dynamics since Lagrange.

With this in mind, it seems worthwhile to give some quotations from Jacobi's posthumously published lectures on dynamics, starting with an extract from the sixth lecture.[5]

> We now come to a new principle which, unlike earlier ones, does not give an integral. This is the "principle of least action" incorrectly called that of least work. Its importance lies first in the form in which it presents the differential equations of motion, and second in that it gives a function which is minimized when these differential equations are satisfied. Indeed, such a minimum does exist in all examples, but the reason for this is unknown. Whereas the interest of this principle consists precisely in the fact that one can generally *construct* a minimum, formerly too much importance was attributed to the existence of such a minimum. An example of the principle under consideration comes from Euler's "de motu projectorum". After ...proving the principle for attraction to fixed centers, he did not succeed in extending it to the n-body problem, for which he did not know the principle of kinetic energy; he contented himself with stating that the computations were very lengthy. But Euler said that the principle of least work had to be valid also here, since the fundamental results of a sound metaphysics revealed that forces in nature always do the least work.

> However, neither a sound nor any metaphysics shows this, and indeed Euler was led to this expression only through misunderstanding of the name "least work." Maupertuis meant that nature achieved her work with the least expenditure of force, and this is the true meaning of the "principle of least action."

> In my opinion, this principle is presented incomprehensibly in all textbooks, even in the best, those by Poisson, Lagrange, and Laplace. Namely, it is stated that the integral $\int \sum m_j v_j ds_s$ (where $v_j = \frac{ds_j}{dt}$ is the velocity of the point m_j,) is minimized, when taken from one position of the system to another. To be sure, this is only stated to be valid for conservative systems, but it is forgotten that one must eliminate time from the above integral and reduce everything to space elements. Moreover, this integral must be understood to be a minimum for given initial and final configurations and all possible paths joining them.

In his eighth lecture, Jacobi obtained the Lagrangian form of the equations of dynamics, and wrote

---

[5]They were published in 1866. Compare the extract in the Birkhoff *Source Book*, 374–379, which gives some of the technical material as well.

In place of the Principle of Least Work, one can substitute another principle, which also consists in the vanishing of a first variation, and which can be derived from the differential equations of motion even more simply than from the Principle of Least Work. This variational principle seems to have been unnoticed previously, because – in contrast to the Principle of Least Work – it does not correspond to a minimum principle. Hamilton was the first to use this principle as a point of departure. We will use it to set down the equations of motion in the form given by Lagrange in his *Mécanique analytique*.

Now let X_j, Y_j, Z_j be the partial derivatives of a function $\tilde{U}$, and let T be half the living force [kinetic energy], that is,

$$T = \frac{1}{2} \sum_j m_j v_j^2 = \frac{1}{2} \sum_j m_j \left(\left(\frac{dx_j}{dt}\right)^2 + \left(\frac{dy_j}{dt}\right)^2 + \left(\frac{dz_j}{dt}\right)^2 \right),$$

then the new principle is

$$\delta \int (T + \tilde{U})dt = 0. \quad (1)$$

This principle is equivalent to the Principle of Least Work, but more general in that t may enter $\tilde{U}$ explicitly, which is excluded in the previous Principle [of Least Work], because in the latter time must be eliminated using the theory of living force [conservation of energy], and this holds only when t does not enter $\tilde{U}$ explicitly. We will use equation (1) to derive the differential equations of motion from a first-order partial differential equation. As Hamilton showed, one can decompose the variation in (1), using integration by parts, into two parts, of which one is outside and the other inside the integral sign, and both must vanish. In this way, the integrand, which equals zero, gives the differential equations of the problem, and the expression outside the integral sign gives its integral equations.

The complete statement of the new principle reads as follows: Let the configurations of the system be given at a specified initial time t_0 and final time t_1. Then the actual intervening motion is determined by the equation $\delta \int (T + \tilde{U})dt = 0$ of (1).

Here the integral is taken from t_0 to t_1; $\tilde{U}$ is the force function and can contain the time explicitly, and T is half the living force.

Jacobi then derived the equations

$$\frac{dp_j}{dt} = \frac{\partial(T + \tilde{U})}{\partial q_j}; \quad p_j = \frac{\partial T}{\partial \dot{q}_j} .$$

In his ninth lecture, Jacobi then obtained the Hamiltonian form of the equations of dynamics. He first obtained the equations in a form Poisson had presented them in 1820:

$$\frac{dp_j}{dt} = \frac{\partial(T + \tilde{U})}{\partial q_j} = \frac{\partial T}{\partial q_j} + \frac{\partial \tilde{U}}{\partial q_j},$$

in which $\frac{\partial \tilde{U}}{\partial q_j}$ depends only on the q_j and $\frac{\partial T}{\partial q_j}$ is a homogeneous quadratic function of the $\dot{q}_j$ and therefore of the p_j. This yields the equations

$$\frac{dp_j}{dt} = P_j, \quad \frac{dq_j}{dt} = Q_j,$$

about which he remarked

> This is Poisson's form of the equations of motion, where the Q_j and P_j contain no variables
> other than the ps and the qs. This system of 2k equations has the following noteworthy
> properties:
>
> $$\frac{\partial Q_j}{\partial p_k} = \frac{\partial Q_k}{\partial p_j}, \quad \frac{\partial Q_j}{\partial q_k} = -\frac{\partial P_k}{\partial p_j}, \quad \frac{\partial P_j}{\partial q_k} = \frac{\partial P_k}{\partial q_j} \quad (2).$$
>
> Of these, Poisson (loc. cit.) writes down those of the first group, while the rest can be written
> down directly from his results. Equations (2) show that the quantities Q_j and P_j are to be
> recognized as the partial derivatives of a *single* function of the p_j and $-q_j$. This observation,
> which comes directly from equations (2), Poisson does not make; still less does he try to
> use this function. It was Hamilton who first expressed it, and through the introduction of
> his characteristic function the whole reformulation is made extraordinarily easier. One can
> reach the same conclusion almost by one's self, if one derives the kinetic energy theorem
> from the second Lagrange form of the differential equations given in formula (9) [of the
> Eighth Lecture].

24.4 First-Order Partial Differential Equation Theory

None of this would be much use unless the actual equations could be solved, and many
authors comment explicitly on the utility of the Hamilton–Jacobi equation. After
all, one's experience is that partial differential equations are usually hard to solve,
whereas ordinary differential equations are usually easier. So, given a first-order
partial differential equation, one looks for the system of characteristic equations,
which are ordinary differential equations, and hopes to solve them. But, as Hilbert
and Courant commented (Vol. 2, 107)

> Hamilton and Jacobi achieved a major success by recognizing that this relationship may be
> reversed. To be sure, the integration of a partial differential equation is usually considered
> as a problem more difficult than that of a system of ordinary differential equations. In
> mathematical physics one is often led, however, to a system of ordinary differential equations
> in canonical form. These equations may be difficult to integrate by elementary methods, while
> the corresponding partial differential equation is manageable; in particular, it may happen
> that a complete integral is easily obtained, e.g., with the help of the separation of variables
> (cf. Ch. I §3). Knowing the complete integral, one can then solve the corresponding system of
> characteristic ordinary differential equations by processes of differentiation and elimination.
> This fact, which is contained in the earlier results of §4 and §8, can be formulated in a
> particularly simple way for the case of canonical differential equations and can be verified
> analytically, independently of the motivation [...].

Now, the Hamilton–Jacobi equation is a first-order partial differential equation
in n variables $q_1, \ldots q_n$ and we expect to find a system of n first-order ordinary
differential equations for it which are the characteristic curves. In fact, these equations
are precisely the equations $\dot{q}_j = H_{p_j}$ that we took on trust earlier. So we see a very
strong connection between the theory of first-order partial differential equations and
the Hamilton–Jacobi theory of dynamics. Curiously, it is not clear how much of
Cauchy's theory was known to either of these men at the time.

A comment of the historian Tom Hawkins provides a helpful conclusion to this difficult mathematics[6]:

> For Hamilton the importance of the equivalence lay in the direction of replacing the equations of motion by the two partial differential equations so that thereby the difficulty of determining the motion of a system of masses "is at least transferred from the integration of many equations of one class to the integration of two of another" [1834]. Jacobi realized that, at least in terms of the integration theory of first-order partial differential equations, which Hamilton does not appear to have had in mind, "reduction" [...] is hardly an advance. As Jacobi explained in a letter to the secretary of the mathematics and physics section of the Berlin Academy: "Little would seem to be gained by this reduction to a partial differential equation since according to Pfaff's method ...– and for more than three variables till now nothing further was known about the integration of partial differential equations of the first order – the integration of the one partial differential equation to which the dynamical problem is reduced is much more difficult than integration of the directly given system of ordinary differential equations of motion." He went on to explain, however, that "if Hamilton's investigations are extended to all first-order partial differential equations, as can be done without difficulty, it is on the other hand a significant discovery in the theory of first-order partial differential equations that they can always be reduced to a single system of ordinary differential equations, which previously according to the Pfaffian method was insufficient" (1837a: 50–51).

> Here Jacobi was referring to his discovery that the problem of determining a complete solution to the general first-order partial differential equation

$$F(x_1, \ldots, x_n, z, p_1, \ldots, p_n) = 0, \quad p_j = \frac{\partial z}{\partial x_j}, \ j = 1, \ldots, n$$

> reduces to the complete integration of a single system of ordinary differential equations [...] of order $2n - 1$, which is in fact the first system that arises in the application of Pfaff's method (1837b: 101–102).

24.5   Exercises

Questions

1. An abundance of daily experience suggests that Euclidean geometry is true, and adequately axiomatised. What would have to be done—a question Jacobi asked when giving his courses on mechanics—to establish an axiomatic account of everyday mechanics?

---

[6]See Hawkins ([139], 206).

Chapter 25
Geometrical Interpretations of Mechanics

25.1 Introduction

One of the most important developments in the application of the calculus of variations was the exploration of the close analogy between Hamiltonian dynamics and Gaussian differential geometry, and it is to this that we now turn. These connections have been carefully explored by the historian Jesper Lützen in his [193], and we follow his account here.[1] The chain of ideas is as follows:

- Gauss introduced the idea of geodesics on a surface and coordinate systems largely made up of geodesics (compare latitude and longitude on a sphere).
- Liouville mimicked these arguments in the context of mechanics but without imposing a geometrical interpretation on mechanics.
- Lipschitz interpreted a mechanical trajectory as a geodesic on a surface (or higher dimensional analogue) and thinks of mechanics in geometric terms
- Darboux wrote all this up carefully and lucidly: mechanics can be studied geometrically.

25.2 Gaussian Curvature

In his *Disquisitiones circa superficies curvas* (or, *General investigations of curved surfaces*) [116] that created the subject of intrinsic differential geometry, Gauss introduced the idea of a surface as either the image of a map from a patch of $\mathbb{R}^2$ to $\mathbb{R}^3$ or a domain with coordinates (p, q) and a metric

$$ds^2 = E dp^2 + 2F dp dq + G dq^2,$$

where E, F, and G are functions of p and q.

---

[1] See also Lützen ([192], Chaps. XVI, XVII).

© The Author(s), under exclusive license to Springer Nature Switzerland AG 2021
J. Gray, *Change and Variations*, Springer Undergraduate Mathematics Series,
https://doi.org/10.1007/978-3-030-70575-6_25

He then imposed a coherent logic on what seemed at times to be a sprawl of formulae related by arbitrary changes of variable. In particular, he defined a concept of curvature of a surface and showed that it was intrinsic to the surface. That is, it could be defined entirely without reference to any ambient space in which the surface lived (such as, say, a normal to the surface implies). As Gauss puts it, if one surface can be mapped isometrically onto another then the values of the curvature agree at corresponding points. This was an entirely novel and unexpected idea (Gauss called the corresponding theorem the exceptional theorem, Theorema egregium), and it gradually transformed the subject of differential geometry.

Gauss defined his measure of curvature by means of the Gauss map (see Sect. 23.4). He defined the (Gaussian) curvature at the point P as the limit

$$\lim_{S \to P} \frac{area\ of\ S'}{area\ of\ S},$$

where S is a region about the point P and S' is its image under the Gauss map. If the surface S is a plane then all the normals point in the same direction and the Gauss map sends the entire plane to a point: the plane has Gaussian curvature zero. The Gauss map sends a circular cylinder to a line, so the curvature of a cylinder is also zero. The Gaussian curvature of a sphere of radius R is easy to find: directions on the domain sphere are scaled by a factor of $1/R$ by the Gauss map, so the Gaussian curvature is $1/R^2$. Lastly, saddle-shaped regions have negative curvature, as can be seen from the figure of the catenoid, Fig. 23.1.

Gauss showed that the value of the (Gaussian) curvature at a point was always the product of the extremal radii of curvature at that point. So, writing K for the Gaussian curvature and k_1, k_2 for the radii of curvature at a point, one has $K = k_1.k_2$. In particular, if the surface is a minimal surface, its mean curvature vanishes and the radii of curvature at a point are $\pm k$ (they will vary with P) then one has $K = -k^2$. So a minimal surface has negative Gaussian curvature.

Gauss's reformulation of differential geometry spread slowly across the mathematical community, and because its deepest discovery concerned the existence of intrinsic properties of surfaces applications of it came only slowly to the study of minimal surfaces, which belong in extrinsic geometry.

To study geodesics on such a surface, Gauss noted that one can impose an analogue of polar coordinates on a surface, by choosing an arbitrary point O as origin, an arbitrary geodesic as the base line, and assigning the coordinates (r, φ) to the point that is a distance r along the geodesic that meets the base line at the angle φ. He then proved that the curve defined by the points a given distance r_0 from O is everywhere at right angles to the geodesics through O. In a system of polar coordinates, $E = 1$, $F = 0$, and G must be positive so that the metric is positive definite.

He then investigated how to change a given coordinate system to a system of polar coordinates, and obtained the equations

$$EG - F^2 = E \left(\frac{\partial r}{\partial q} \right)^2 - 2F \frac{\partial r}{\partial p} \frac{\partial r}{\partial q} + G \left(\frac{\partial r}{\partial p} \right)^2, \tag{25.1}$$

from which r can be determined, and

$$\left(E\frac{\partial r}{\partial q} - F\frac{\partial r}{\partial p}\right)\frac{\partial \varphi}{\partial q} = \left(F\frac{\partial r}{\partial q} - G\frac{\partial r}{\partial p}\right)\frac{\partial \varphi}{\partial p} \qquad (25.2)$$

from which, when r is known, φ can be found. He observed that these equations can certainly be solved, and indicated that power series solutions would be particularly interesting. In this way, geodesics are found that are the orthogonal trajectories to the curves $r = const$.

25.2.1 Liouville's Contributions

Gauss's ideas about differential geometry were brought to France by Liouville and Bonnet. Liouville was a particularly versatile mathematician and an editor of a journal he had founded, which put him in a good position to know what was going one generally, and one of his contributions was to find an important way in which differential geometry connected to the study of partial differential equations.

In the 1840s, one of Liouville's concerns was to acquaint his fellow French mathematicians with the work of Gauss on differential geometry.[2] One of the ways he did this was by re-issuing Monge's *Application d'analyse à la géométrie*, to which he added a series of notes, remarking that the notes

> deal with those points …for which Mr. Gauss has opened new ways; besides our aim is to indicate to young people the sources where they can find information rather than giving them regular lectures.

One result rather sketchily proved by Gauss—although one can legitimately wonder about Liouville's argument, as we shall see—was that a surface given in the form $(x(u, v), y(u, v), z(u, v))$ and with a metric of the form $ds^2 = Edu^2 + 2Fdudv + Gdv^2$ can be given a new coordinate system α and β for which the metric takes the form $ds^2 = \lambda(\alpha, \beta)(d\alpha^2 + d\beta^2)$.

Such a system of coordinates is called isothermal, and its advantage is not only that it is easier to calculate with but that the coordinate curves $\alpha = const.$ and $\beta = const.$ meet everywhere at right angles. This condition on two families of curves occurs naturally in differential geometry, for example, the principal curves on a surface meet at right angles.[3]

---

[2] This account follows [192], the definitive biography of Liouville, see pp. 739–747.

[3] Away, that is, from what are called umbilic points.

To prove this claim, Liouville formally factorised the metric as a product

$$\left(du\sqrt{E} + dv\frac{1}{\sqrt{E}}(F + i\sqrt{EG - F^2})\right)\left(du\sqrt{E} + dv\frac{1}{\sqrt{E}}(F - i\sqrt{EG - F^2})\right).$$

Liouville now supposed that there was an integrating factor $\mu + i\nu$ that makes the first factor an exact differential, which he wrote as $d(\alpha + i\beta)$. Multiplying the second factor by the integrating factor $\mu - i\nu$ makes it exact and equal to $d(\alpha - i\beta)$, and multiplying the two factors yields

$$(\mu^2 + \nu^2)ds^2 = d\alpha^2 + d\beta^2,$$

which Liouville wrote as

$$ds^2 = \lambda(\alpha, \beta)(d\alpha^2 + d\beta^2), \quad \lambda(\alpha, \beta) = (\mu^2 + \nu^2)^{-1}.$$

The weak point in this argument is the existence of an integrating factor, which had been proved in the real case by an argument that does not extend to the complex case that Liouville was dealing with without more thought than he gave it. In fact, that point had been dealt with by Cauchy in 1819, but his result was not well known and it can seem that even Cauchy had forgotten it; in any case, Liouville did not mention it.

Gauss's great discovery had been the intrinsic nature of the curvature K of a surface. Liouville offered what he believed was a simpler proof of this result, by showing that K satisfies a partial differential equation in terms of λ, which is an intrinsic quantity:

$$K = -\frac{2}{\lambda}\left(\frac{\partial^2 \log \lambda}{\partial \alpha^2} + \frac{\partial^2 \log \lambda}{\partial \beta^2}\right).$$

It would take us too far afield to give his proof, but we can observe the uses of this result. For example, Liouville showed that a surface of zero curvature can be mapped isometrically onto a plane by solving the equation

$$\frac{\partial^2 \log \lambda}{\partial \alpha^2} + \frac{\partial^2 \log \lambda}{\partial \beta^2} = 0.$$

As for surfaces of non-zero curvature, those of constant curvature stand out as the first to analyse. If the curvature of such a surface is, say, $\pm\frac{1}{a^2}$ then λ satisfies the partial differential equation

$$\frac{\partial^2 \log \lambda}{\partial \alpha^2} + \frac{\partial^2 \log \lambda}{\partial \beta^2} \pm \frac{2\lambda}{a^2} = 0,$$

an equation that has become known as Liouville's equation. It is more often written today in the form

$$\frac{\partial^2 z}{\partial \alpha^2} + \frac{\partial^2 z}{\partial \beta^2} = \mp \frac{2e^z}{a^2},$$

where $e^z = \lambda$.

Liouville tackled this equation by introducing complex coordinates

$$\alpha + i\beta = u, \quad \alpha - i\beta = v,$$

which enabled him to write the equation as

$$\frac{\partial^2 \log \lambda}{\partial u \partial v} \pm \frac{\lambda}{2a^2} = 0.$$

In Note 4 of the re-edited book by Monge, Liouville merely stated the solution he had found, offering as the complete integral depending on two arbitrary functions $\varphi(u)$ and $\psi(u)$ the expression

$$\lambda = \frac{4a^2 \varphi'(u)\psi'(v)e^{\varphi(u)+\psi(u)}}{(1 \pm e^{\varphi(u)+\psi(u)})^2}.$$

Further work by Liouville enabled him to obtain as a surface of constant negative curvature the curve obtained by rotating a tractrix about its axis—the surface later known as the pseudosphere.

25.3 Geometrising Mechanics

In a paper [187], Liouville observed that the kinetic energy of a conservative system is

$$\sum mv^2 = 2(U + K),$$

where K is a constant. From the definition $v = \frac{ds}{dt}$, he obtained

$$dt = \sqrt{\frac{\sum m ds^2}{2(U + K)}}'$$

and so the action integrand can be written as the square root of

$$2(h + U) \sum_{j,k=1}^{m} q_{jk} dq_j dq_k,$$

where $h = T - U$, the sum of the kinetic energy T and the (negative of what we would call the) potential energy U is a constant (the total energy) and

$$\sum_{j=1}^{m} m_j ds_j^2 = \sum_{j,k=1}^{n} q_{jk} dq_j dq_k.$$

Liouville now observed that because the quadratic differential form is positive definite it can be written as a sum of squares in a new system of coordinates, so

$$\sum_{j,k=1}^{n} q_{jk} dq_j dq_k = \sum_{j=1}^{n} l_j^2,$$

where

$$l_j = \sum_{k=1}^{n} P_{jk} q_k.$$

We can think of the array (P_{jk}) as a matrix, and the array (P^{jk}), which we shall shortly meet, as its inverse. Those ideas had not yet fully come into mathematics, and Liouville had to write everything out in full.

He then wrote

$$d\theta = \sum_{j=1}^{n} n_j l_j,$$

where $\theta = \theta(q_1, \ldots, g_n)$ is a function yet to be determined. It follows that

$$n_j = \sum_{j=1}^{n} P^{jk} \frac{\partial \theta}{\partial q_j}.$$

Liouville now demanded that θ satisfies the first-order partial differential equation

$$\sum_{j=1}^{n} \left(\sum_{k=1}^{n} P^{jk} \frac{\partial \theta}{\partial q_j} \right)^2 = 2(h - U).$$

A little more work (here omitted) allowed Liouville to write the action integral as

$$A = \int_0^1 \left((d\theta)^2 + \sum_{j>k} (n_j l_k - n_k l_j)^2 \right)^{1/2}.$$

It is clear that the integral takes a minimum value when $\sum_{j>k} (n_j l_k - n_k l_j)^2 = 0$, which occurs when

$$\frac{l_1}{n_1} = \frac{l_2}{n_2} = \cdots = \frac{l_n}{n_n}$$

and along this trajectory $A = \theta(1) - \theta(0)$.

Thus, as Lützen ([193], 33) points out, Liouville rederived the Hamilton–Jacobi formalism in an elegant way. But it seems that he did not draw any analogy with the differential geometry that we are tempted to read in to his work, and most likely did not appreciate it.

The posthumous publication of Riemann's Habilitation lecture on geometry in 1867 slowly provoked mathematicians to investigate differential geometry in n dimensions. Importantly, Beltrami's two papers [5, 6] showed how to rigorously define intrinsic non-Euclidean geometry in any number of dimensions. In his [188], Rudolf Lipschitz showed that Hamiltonian mechanics was possible in such a setting, thus keeping alive the idea that these new geometries are candidates for physical space.[4] Lipschitz had already written on mechanics with an eye to geometry in his [188] and in a later French summary of his work ([189], 297–298) he wrote[5]

> One of the principal aims of the research that we shall analyze here was the profound study of Gauss's measure of curvature. In addition to the approaches which have hitherto led to this goal one may chose an approach which consists in presenting, in a general way, all the fundamental concepts related to the curvature of a surface and then deduce from them the concept of the measure of curvature. With this in mind, the idea is to find a definition of the radius of curvature which will lend itself to a natural extension. Here the principles admitted in ordinary mechanics leads to the following theorem: *When a material point which is not influenced by any accelerating forces is bound to move on a given surface, the pressure exerted in each point of the trajectory is inversely proportional to the radius of curvature of this trajectory.* Accordingly, one may define the *reciprocal of the radius of curvature* as a quantity which is directly proportional to the resulting pressure of this motion.

The measure of curvature at issue here is what is called geodesic curvature. Recall that a parameterised curve in space has at each point a tangent vector, a principal normal that measures the rate of change of the tangent vector, and a binormal (with which we shall not be concerned). Just as the tangent vector captures the velocity of a point moving on the curve, the normal vector captures its acceleration. Now suppose that the curve lies on a surface. At each point, the magnitude of the component of the normal to the curve that lies in the tangent plane to the surface is the geodesic curvature. It has this name because when that component vanishes the normal to the curve and the normal to the surface coincide and there is no acceleration in the tangent plane. Therefore, in terms of the intrinsic geometry of the surface, the curve is as straight as it can be—the definition of a geodesic.

---

[4]Earlier Schering, Riemann's successor in Göttingen, had written his [240] on potential theory in non-Euclidean geometry.

[5]Quoted in Lützen ([193], 36).

25.4   The Connection to Hamilton–Jacobi Theory

The geometrical interpretation of the solution of the Hamilton–Jacobi equation is interesting and illuminating. It turns out that one can suppose that the S-surface is moving in Q space along curves that can be regarded as geodesics in Q space with a metric determined by the variational principle of the dynamical problem.

We shall suppose that the points $(q_1, \ldots, q_n)$ occupy a simply connected domain Q. We now write the interval $[t_0, t_1] = I$, and consider the space $Q \times I$. We can regard this space as being made up of hypersurfaces $(q_1, \ldots, q_n, t)$ for each fixed value of $t \in I$. As we have seen, problems in dynamics often throw up a different family of hypersurfaces, given by the equations

$$S(q_1, \ldots, q_n, t) = \sigma,$$

where σ is a constant. The idea here is that through each point $(q_1, \ldots, q_n, t_0)$ there passes a curve that leads in $Q \times I$ to the hypersurface $(q_1, \ldots, q_n, t_1)$—we will see how to specify the right curve shortly, it will be an extremal corresponding to a variational principle. We assume that these curves never cross, and so we can suppose that the hypersurface $(q_1, \ldots, q_n, t_0)$ flows along them in the direction of the hypersurface $(q_1, \ldots, q_n, t_1)$. However, we do not assume that the flow is at the same speed on each of the designated curves; as the example of geometrical optics suggests, commonly one imagines light is travelling along these curves in a way that depends on the medium through which it is passing.

Or, which is not very different, one supposes that at each point P_0 of the hypersurface $(q_1, \ldots, q_n, t_0)$ there is a metric on $Q \times I$ and a geodesic in $Q \times I$ that joins P_0 to a unique point P_1 on the hypersurface $(q_1, \ldots, q_n, t_1)$. Now one considers the surfaces that are defined as the points a fixed distance σ from the initial hypersurface $(q_1, \ldots, q_n, t_0)$.

In his [188], Lipschitz generalised Hamilton's principle and deduced conservation of energy in the new setting. Then he noted that in the special case where $n = 2$ and U is a constant his formulation of the Hamilton–Jacobi equation is Gauss's equation (25.1). This suggested to Lipschitz that in any number of variables the Hamilton–Jacobi equation can be considered as a transformation of a metric, and the best case is when the equations of motion (the Euler–Lagrange equations) in the new coordinates $y_1, y_2, \ldots, y_n$ are solved by equations of the form $y_j = const.$

He was led to establish this theorem (here and below $f(dq)$ is shorthand for the quadratic form in the transformed system):

Let $P(q_1, \ldots, q_n, a_1, \ldots, a_n)$ be a complete solution of the Hamilton–Jacobi equation. Fix the values of $a_1, \ldots a_n$ and consider the family of trajectories of the mechanical system that are orthogonal to the $(n-1)$ dimensional manifold $P = A$ with respect to the form $f(dq)$. Then the trajectories are determined by the equations

$$\frac{\partial P}{\partial a_j} = \left(\frac{\partial P}{\partial a_j} \right)_0 ,$$

where $\left(\frac{\partial P}{\partial a_j}\right)_0$ is the value of $\frac{\partial P}{\partial a_j}$ at the intersection point. Moreover, any other $(n-1)$ dimensional manifold $P = B$ will cut the trajectories orthogonally with respect to the form $f(dq)$, and the action integral along the trajectories between the two $(n-1)$ dimensional manifolds $P = A$ and $P = B$ will all have the value $B - A$.

He also established the converse result[6]:

> Let $P(q_1, \ldots, q_n) = A$ denote an $(n-1)$-manifold. Consider the family of trajectories of the mechanical system cutting this manifold orthogonally with respect to $f(dq)$. On each trajectory and on the same side of the $(n-1)$-manifold determine a point such that the action integral V between the $(n-1)$-manifold and this point is equal to $B - A$. Then these points make up an $(n-1)$-manifold which is orthogonal to all the trajectories with respect to $f(dq)$. Moreover if the action integral V along a trajectory from its intersection with $P = A$ to an arbitrary point is considered a function of this latter point, then $R = A + V$ is a solution of the Hamilton–Jacobi equation and the $(n-1)$-manifold $R = A$ will coincide with the original $(n-1)$-manifold $P = A$.

Thus, Lipschitz's work made clear the close analogy between the geometric study of geodesics in a manifold and the dynamical study of trajectories in Hamiltonian mechanics. To be sure, he spoke of a quadratic form, not a metric. But he knew Riemann's work very well, and Beltrami's, and he referred to Gauss's work on geodesics.

It seems that he was unaware of the earlier work of Liouville, and in turn his work was unknown to Thomson and Tait, who had stated Lipschitz's result (for a single particle subject to a force) in vol. 1, p. 353 of their book. But other authors did read Lipschitz's paper, among them the French geometer Gaston Darboux.

Darboux makes an interesting contrast with Felix Klein. Both men saw geometry as the natural way to formulate and solve problems, both were energetic writers of books as well as research articles. Darboux inclined, as a well-educated French mathematician, to the study of differential equations and differential geometry; Klein, as a well-educated German mathematician, to projective geometry, with an original interest in groups.

In Volume 2 of his *Leçons sur la Théorie Générale des Surfaces* [58], Darboux developed the study of curves on surfaces, the ideas of curvature and torsion, and geodesics. Then he turned (in Book 5, Chaps. 6–8) to what he saw as the close analogy between Gauss's theory of geodesics and Jacobi's theory of analytical mechanics. In this spirit, he mentioned the work of Thomson and Tait, introduced Hamilton's principle, and took his readers through Lagrangian and Hamiltonian dynamics in the manner of Liouville and Lipschitz. With his lucid exposition the theories of classical mechanics and differential geometry were united.

Indeed, as Darboux puts it (§569) the work of Liouville and Lipschitz

> establishes the principle of least action without the use of the calculus of variations, and by methods that are entirely algebraic.

Lützen has drawn attention to a particularly attractive point in Darboux's account (Chap. VIII, §§571–577). Darboux formally eliminated the time variable from the action integral by noting that conservation of energy implies that

---

[6]This and the earlier result are quoted in Lützen ([193], 43–44).

$$dt = \sqrt{\frac{T}{U+h}},$$

where

$$2T = \sum_{j,k} a_{jk} dq_j dq_k.$$

Therefore, Lagrange's equations can be written in the form (in Darboux's disturbing notation)

$$d\frac{\partial}{\partial dq_j}\sqrt{(U+h)T} - \frac{\partial}{\partial q_j}\sqrt{(U+h)T} = 0.$$

But this is what is found by saying that the first variation of the integral

$$\int_0^1 \sqrt{(U+h)T} = \frac{1}{2}\int_0^1 \sqrt{(U+h)T}\sum_{j,k} a_{jk} dq_j dq_k$$

vanishes.

Darboux therefore set

$$ds^2 = 2(U+h)\sum_{j,k} a_{jk} dq_j dq_k.$$

He called ds the elementary action, and remarked that the general problem of mechanics is reduced to the study of the extrema of the integral $\int_1^2 ds$. But

> This is what the principle of least action consists of; and one sees immediately, thanks to this principle, that the general problem of mechanics is only an extension to any number of variables of the problem of studying geodesic curves.

(All that is missing is the fully Riemannian idea that any number of variables together with a metric define a geometric space.)

The final detail was contributed by Heinrich Hertz. His expertise was in physics, and he fulfilled his initial promise by being the first person to confirm Maxwell's prediction that electro-magnetic waves travel at the speed of light and that light is itself an electro-magnetic wave. The book in which he set out his most fundamental ideas about mechanics, his *Die Prinzipien der Mechanik in neuen Zusammenhangen dargestellt*, was published posthumously in the year of his death, 1894.

Hertz had a dislike of the concept of force—there was in fact a long tradition of mathematicians and physicists for whom the word covered up a lack of understanding and needed to be replaced. He also disliked the concept of energy, and his own way of doing without these concepts was to re-interpret the Lagrangian equations of dynamics in terms of some new, "hidden" masses. This seems to have convinced no one, but on the way he came up with a thorough-going geometrical reading of dynamics that is close to that of Lipschitz and Darboux, but which went one step further by calling the quadratic form a metric. Surprisingly, it seems from the definitive study of Hertz's mechanics that Hertz came to these ideas initially unaware of this earlier work ([194], 160); however, much of his later acquaintance with them may have affected his exposition.

It was now possible for mathematicians to say explicitly what Darboux had said at length but without the concept of a general space up front: Trajectories in Hamiltonian dynamics evolve in phase space along trajectories that are geodesics with respect to the metric that is the quadratic form in the energy functional.

25.5 Exercises

[In defiance of my earlier proscription!]

1. Given a straight line (a geodesic) in the plane and a family of geodesics at right angles to it and of equal length, what is the curve formed by the end points of these geodesics?
2. The same question, but for geodesics on the sphere.
3. The same question, but for geodesics in the disc with a metric of constant negative curvature (the non-Euclidean or hyperbolic disc).

Questions

1. The extrema of the integral of a Lagrangian with fixed times for the lower and upper end points determine the curves along which a dynamical system evolves. By studying what happens when the upper end point or time is allowed to vary, Hamilton studied how those trajectories evolve. The time-dependent Lagrangian satisfies a partial differential equation with respect to which these trajectories are the characteristic curves. These curves can be seen as geodesics with respect to metric determined by the Lagrangian. If you know about general relativity, this is the bridge between Einstein's approach and Hilbert's; see what you can find out about their competitive rivalry in 1915, starting with Corry [48].

Chapter 26
The Calculus of Variations in the nineteenth Century

26.1 Introduction

Lagrange's theory of the calculus of variations was successful, influential, and, just like the early calculus itself, hard to explain. As it was steadily improved, first by Legendre and then by Jacobi, it also became clear that it reflected an eighteenth century naivety about the nature of functions and had not properly considered the range of possibilities for candidate curves. In particular, functions were generally taken to be infinitely differentiable. The best nineteenth century theory for this was presented by Adolf Kneser towards the end of the century, building on earlier ideas of Weierstrass.[1] The subject also forms the last of the famous Hilbert Problems.

26.2 After Lagrange

Lagrange's elegant enrichment of Euler's insights created a workable calculus of variations, but it rested on a mysterious theory, and as Lagrange's personal confidence in an algebraic foundation for the calculus as a whole found few who shared it, it was natural for others to investigate and extend the theory more carefully.

It was clear, for example, that the solutions found were generally either maxima or minima, and that this issue could usually be decided from the context, but it would be better to have a way of deciding it on theoretical grounds. Legendre initiated such an investigation in a paper of 1788, when he studied what is called the second variation of the integral. (The name derives from the analogy with the calculus of functions $y = y(x)$: if at a given point $\frac{dy}{dx} = 0$ then the sign of $\frac{d^2y}{dx^2}$ can determine if the point is a local maximum or minimum of $y(x)$.)

---

[1] An excellent historical account, going into more detail than is possible here, is Fraser [107].

© The Author(s), under exclusive license to Springer Nature Switzerland AG 2021
J. Gray, *Change and Variations*, Springer Undergraduate Mathematics Series,
https://doi.org/10.1007/978-3-030-70575-6_26

Legendre took the integral

$$I = \int_a^b f(x, y, y')dx$$

and assumed that the function y is an extremal for the integral I. He then considered a function $w(x)$ which varied $y(x)$, writing $\delta y = w(x)$, where $w(a) = 0 = w(b)$. The first variation of I is

$$I_1 = \int_a^b \left(\frac{\partial f}{\partial y} w + \frac{\partial f}{\partial y'} w' \right) dx,$$

and the second variation of I is

$$I_2 = \int_a^b \left(\frac{\partial^2 f}{\partial y^2} w^2 + 2 \frac{\partial^2 f}{\partial y \partial y'} w w' dx + \frac{\partial^2 f}{\partial y'^2} w'^2 \right) dx.$$

The total variation in I is given by

$$\Delta I = I_1 + \frac{1}{2} I_2 + \cdots .$$

On the reasonable assumption that I_1 dominates this expression, for an extremal y it is clear that I_1 must vanish. This ensures the validity of the Euler–Lagrange equations for y. To determine whether the extremal is a maximum or a minimum, Legendre looked at the sign of I_2. After some work, here omitted, he concluded that the extremal will be a minimum if

$$\frac{\partial^2 f}{\partial y'^2} > 0.$$

This criterion is easy to use, but its derivation was troubling. Legendre produced a transformation of the integral I_2 that converted it into an the integral

$$I_2 = \int_a^b \left(\frac{\partial^2 f}{\partial y'^2} \right) (h(x, y, y'))^2 dx,$$

from which the conclusion follows immediately. But the function h is found by solving the differential equation

$$\left(\frac{\partial^2 f}{\partial y'^2} \right) \left(\frac{\partial^2 f}{\partial y^2} + v' \right) = \left(\frac{\partial^2 f}{\partial y \partial y'} + v \right)^2 \tag{26.1}$$

for the unknown function $v = v(x)$. However, Legendre had produced no general method for solving this non-linear equation, and Lagrange was able to show that solutions may not exist on the whole interval $[a, b]$.

The next advance was made by Jacobi, in his major paper [150]—almost 50 years after Legendre's. In it, he showed how to use a solution to the Euler–Lagrange equations for y to obtain a solution to Legendre's differential equation. In particular, he was able to show that the existence of a solution to the Euler–Lagrange equations stands or falls with the existence of solutions to a transformed version of Legendre's differential equation that never vanish in the interval (a, b). But, for whatever reason, Jacobi gave no proofs of his results in this paper, and it, therefore, generated a considerable amount of research by German mathematicians of the next generation, chiefly Otto Hesse, Rudolf Clebsch, and Adolph Mayer, before its conclusions were considered fully established in the mid-1850s.

26.3   Weierstrass's Theory

Such were the difficulties inherent in the calculus of variations, and perhaps also such was the naiveté about the varied nature of functions throughout much of the nineteenth century, that it is only with Weierstrass's lectures at the University of Berlin in 1879 and thereafter that some fundamental issues were confronted for the first time (Fig. 26.1).

One concerns the very idea of a variation, or more precisely, a *small* variation. If one attempts with goodwill to copy the graph of a function, say for definiteness $y = \sin x$ in the interval $[0, 2\pi]$, one surely draws a smooth curve that looks very like the original even though it may agree with it at very few points. However, it is easy to convince oneself intellectually that there is, for example, a very wriggly

Fig. 26.1 Karl Weierstrass (1815–1897) by Conrad Fehr 1895

curve that crosses the sine curve very many times and is always closer to the sine curve than the first smooth approximation that was drawn. Which of these curves should be considered as the smaller variation of the sine curve?

As put, Lagrange's theory gives no reason to hesitate: the comparison is made only between curves of the first kind, whose values and slopes differ little. But as sophistication about curves grew, mathematicians realised that it might be advisable to decide whether rapidly oscillating curves were close to slowly oscillating ones, or if the marked variation in corresponding values of $y'(x)$ should be taken into account. In particular, the English mathematician Isaac Todhunter was worried by the possibility of a few abrupt changes in values of $y'(x)$, but he did not produce a systematic theory to cope with the difficulties that he had identified.

Weierstrass, who was alert to the great difference between differentiable and merely continuous functions, went much further. Given an integral I of the form $I = \int_a^b f(x, y, y')dx$ and a function $\bar{y}(x)$ that is a solution of the corresponding Euler–Lagrange equation, Weierstrass considered the curve described by $\bar{y}$ and a comparison curve $y(x)$ that agreed with $\bar{y}$ at $x = a$ and again at a point P where $x = c, a < c < b$ where they crossed.[2] So their slopes will differ at $x = a$. He considered the excess function

$$E(x, \bar{y}, \bar{y}', y') = f(x, \bar{y}, y') - f(x, \bar{y}, \bar{y}') - \frac{\partial f}{\partial \bar{y}'}(x, y, y')(y' - \bar{y}).$$

This allowed him to consider comparison curves with different slopes, and he was able to show that it is necessary and (he believed) also sufficient for the curve $\bar{y}$ to be a minimum that

$$E(x, \bar{y}, \bar{y}', y') \geq 0.$$

Weierstrass's analysis began to clarify the nature of the comparison curves, because it opened the way to curves that oscillate much more than the extremal. Later, Adolf Kneser was able to show that if a minimiser of an integral is sought only among curves that differ from an extremal only slightly in respect of both their values and the values of their derivatives, then Weierstrass's necessary and sufficient conditions for a minimum are correct. But if comparison curves may differ in the values of the derivatives, then Weierstrass's condition is not sufficient. He called the first case the "weak" variation and the second case the "strong" variation.

In his lectures at Berlin, Weierstrass developed his theory in the formalism of parameterised curves. This had the effect of making it harder to use in any but geometrical problems, but that has the incidental effect of making it easier to describe in general terms.

At issue is the search for a curve in the plane that joins the points (a_0, b_0) and (a_1, b_1) and minimises a given integral J. The curves through these points will be considered in the form $(x(t), y(t))$, $0 \leq t \leq 1$. Given one such curve a nearby curve will be written as

---

[2]The bar notation for the extremal is due to Kneser; it was not used by Weierstrass.

$$(x(t) + \xi(t), y(t) + \eta(t)), \ 0 \le t \le 1,$$

where $\xi(0) = \xi(1) = 0 = \eta(0) = \eta(1)$. (Precisely what conditions can be imposed on the curves and their variations soon became a topic of research—here we shall assume continuously twice differentiable.)

The variation in the integral J can be written down and analysed, and the conclusion is that δJ must vanish for all admissible functions ξ and η. This implies Weierstrass's form of the Euler–Lagrange equations:

$$F_x - \frac{d}{dt}F_{x'} = 0, \quad F_y - \frac{d}{dt}F_{y'} = 0.$$

These equations are not independent, however, and are equivalent to the differential equation

$$F_{xy'} - F_{x'y} + F_1(x'y'' - x''y') = 0,$$

where $F_1 = F_1(x, y, x', y')$ satisfies

$$F_{x'x'} = y'^2 F_1, \quad F_{x'y'} = -x'y'F_1, \quad F_{y'y'} = x'^2 F_1.$$

This function, which Weierstrass showed exists, is "infinite" when x' and y' vanish simultaneously. Curves $(x(t), y(t))$, $0 \le t \le 1$ satisfying this differential equation are extremals for the problem. A further equation is needed to determine the functions $x(t)$ and $y(t)$ precisely, should that be necessary.

The method works well for finding shortest curve on a surface and joining two given points when the surface is given as a graph over a region of the plane, and so in the form $z = f(x, y)$. The integral to be minimised is

$$J = \int_0^1 (Eu'^2 + 2Fu'v' + Gv'^2)^{1/2} dt,$$

where $u = \xi(t)$ and $v = \eta(t)$. Here, as usual in differential geometry, with $\mathbf{r}(u, v) = (u, v, z(u, v))$,

$$E = \mathbf{r}_u . \mathbf{r}_u, \quad F = \mathbf{r}_u . \mathbf{r}_v, \quad G = \mathbf{r}_v . \mathbf{r}_v.$$

As is also usual in differential geometry, the resulting differential equation looks intimidating at first but has a conceptually simple interpretation: the acceleration of the minimising curve with respect to parameterisation by arc length (the principal normal to the curve) is normal to the surface. This means that in terms of the intrinsic geometry of the surface there is no acceleration, and so the curve is as "straight" as it can be, which makes it a geodesic as required.

Weierstrass's method extended to the second variation, and encompassed both Legendre's and Jacobi's conclusions, and resulted in necessary and sufficient conditions for both weak and strong minima. The method shows, for example, that in the above example the curve is indeed of shortest length between the given end points.

26.3.1   Two Examples

I take these examples from Bolza ([19], 128–129, 210–211). The first requires the above characterisation of geodesics, and the second one is straightforward.

a) Example XI: To determine the curve of shortest length which can be drawn on a given surface between two given points.

If the rectangular coordinates x, y, z of a point of the surface are given as functions of two parameters u, v and the curves on the surface are expressed in parameter-representation

$$u = \phi(t), v = \psi(t) \quad (27)$$

the problem is to minimise the integral

$$j = \int_0^1 \sqrt{Eu'^2 + 2Fu'v' + Gv'^2}\, dt,$$

where

$$E = \sum x_u^2, \quad F = \sum x_u x_v, \quad G = \sum x_v^2,$$

the summation sign referring to a cyclic permutation of x, y, z.

The curves must be restricted to such a portion S of the surface that the correspondence between S and its image T in the u-, v-plane is a one-to-one correspondence. We further suppose that E, F, G are of class C'' in T and that S is free from singular points, i.e.

$$EG - F^2 > 0.$$

α) If we use Weierstrass's form (I) of *Euler's equation*, and denote by $\Phi(F)$ the differential expression

$$\Phi(F) = F_{xy'} - F_{yx'} + F_1(x'y'' - x''y'),$$

we obtain easily

$$\Phi(\sqrt{Eu'^2 + 2Fu'v' + Gv'^2}) = \frac{\Gamma}{\sqrt{Eu'^2 + 2Fu'v' + Gv'^2}} \quad (28),$$

where

$$\Gamma = (EG - F^2)(u'v'' - u''v') + (Eu' + Fv')((F_u - \frac{1}{2}E_v)u'^2 + G_u u'v' + \frac{1}{2}G_v v'^2)$$

$$- (Fu' - Gv')(\frac{1}{2}E_u u'^2 + E_u u'v' + (F_v - \frac{1}{2}G_u)v'^2). \quad (29)$$

The extremals satisfy, therefore, the differential equation

$$\Gamma = 0 \quad (29a).$$

This differential equation admits of a simple geometrical interpretation: The geodesic curvature of the surface (27) at the point t is given by the expression

$$\frac{1}{\rho_g} = \frac{\Gamma}{\sqrt{EG - F^2}\sqrt{Eu'^2 + 2Fu'v' + Gv'^2}^3}.$$

Hence the curve of shortest length has the characteristic property that *its geodesic curvature is constantly zero*, i.e. it is a *geodesic*.

For the second example I quote without proof that extremals of an integral

$$J = \frac{1}{2}\int_{t_0}^{t_1}(xy' - x'y)dt$$

subject to the condition

$$K = \int_{t_0}^{t_1}\sqrt{x'^2 + y'^2}dt$$

is constant are found by defining $H = F + \lambda G$ and solving the equation

$$H_{xy'} - H_{x'y} - H_1(x'y'' - x''y') = 0,$$

where

$$H_1 = H_{x'x'}/y'^2 = -H_{x'y'}/x'y' = H_{y'y'}/x'^2.$$

b) Example XIII: Among all curves of given length joining two given points A and B, determine the one which, together with the chord AB, bounds the maximum area. [This is Dido's problem.]

Taking the straight line join A and B for the x-axis, with BA for positive direction, we have to maximise the integral

$$J = \frac{1}{2}\int_{t_0}^{t_1}(xy' - x'y)dt,$$

while

$$K = \int_{t_0}^{t_1}\sqrt{x'^2 + y'^2}dt$$

has a given value, say l, which we suppose greater than the distance AB. Since

$$H = \frac{1}{2}(xy' - x'y) + \lambda\sqrt{x'^2 + y'^2}$$

we get

Fig. 26.2 David Hilbert
(1862–1943) c. 1886

$$H_1 = + \frac{\lambda}{\sqrt{x'^2 + y'^2}^3} \, ,$$

and therefore [...]

$$\frac{x'y'' - x''y'}{\sqrt{x'^2 + y'^2}^3} = -\frac{1}{\lambda}.$$

Hence the radius of curvature of the maximising curve is constant and has the value $|\lambda|$, while its direction is determined by the sign of λ.

Again, since H_1 never vanishes, there can be no corners, and therefore the curve must be an *arc of a circle of radius* $|\lambda|$. The centre and the radius of the circle are determined by the conditions that the arc shall pass through the two given points and shall have length l. There are two arcs satisfying these conditions, symmetrical with respect to the x-axis.

26.4   Hilbert's Problem 23 and the Theory of the Calculus of Variations

By 1900 David Hilbert (Fig. 26.2) was the agreed new leader of mathematics in Germany. He was the most powerful figure in the growing collection of highly talented mathematicians that Felix Klein was bringing together in Göttingen, an authority at successive stages in his career on algebraic invariant theory, algebraic number theory, and plane geometry—a surprising choice that allowed him to produce a new branch of mathematics, the study of axiom systems.[3]

---

[3] As Adolf Hurwitz, his former student, remarked.

Also in 1900, the French had managed to stage the largest celebration of the new century: 6 months of Congresses on various topics. Philosophy, physics, and mathematics had week-long meetings in August; Poincaré spoke at all three. But the Congress of Mathematicians is famous for Hilbert's contribution. Hilbert's close friend Hermann Minkowski had suggested to Hilbert that he speak on the future of mathematics as a particularly appropriate subject, and that is what Hilbert did. He offered a general panorama on the dialogue between problems and theory, giving Fermat's last theorem and the brachistochrone problem as his key examples of provocative problems that had produced important new theories. Then he presented some 23 problems (10 in his address at the congress, all 23 in his published paper). They were on a great variety of topics, and over the years, helped by the prestige of Göttingen, they became celebrated, and reputations were made for solving one.

The last five were on analysis, indicative of Hilbert's shift to a new interest (he was occupied with functional analysis for most of the next few years, research that provides the origin of "Hilbert space"). Hilbert's 23rd and final problem in his list in Paris connected to the opening words of his lecture, where he spoke of the importance of the brachistochrone problem and went on to remark that[4]

> ...for example, the problem of the shortest line plays a chief and historically important part in the foundations of geometry, in the theory of lines and surfaces, in mechanics and in the calculus of variations.

A little later in his address he returned to the topic and made the striking remark that

> It is an error to believe that rigor in the proof is the enemy of simplicity. On the contrary, we find it confirmed by numerous examples that the rigorous method is at the same time the simpler and the more easily comprehended. The very effort for rigor forces us to discover simpler methods of proof. It also frequently leads the way to methods which are more capable of development than the old methods of less rigor.

He offered some examples and then continued

> But the most striking example of my statement is the calculus of variations. The treatment of the first and second variations of definite integrals required in part extremely complicated calculations, and the processes applied by the old mathematicians lacked the necessary rigor. Weierstrass showed us the way to a new and sure foundation of the calculus of variations. By the examples of the simple and double integral I will show briefly, at the close of my lecture, how this way leads at once to a surprising simplification of the calculus of variations. For in the demonstration of the necessary and sufficient criteria for the occurrence of a maximum and minimum, the calculation of the second variation and in part, indeed, the tiresome reasoning connected with the first variation may be completely dispensed with to say nothing of the advance which is involved in the removal of the restriction to variations for which the differential coefficients of the function vary only slightly.

Then, at the end of his lecture Hilbert outlined the problems posed by the calculus of variations in these terms. What he says is not easy to follow, and it will be enough

---

[4]For accounts of Hilbert's Paris problems, their origins and their influence down the twentieth century, see Gray [125] and Yandell [275].

to go straight to his conclusion; note only that he presented a rigorous and ingenious argument to his conclusion. In fact, although his reputation has become that of a highly abstract thinker, his contemporaries regarded him as a problem-solver first and foremost, and without getting into the details of his new presentation of the calculus of variations you can see that his definition of J^* exhibits the cunning of a problem-solver more than the sweep of a theorist.

His approach soon became an accepted part of the theory. The connection to Hamilton–Jacobi theory was appreciated, but so was the simplicity of the derivation. It can be found, for example, in Osgood's paper [208], Bolza's book ([19], 92) and in the 15-page article by Ernest Zermelo and Hans Hahn on recent further developments in the calculus of variations that they published in the Encyklopädie *der Mathematischen Wissenschaften* in 1904. Hilbert himself gave a more detailed account in his [144].

23. FURTHER DEVELOPMENT OF THE METHODS OF THE CALCULUS OF VARIATIONS.

So far, I have generally mentioned problems as definite and special as possible, in the opinion that it is just such definite and special problems that attract us the most and from which the most lasting influence is often exerted upon science. Nevertheless, I should like to close with a general problem, namely with the indication of a branch of mathematics repeatedly mentioned in this lecture – which, in spite of the considerable advance Weierstrass has recently given it, does not receive the general appreciation which, in my opinion, is its due – I mean the calculus of variations.

The lack of interest in this is perhaps due in part to the need of reliable modern text books. So much the more praiseworthy is it then that A. Kneser, in a work published very recently, has treated the calculus of variations from the modern points of view and with regard to the modern demand for rigor.

The calculus of variations is, in the widest sense, the theory of the variation of functions, and as such appears as a necessary extension of the differential and integral calculus. In this sense, Poincaré's investigations of the three body problem, for example, form a chapter in the calculus of variations, in so far as Poincaré derived from known orbits by the principle of variation new orbits of similar character.

I add here a short justification of the general remarks upon the calculus of variations made at the beginning of my lecture.

The simplest problem in the calculus of variations proper is known to consist in finding a function y of a variable x such that the definite integral

$$J = \int_a^b F(y', y; x)dx, \quad y' = \frac{dy}{dx}$$

assumes a minimum value compared with the values it takes when y is replaced by other functions of x with the same initial and final values.

The vanishing of the first variation in the usual sense

$$\delta J = 0$$

gives for the desired function y the well-known differential equation

$$\frac{dF_{y_x}}{dx} - F_y = 0, \tag{26.2}$$

$[F_{y_x} = \frac{\partial F}{\partial y_x}, \quad F_y = \frac{\partial F}{\partial y}.]$

In order to investigate more closely the necessary and sufficient criteria for the occurrence of the required minimum, we consider the integral

$$J^* = \int_a^b (F + (y_x - p)F_p)dx,$$

$[F = F(p, y, x), F_p = \frac{\partial F(p,y,x)}{\partial p}.]$

Now we inquire how p is to be chosen, as function of x, y in order that the value of this integral J^* shall be independent of the path of integration, i. e., of the choice of the function y of the variable x. The integral J^* has the form

$$J^* = \int_a^b (Ay_x - B)dx,$$

where A and B do not contain y_x, and the vanishing of the first variation

$$\delta J^* = 0,$$

in the sense which the new question requires, gives the equation [see below]

$$\frac{\partial A}{\partial x} + \frac{\partial B}{\partial y} = 0,$$

i. e. we obtain for the function p of the two variables x, y the partial differential equation of the first order

$$\frac{\partial F_p}{\partial x} + \frac{\partial (pF_p - F)}{\partial y} = 0. \tag{26.3}$$

The ordinary differential equation of the second order (26.2) and the partial differential equation (26.3) stand in the closest relation to each other. This relation becomes immediately clear to us by the following simple transformation

$$\delta J^* = \int_a^b \left(F_y \delta y + F_p \delta p + (\delta y_x - \delta p)F_y + (y_x - p)\delta F_p \right) dx$$

$$= \int_a^b \left(F_y \delta y + \delta y_x F_p + (y_x - p)\delta F_p \right) dx$$

$$= \delta J + \int_a^b (y_x - p)\delta F_p dx.$$

We derive from this, namely, the following facts: If we construct any *simple* family of integral curves of the ordinary differential equation (26.2) of the second order and then form an ordinary differential equation of the first order

$$y_x = p(x, y) \tag{26.4}$$

which also admits these integral curves as solutions, then the function $p(x, y)$ is always an integral of the partial differential equation (26.3) of the first order; and conversely, if $p(x, y)$ denotes any solution of the partial differential equation (26.3) of the first order, all the non-singular integrals of the ordinary differential equation (26.4) of the first order are at the same time integrals of the differential equation (26.2) of the second order, or in short if $y_x = p(x, y)$ is an integral equation of the first order of the differential equation (26.2) of the second order, $p(x, y)$ represents an integral of the partial differential equation (26.3) of the first order and conversely; the integral curves of the ordinary differential equation of the second order are therefore, at the same time, the characteristics of the partial differential equation (26.3) of the first order.

In the present case we may find the same result by means of a simple calculation [see below]; for this gives us the differential equations (26.2) and (26.3) in question in the form

$$y_{xx} F_{y_x y_x} + y_x F_{y_x y} + F_{y_x x} - F_y = 0,$$

$$(p_x + p p_y) F_{pp} + p F_{py} + F_{px} - F_y = 0,$$

where the lower indices indicate the partial derivatives with respect to x, y, p, y_x. The correctness of the affirmed relation is clear from this.

The close relation derived before and just proved between the ordinary differential equation (26.2) of the second order and the partial differential equation (26.3)) of the first order, is, as it seems to me, of fundamental significance for the calculus of variations. For, from the fact that the integral J^* is independent of the path of integration it follows that

$$\int_a^b (F(p) + (y_x - p) F_p(p)) dx = \int_a^b F(\bar{y}_x) dx \qquad (26.5)$$

if we think of the left hand integral as taken along any path y and the right hand integral along an integral curve $\bar{y}$ of the differential equation

$$\bar{y}_x = p(x, \bar{y}).$$

With the help of Eq. (26.5) we arrive at Weierstrass's formula

$$\int_a^b F(y_x) dx - \int_a^b F(\bar{y}_x) dx = \int_a^b E(y_x, p) dx, \qquad (26.6)$$

where E designates Weierstrass's expression, depending upon y_x, p, y, x,

$$E(y_x, p) = F(y_x) - F(p) - (y_x - p) F_p(p).$$

Since, therefore, the solution depends only on finding an integral $p(x, y)$ which is single valued and continuous in a certain neighborhood of the integral curve $\bar{y}$, which we are considering, the developments just indicated lead immediately – without the introduction of the second variation, but only by the application of the polar process to the differential equation (26.2) – to the expression of Jacobi's condition and to the answer to the question: How far this condition of Jacobi's in conjunction with Weierstrass's condition $E > 0$ is necessary and sufficient for the occurrence of a minimum.

Hilbert ended his discussion of this problem with some remarks about Kneser's approach to Weierstrass's theory, which led to a partial differential equation that could be considered a generalisation of the Hamilton–Jacobi equation.

I add a few lines on the derivation of the equation

$$\frac{\partial A}{\partial x} + \frac{\partial B}{\partial y} = 0.$$

If we set $G = A y_x - B$ then $G_y = A y_{y_x} - B_y$ and $G_{y'} = A$, so

$$G_y - \frac{d}{dx} G_{y'} = A y_{y_x} - B_y - A_x - A y_{y_x},$$

and so (note that $y' = y_x$)

$$\delta J^* = A_x + B_y,$$

as Hilbert said.

I also add a few lines on the derivation of the equation

$$(p_x + pp_y)F_{pp} + pF_{py} + F_{px} - F_y = 0.$$

This comes about through a change of variable argument, in which the variables x and y are replaced by the variables x and p.

26.5   Exercises

Questions

1. I am not entirely sure why Hilbert was so interested in the calculus of variations around 1900; one possibility is that it was the disparity between its importance and its complexity that intrigued him. But it is also possible that it was minimising principles like the law of least action that had caught his attention. In 1898–99 he lectured on mechanics at Göttingen, and topics included "the energy conservation principle, the principle of virtual velocities and the d'Alembert principle, the principles of straightest path and of minimal constraint, and the principles of Hamilton and Jacobi" and their logical and conceptual inter-relations.[5] Can you form an assessment of these principles and the relationships between them?

---

[5]The quote comes from Corry ([48], 93).

Chapter 27
Poincaré and Mathematical Physics

27.1 Introduction

In the nineteenth century the wave equation, the heat equation, and Laplace's equation (the Dirichlet problem) were solved. Or rather, and more accurately, they were solved for a wide range of domains and initial conditions. But there was a fundamental lack of clarity about the initial conditions for these equations, and almost nothing was known of the possibility of standard methods, such as the method of characteristics, for dealing with more general equations. Even the now-standard classification of second-order linear partial differential equations into three types (elliptic, parabolic, hyperbolic) was only established in 1889, in a paper by Paul du Bois-Reymond.

In the 1890s Poincaré shook up the subject of partial differential equations with new methods, and shed light on the question of suitable initial conditions. We shall consider only his account only of the Laplace equation, which was to lead Hadamard to some surprising insights into the existence and uniqueness of solutions of partial differential equations. But Poincaré wrote widely on many aspects of partial differential equations and applied mathematics; his discussions of eigenvalue problems being a notable success there is not room to discuss in this book.[1]

27.2 The Classical Classification of Linear Partial Differential Equations

The now-standard division of linear second-order partial differential equations into elliptic, parabolic, and hyperbolic seems to have been introduced surprisingly late, in a paper Paul du Bois-Reymond published in the *Journal für die reine und angewandte*

---

[1] See Verhulst ([260], Chap. 11).

© The Author(s), under exclusive license to Springer Nature Switzerland AG 2021
J. Gray, *Change and Variations*, Springer Undergraduate Mathematics Series,
https://doi.org/10.1007/978-3-030-70575-6_27

Mathematik on the subject in 1889. By the time the paper appeared he had died, at the age of 57.

He had been introduced to mathematical physics by Franz Neumann, who taught him about fluids at the University of Zürich, and the subject became the topic of his Ph.D. at Berlin in 1859. He then pursued a career in mathematics, and eventually became a Professor at the Technical University in Berlin in 1884. Although some criticised his work for lack of rigour, he made a number of interesting discoveries about Fourier series representations, and about the growth of functions and infinite numbers.

The non-degenerate second-order linear partial differential equation with constant coefficients for an unknown function u in two variables x and y can be reduced to one of three forms:

$$u_{xx} + u_{yy} + 2au_x + 2bu_y + c = d;$$

$$u_{xx} - u_{yy} + 2au_x + 2bu_y + c = d;$$

$$u_{xx} + u_y + 2au_x + c = d,$$

where a, b, c, d are constants. It is also easy to see that introducing the new variable

$$u = e^{-dx-ey}v$$

further reduces these equations to

$$v_{xx} + v_{yy} + kv = f(x, y);$$

$$v_{xx} - v_{yy} + kv = f(x, y);$$

$$v_{xx} + v_y = f(x, y),$$

where k is a constant.

Du Bois-Reymond wrote the general partial differential equation of the type he was considering in the form

$$F(z) \equiv Rr + Ss + Tt + Pp + Qq + Zz = 0,$$

where p, q, r, s, t, have their usual meanings as the various first (p, q) and second (r, s, t) derivatives of z, and R, S, T, P, Q, Z are sufficiently differentiable functions of x and y.

Differentiation gave him equations such as

$$dp = rdx + sdy, \quad dq = sdx + tdy,$$

and so on, and du Bois-Reymond deduced that

$$Rdpdy + Tdqdx - s(Rdy^2 - Sdxdy + Tdx^2) + \Pi dxdy = 0, \qquad (27.1)$$

where $\Pi = Pp + Qq + Zs$.

He applied these equations to an arbitrary curve C on a solution surface, which necessarily satisfied these equations:

$$dz = pdx + qdy, \ dp = rdx + sdy, \ dq = sdx + tdy, \ F = 0,$$

and observed that two of z, p, q, r, s, t remain arbitrary on C but once they are given the other four can be found by integration. The simplest assumption to make, he said, is that z and one of p or q is given arbitrarily, and that z and the tangent plane to the surface are given along a curve C. More precisely, he said, no more than two of z, p, q can be arbitrary, and for curves that project to a given plane curve only one is at our disposal.

Consider, however, when the equation of the curve is such that

$$Rdy^2 - Sdxdy + Tdx^2 = 0. \qquad (27.2)$$

This equation can be written in the form

$$(dy - N_+dx)(dy - N_-dx) = 0,$$

where $N_\pm = \frac{1}{2R}\left(S \pm \sqrt{S^2 - 4RT}\right)$, and, he remarked, the solutions to the equations

$$dy - N_+dx = 0, \ dy - N_-dx = 0$$

define two families of curves that cross at every point of the (x, y)-plane when $S^2 - 4RT > 0$. These curves are the projections onto the (x, y)-plane of the characteristics of the partial differential equation $F = 0$.

The situation on an arbitrary curve and a characteristic curve are very different, du Bois-Reymond pointed out, in terms of what is known along them, and this is connected, he went on, to the question of what has to be given before an integral surface is determined. He had investigated this matter in an earlier paper, he said, and here he preferred to draw attention to "the most important point" (p. 245), which was how widely different the boundary conditions are for real and imaginary characteristics. In the case of positive characteristics (real characteristic curves) a small change in the curve C changes the surface inside the region bounded by C and the characteristics at its end points (du Bois-Reymond thought of these three curves as forming a triangle). But with imaginary characteristics, when $S^2 - 4RT < 0$, the entire solution surface is determined by an arbitrarily small part of the curve C. In the case of real characteristics du Bois-Reymond noted that if one attempts to define the surface initially along a characteristic, then one needs an extra arbitrary function that is not required when starting with a curve that is not a characteristic (one might say that the moral is that characteristics make bad boundary curves).

After discussing a number of other topics, du Bois-Reymond turned in Chap. 4 of his paper to the reduction of second-order linear partial differential equations to canonical form. He showed that a linear change of variables x and y can have these consequences:

1. When $S^2 = 4RT$, if one of R or T vanishes then so does S, and the equation can be made to take the form

$$Rr + Pp + Qq + Zz = 0, \text{ or } Tt + Pp + Qq + Zz = 0.$$

2. When $S^2 - 4RT > 0$, then either S or both R and T can be made to vanish, and the equation can be made to take the form

$$Ss + Pp + Qq + Zz = 0, \text{ or } Rr - Tt + Pp + Qq + Zz = 0.$$

3. When $S^2 - 4RT < 0$, then S can be made to vanish, but not R and T, and when S vanishes, the new R and T will have the same sign, and the equation can be made to take the form

$$Rr + Tt + Pp + Qq + Zz = 0.$$

At this point he wrote (1889, 265):

> I shall call the differential equations in the two first forms parabolic, the second two hyperbolic, and the third form elliptic.

So any linear, second-order partial differential equation can be reduced to one of these forms, provided the condition on R, S, and T is satisfied, and the task of the theory of such equations is to study the solutions appropriate to each form.

In Chap. 6 of the paper, du Bois-Reymond offered a proof that a hyperbolic (linear, second-order) partial differential equation can be solved when one is given an arc $P_2 P_4$ that meets no characteristic more than once, and along which z and p or z and q (and thus z, p, and q are given), and the solution holds in the region bounded by the arc and the characteristics through P_2 and P_4. He argued that the equation can be taken in the form

$$F(z) = s + up + vq + wz = 0,$$

when the characteristics are $x = const.$ and $y = const..$ In this case, the key result is that the solution of the partial differential equation is known in a rectangle when it is known on two adjacent sides, and the data is continuous at the common corner. Discontinuities on either edge will propagate into the rectangle. But he admitted that his argument was intuitive and inconclusive, and that a rigorous proof would be hard to find. All he had was a power series argument for which a good convergence result was lacking.

He ended his paper several pages later, remarking that

In this article, I have pulled together the few cases where, so far as I can see, the principal integral can be written down at once. Leaving aside a few particular cases, such as Riemann's equation for the propagation of sound, it seems that from now on new and more general methods must be found, that I will explain in future papers.

27.3   Poincaré and the Dirichlet Problem

In 1890 Poincaré wrote a paper [216] on the partial differential equations of mathematical physics that became much quoted. He began by observing that a number of problems in physics—electrostatics, electrodynamics, the propagation of heat, optics, elasticity, and hydrodynamics—all lead to the same family of partial differential equations. Among these is Laplace's equation, which raises the Dirichlet problem, where many different boundary conditions can be handled by the method of Green's functions, but there are several others. After surveying them, he went on

Unfortunately, the first property common to all these problems is their extreme difficulty. Not only can one often not solve them completely, but it is only at the price of the greatest effort that one can rigorously prove the possibility.

So, he asked himself, is all this hard work necessary? After all, most physicists do very well, guided by their experiments. But, he said, analysis ought to be able to do it, and a rigorous proof that a problem can be solved may be quite unsuited to providing numerical estimates but it teaches us something. Should we nonetheless relax the demand for rigour, on the grounds that the differential equations themselves have often been established by less than rigorous arguments, and experimental results are necessarily approximate? He rejected this too: how can one decide if a less than rigorous argument is valid? Who has the right to say that an argument insufficient for a mathematician is good enough for a physicist? Moreover, he concluded, it is hard to give up a problem that has not been completely solved, and some of these equations also play a role in pure analysis (in Riemann's work, for example).

   Poincaré then set out a new method for solving the Dirichlet problem; the physical context made him cast the problem in three dimensions, which was a significant mathematical advance. He remarked that the problem was known to always admit a solution, and that Riemann had proved this—remarks that no German contemporary in the field would have accepted, even in spirit. Then he noted, more securely, that solutions had been provided by Schwarz and Carl Neumann in Germany and Robin in France. After that, he presented his own solution, known as the method of "sweeping out" ("balayage" in French), which has something in common with Schwarz's alternating method.

   Poincaré used Green's theorem, which says that a solution to a Dirichlet problem is obtained by finding a suitable Green's function. He also used a rigorous version of another of Green's theorems to show that given a positive electric charge at a point inside a virtual sphere one can arrange for a charge distribution on the sphere with the property that the potential functions of the two distributions agree outside the virtual sphere.

We can give an informal argument to this effect. It is trivially true of a point charge at the centre of a virtual sphere: the potential outside the sphere is the same as that of a uniform charge distribution on the sphere. If we now move the point, keeping it inside the virtual sphere, we expect that the charge distribution on the sphere can vary in such a way as to produce a new potential function equal to that of the charge in its new position.

Poincaré gave the formulae for this, but left it to the reader to check; in fact this had been done first by Thomson and then repeated by Maxwell in his great book on electromagnetic theory.[2]

More specifically, Poincaré showed that a Green's function u relative to a sphere of radius R and concentrated at P is the potential function associated to a unit mass placed at P inside the sphere and a mass of $-\sqrt{\frac{OQ}{OP}}$ placed at Q outside the sphere.[3] The contribution to $\dfrac{dU}{dn}$ of a point M on the sphere is given by

$$\frac{R^2 - OP^2}{R} \frac{1}{MP^3}.$$

If the point P is inside the sphere, the potential of a unit charge at P is equal to $\frac{1}{MP}$ at P.

If, therefore, a unit charge is distributed over the sphere in such a way that the charge density at each point M on the sphere varies inversely with MP^3, then, Poincaré argued that the potential W of this distribution equals $\frac{1}{M'P}$ at points M' outside the sphere and at points M'' inside the sphere the potential is less that $\frac{1}{M''P}$. From this he deduced that a function equal to the Green's function U inside the sphere and zero outside the sphere is harmonic everywhere except at P and on the surface of the sphere, where it is continuous but its normal derivative jumps by $\dfrac{R^2 - OP^2}{R.MP^3}$. This function is equal to the potential function associated to a unit charge at the point P and the potential function associated to a charge density on the sphere given by $-\dfrac{R^2 - OP^2}{4\pi R.MP^3}$, which differs from the formula above only by a change of sign. Therefore the function V that is harmonic inside the sphere and takes prescribed values on the sphere, given by a function V^0, is defined by the equation

$$V(M'') = \int_S \frac{1}{4\pi R} \frac{V^0(R^2 - OP^2)}{M''P^3} d\omega.$$

---

[2]The method involves inversion in spheres. Inversion in a sphere S with centre O and radius R maps a point P to the point Q on the half-line from O through P and such that $OP.OQ = R^2$ (the map is not defined at O). The map switches P and Q, and therefore switches the inside and the outside of the sphere; it is an anti-conformal map (like a reflection). In the plane it is an inversion in a circle (see Chap. F). A harmonic function is transformed by an inversion of its domain to another harmonic function with its singular point somewhere else. Traces of this process are visible in Poincaré's map.

[3]Note that if the sphere is inverted into a plane, P and Q become mirror image points; see Sect. D.1.

Importantly, its maximum and minimum values lie between those of the given V^0, which can be assumed to be positive. Our earlier informal argument could not guarantee this crucial detail, which will be used to guarantee the existence of a lower bound later on.

Poincaré now showed how to solve the Dirichlet problem for the region R outside an isolated charged conductor of an arbitrary shape, provided only that it has a tangent plane at every point and distinct principal curvatures—conditions that enabled Poincaré to establish some convergence arguments. First, he indicated briefly that the region R outside the conductor can be covered by an infinite number of spheres S_j, $j = 1, 2, \ldots$ of various sizes so that each point of the region R lies in at least one of these spheres. Then he proposed to define a harmonic function on the region R that tends to the value 1 at points P on the surface of the conductor and that tends to the value 0 as P moves off to infinity.

To do so, he observed that given a sphere S_j and the electric charge it contained, the charge distribution can be switched with an equivalent one entirely on the surface of the sphere. This has no effect on the potential outside the sphere, which is unchanged, and reduces the potential inside the sphere. He called this operation "sweeping out" the sphere.

He started with a large external sphere Σ that surrounds the conductor and has a uniform charge distribution on it that gives rise to a potential V_0 outside the sphere (the potential goes to zero at infinity) and a constant potential of 1 everywhere inside it, including the conductor. Essentially his argument is that the spheres are swept out, lowering the potential function by a sequence of harmonic functions inside them but not altering the potential outside them. So the potential function continues to take the value 1 on the boundary of the conductor but otherwise drops. It cannot become negative, because all the charges introduced are positive, so by Harnack's theorem—if an increasing or decreasing sequence of harmonic functions is bounded it tends to a limit that is a harmonic function—the limit is everywhere harmonic and it is unaltered on the boundary of the conductor, so it continues to take the value 1 there.

In slightly more detail, at least one of the spheres S_j meets Σ and contains some of the charge distributed over Σ. Poincaré let S_1 be such a sphere and swept it out. The potential function becomes V_1, and there is no charge inside S_1. He now swept out S_2, an operation that can put some charge back inside S_1. Now he swept out S_1 and S_2. Then he turned to S_3, and so on. To sweep out every sphere infinitely often, Poincaré swept them out in this order

$$S_1, S_2; S_1, S_2, S_3; S_1, S_2, S_3, S_4; \ldots.$$

If the nth sweeping out operation empties the sphere S_k the potential resulting function V_n agrees with the preceding one, V_{n-1}, outside S_k and inside S_k it is less: $V_n \leq V_{n-1}$. So everywhere one has $V_n \leq V_{n-1}$. Because a negative charge never occurs, the decreasing sequence of V_ns is bounded below at every point and so tends to a limit, a function Poincaré called V.

Consider now the jth sphere S_j, which is swept out infinitely often, say at times $\alpha_k, k = 1, 2, \ldots$. Each time there is no charge in its interior and the corresponding potential function V_{α_k} is harmonic. Now, the sequence of values of the V_{α_k} tends to a limit and so Poincaré used Harnack's theorem from 1887 to deduce that the limit function V is also harmonic. But every point of the region R lies in at least one sphere so, said Poincaré, there is a harmonic function defined everywhere on R. Because everywhere one has $V_0 > V > 0$, and V_0 tends to zero at points arbitrarily far from the conductor, it follows that V tends to zero at points arbitrarily far from the conductor.

To show that the potential function $V(P)$ tends to 1 as the point P tends to a point M, say, on the conductor, Poincaré invoked his assumptions on the shape of the conductor to allow him to define a sphere that touches the conductor at M and otherwise lies entirely inside the conductor. A limiting argument allowed him to show that the potential function on a sequence of points tending to M from outside the conductor tends to 1, as required. He concluded that the Dirichlet principle had been established; it would have been more accurate to say that the Dirichlet problem had been solved.

Poincaré then spent some time weakening the conditions he had imposed on the boundary of the conductor, so that it could, for example, have a finite number of cone points.

Then he defended having introduced a new method into an already popular field, although it was no better than those of Robin or Neumann and in some cases actually slower. But he argued that no known method allowed one to go beyond the first approximation without calculations that were too repellent, and so the skilled analyst will welcome a new method, and his, he remarked, was particularly elastic ("if I may use the term"—[216], 231). He then proceeded to show how it can be adapted in various ways.

27.4   Exercises

Questions

1. Du Bois-Reymond's classification of linear second-order partial differential equations is the formal face of a fundamental division. His presentation emphasises the varying nature and therefore role of the associated characteristic curves. Reach the same classification by looking at boundary or initial data for these equations.

Chapter 28
Elliptic Equations and Regular Variational Problems

28.1 Introduction

This chapter, and Chap. 29 on hyperbolic equations, concludes this history of differential equations. Topics that emerge of considerable importance are the regularity of the solutions of elliptic equations—this was a particular interest of David Hilbert's—and the introduction of more rigorous methods in potential theory.

28.2 Picard on Second-Order Linear Elliptic Equations

In a paper in the *Journal de Mathématiques* for 1890, Picard considered the linear second-order partial differential equation

$$A\frac{\partial^2 u}{\partial x^2} + 2B\frac{\partial^2 u}{\partial x \partial y} + C\frac{\partial^2 u}{\partial y^2} = F\left(u, \frac{\partial u}{\partial x}, \frac{\partial u}{\partial y}, x, y\right),$$

in which the coefficients are functions of x and y in a domain for which $B^2 - AC < 0$. He showed that the solutions are determined by their values on the boundary of the domain, provided the domain is suitably small (a condition that ensures that the solution is single-valued and is therefore a function of x and y).

He took the equation in the form

$$\Delta u = u_{xx} + u_{yy} = F(u, u_x, u_y, x, y).$$

To solve it, he took an arbitrary function $u_1(x, y)$ and formed the equation

$$\Delta u_2 = F(u_1, u_{1x}, u_{1y}, x, y),$$

© The Author(s), under exclusive license to Springer Nature Switzerland AG 2021
J. Gray, *Change and Variations*, Springer Undergraduate Mathematics Series,
https://doi.org/10.1007/978-3-030-70575-6_28

which he supposed had the solution $u_2(x, y)$. He then formed the equation

$$\triangle u_3 = F(u_2, u_{2x}, u_{2y}, x, y),$$

which he supposed had the solution $u_3(x, y)$, and so on. Each solution is fixed by its values on the boundary of the domain. If the sequence of functions $u_1, u_2, u_3, \ldots$ converges to a function $u(x, y)$ then the limit function would be a solution of the equation

$$\triangle u = F(u, u_x, u_y, x, y),$$

as required. The issue therefore is to find conditions that guarantee the convergence of the sequence of functions.

Picard began with the case where

$$F(u, u_x, u_y, x, y) = au_x + bu_y + c,$$

where a, b, and c are functions of x and y. Then he dealt with the general case, and then with the special case where F is a function of u, x, and y alone, and is also an increasing function of u, when he showed that no restrictions on the domain are necessary. His method was to establish the theorem for small contours, and then extend it to arbitrary ones by looking at overlapping contours, as in Schwarz's alternating method, to which he explicitly referred.

This case included the partial differential equation

$$u_{xx} + u_{yy} = A(x, y)e^u,$$

to which Picard paid special attention. As Liouville had showed, this is the equation for a surface with metric $E = G = e^u$, $F = 0$ to have its curvature given by the function $A(x, y)$.

Finally, Picard observed that the same method of successive approximations could be used to prove the existence of solutions to ordinary differential equations.[1]

We can look a small way past the introduction and glimpse the subtleties of the problem. One case that Picard looked at was the equation

$$\frac{\partial^2 u}{\partial x^2} + \frac{\partial^2 u}{\partial y^2} = f(x, y).$$

In this case, the solution of the partial differential equation is

$$u(\xi, \eta) = -\frac{1}{2\pi} \int\!\int f(x, y)G(x, y, \xi, \eta)dxdy,$$

---

[1] The method had been used earlier by Liouville in connection with Sturm–Liouville theory, see Lützen ([192], Chap. X).

where the double integral is taken over the given surface and G is a Green's function that becomes infinite at the point (ξ, η) of the surface like $\log(1/r)$ and vanishes on the boundary.[2]

Picard investigated this solution and found that it was necessary to ensure an upper bound on $\frac{\partial u}{\partial x}$ and $\frac{\partial u}{\partial y}$. This he could do when the boundary was either a circle or a curve analytically equivalent to a circle.

For the more complicated equation

$$\frac{\partial^2 u}{\partial x^2} + \frac{\partial^2 u}{\partial y^2} = au_x + bu_y + c,$$

where $au_x + bu_y + c$ is continuous, it was not possible to write down the solution, and so an iterative approach had to be used. Picard now showed that the sequence of approximations converged provided the boundary curve was small, by which he meant that the solution function $u(x, y)$ has no chance to grow uncontrollably and thereby cease to be single-valued.

In a paper he then published in the *Journal de l'École Polytechnique* in the same year, 1890, Picard now imposed the condition that the coefficients are *analytic* functions of x and y and were able to show that in this case the solution is also analytic.[3] His method was the method of successive approximations.

28.3 Hilbert's Problems 19 and 20

This and the next extract are taken from the published version of Hilbert's address on the problems of mathematics at the ICM in Paris in 1900, ([143]).

19. ARE THE SOLUTIONS OF REGULAR PROBLEMS IN THE CALCULUS OF VARI-ATIONS ALWAYS NECESSARILY ANALYTIC ?

One of the most remarkable facts in the elements of the theory of analytic functions appears to me to be this: That there exist partial differential equations whose integrals are all of necessity analytic functions of the independent variables, that is, in short, equations susceptible of none but analytic solutions. The best known partial differential equations of this kind are the potential equation

$$\frac{\partial^2 f}{\partial x^2} + \frac{\partial^2 f}{\partial x^2} = 0$$

and certain linear differential equations investigated by Picard [*J. Ec Poly* 1890]; also the equation

$$\frac{\partial^2 f}{\partial x^2} + \frac{\partial^2 f}{\partial x^2} = e^f,$$

---

[2]Later, in §4 of his paper, he showed that u will exist if f is not even required to be continuous, but for u to be twice differentiable it is necessary that f be continuously once differentiable or at least satisfy some sort of Hölder condition (not to be discussed here).

[3]He defined an analytic function of two variables to be one that can be written as a convergent power series in the variables.

the partial differential equation of minimal surfaces, and others. Most of these partial differential equations have the common characteristic of being the Lagrangian differential equations of certain problems of variation, viz. , of such problems of variation

$$\iint f(p, q, z; x, y) dx dy = \text{minimum}$$

$$[p = \frac{\partial z}{\partial x}, \; q = \frac{\partial z}{\partial y}],$$

as satisfy, for all values of the arguments which fall within the range of discussion, the inequality

$$\frac{\partial^2 f}{\partial p^2} \frac{\partial^2 f}{\partial q^2} - \left(\frac{\partial^2 f}{\partial p \partial q} \right)^2 > 0,$$

f itself being an analytic function. We shall call this sort of problem a regular variation problem. It is chiefly the regular variation problems that play a rôle in geometry, in mechanics, and in mathematical physics; and the question naturally arises, whether all solutions of regular variation problems must necessarily be analytic functions. In other words, *does every Lagrangian partial differential equation of a regular variation problem have the property of admitting analytic integrals exclusively?* And is this the case even when the function is constrained to assume, as, e. g., in Dirichlet's problem on the potential function, boundary values which are continuous, but not analytic?

I may add that there exist surfaces of constant negative Gaussian curvature which are representable by functions that are continuous and possess indeed all the derivatives, and yet are not analytic; while on the other hand it is probable that every surface whose Gaussian curvature is constant and positive is necessarily an analytic surface. And we know that the surfaces of positive constant curvature are most closely related to this regular variation problem : To pass through a closed curve in space a surface of minimal area which shall enclose, in connection with a fixed surface through the same closed curve, a volume of given magnitude.

Hilbert then went on

20. THE GENERAL PROBLEM OF BOUNDARY VALUES.

An important problem closely connected with the foregoing is the question concerning the existence of solutions of partial differential equations when the values on the boundary of the region are prescribed. This problem is solved in the main by the keen methods of H. A. Schwarz, C. Neumann, and Poincaré for the differential equation of the potential. These methods, however, seem to be generally not capable of direct extension to the case where along the boundary there are prescribed either the differential coefficients or any relations between these and the values of the function. Nor can they be extended immediately to the case where the inquiry is not for potential surfaces but, say, for surfaces of least area, or surfaces of constant positive Gaussian curvature, which are to pass through a prescribed twisted curve or to stretch over a given ring surface. It is my conviction that it will be possible to prove these existence theorems by means of a general principle whose nature is indicated by Dirichlet's principle. This general principle will then perhaps enable us to approach the question : *Has not every regular variation problem a solution, provided certain assumptions regarding the given boundary conditions are satisfied* (say that the functions concerned in these boundary conditions are continuous and have in sections one or more derivatives), *and provided also if need be that the notion of a solution shall be suitably extended?*[Cf. my lecture on Dirichlet's principle in the *Jahresbericht der Deutschen Math.-Vereinigung*, vol. 8 (1900), p. 184.]

Hilbert then published two papers on the Dirichlet problem. The first is an indication of what he developed at length in the second. It would take us too far afield to follow him and to sort out what he did, and did not, achieve with them, but we can see how he intended to elucidate his programmatic remarks in Paris by looking at an extract from the first paper ([142]).

> The Dirichlet principle is a method that Dirichlet, drawing on an idea of Gauss, used to solve the so-called boundary value problem, and which can be briefly characterised in the following way. One erects verticals at the points of the given boundary curve in the (x, y)-plane and gives them the corresponding boundary values. On the surface $z = f(x, y)$ that is bounded by this curve one looks for a surface the minimises the value of the integral
>
> $$ J(f) = \iint \left(\left(\frac{\partial f}{\partial x} \right)^2 + \left(\frac{\partial f}{\partial y} \right)^2 \right) dxdy. $$
>
> This surface, as one can easily see using the calculus of variations, is necessarily a potential surface. With the use of considerations of this kind Riemann gave a proof of the existence of the solution to the boundary value problem and then immediately based his great theory of Abelian functions upon it.
>
> It was first recognised by Weierstrass that this use of the method of the Dirichlet principle is not sound; indeed, if only a finite number of numerical values are given one can conclude without further ado that there must be a least numerical value among them; from an unbounded number of numerical values one cannot conclude that there is least one; rather, it requires a proof that in the given case there is a surface $z = f(x, y)$ which gives the least value of the integral $J(f)$.
>
> The important researches of C. Neumann, H.A. Schwarz and H. Poincaré have shown that under certain very general assumptions about the nature of the boundary curve and the boundary values the boundary value problem is solvable and therefore the existence of a minimal function $f(x, y)$ is assured.
>
> The Dirichlet principle owes its fame to the attractive simplicity of its fundamental mathematical idea, to the undeniable richness of its possible applications to pure and to physical mathematics and its intrinsic plausibility. But since Weierstrass's critique the Dirichlet principle became of historical value only and seemed to lose its ability to lead to solutions of the boundary value problem. C. Neumann spoke regretfully that the so beautiful and so much-used Dirichlet principle would now always decline; only A. Brill and M. Noether called for new hope to grow in us and expressed the conviction that the Dirichlet principle, present in nature, could once more enjoy a revival, perhaps in a modified fashion.
>
> The following is an attempt at a revival of the Dirichlet principle.
>
> Inasmuch as we think of the Dirichlet problem as only a particular problem in the calculus of variations, we have been led to express it in the following more general form. Every regular problem in the calculus of variations has a solution provided suitable restrictions are imposed on the given boundary conditions and if necessary the idea of a solution is suitably extended.[4]
>
> How this principle can be used as a guide to the discovery of rigorous and simple existence proofs will be shown by the following two examples:
>
> I. Draw the shortest curve between two given points P and P_1 on a given surface $z = f(x, y)$.
>
> Let ℓ be the lower bound on all curves on the surface between the two points. From the totality of all connecting curves we look for those curves $C_1, C_2, C_3, \ldots$ whose lengths

---

[4]Hilbert here footnoted his Paris address and the papers ([14, 15]).

$L_1, L_2, L_3, \ldots$ respectively approach the limit ℓ. On C_1 we draw from P a length $\frac{1}{2}L_1$ and obtain on C_1 the point $P_{1,\frac{1}{2}}$; on C_2 we draw from P a length $\frac{1}{2}L_2$ and obtain on C_2 the point $P_{2,\frac{1}{2}}$; on C_3 we draw from P a length $\frac{1}{2}L_3$ and obtain on C_3 the point $P_{3,\frac{1}{2}}$, and so on. The points $P_{1,\frac{1}{2}}, P_{2,\frac{1}{2}}, P_{3,\frac{1}{2}}, \ldots$ have an accumulation point $P_{\frac{1}{2}}$ which is also a point of the surface $z = f(x, y)$.

This procedure, which we have applied to P and P_1 alike, and has led to the point $P_{\frac{1}{2}}$ we now apply to the points P and $P_{\frac{1}{2}}$ and obtain in this way the point $P_{1/4}$ on the given surface, and also the point $P_{3/4}$ when we apply the procedure to $P_{\frac{1}{2}}$ and P_1. In the same way we find the points $P_{1/8}, P_{3/8}, P_{5/8}, P_{7/8}, P_{1/16}, \ldots$. All these points and their accumulation points taken together form a continuous curve that is the sought-for shortest curve.

The proof of this fact is easily found when one thinks of the length of a curve as defined as the limiting value if the lengths of inscribed polygons. As we see at the same time, it is necessary for this approach that we assume that the given function $f(x, y)$ and its first differential quotients with respect to x and y are continuous.

Hilbert's second example was that of the Dirichlet problem itself. Unfortunately, his argument is too long to reproduce here.

28.4 Exercises

Questions

1. Hilbert's interest in the regularity of solutions to partial differential equations is perhaps a pure mathematician's attitude. Do you agree? What are the implications for physics of his conjecture (when it is proved, as it shortly was)?

Chapter 29
Initial Value Conditions for Hyperbolic Partial Differential Equations

29.1 Introduction

By the late nineteenth century it was becoming clear that solutions to hyperbolic partial differential equations have a particular kind of relation to the initial conditions that can be imposed on them. This had become increasingly clear in the later decades of the nineteenth century, as the work of Picard and others show; the person who cleared this up decisively was the French mathematician Jacques Hadamard in the early years of the twentieth century, who made a powerfully provocative study of the relation between elliptic and hyperbolic partial differential equations and between boundary and initial conditions.

29.2 Picard on Second-Order Linear Hyperbolic Equations

In the same paper in the *Journal de Mathématiques* for 1890 that we have already looked at, Picard showed how to use the method of successive approximations to solve hyperbolic second-order partial differential equations. One of his examples, the equation

$$\frac{\partial^2 z}{\partial x \partial y} = a \frac{\partial z}{\partial x} + b \frac{\partial z}{\partial y} + cz, \tag{29.1}$$

where a, b, c are functions of x and y, is reproduced in Sect. 31.8, but here it is more worth repeating his analysis of why his method for solving hyperbolic partial differential equations would not work for elliptic ones. He has posed the problem of solving the partial differential equation for which the solution is to take a prescribed value at a point on an arc AB and its first partial derivatives are to take prescribed values along the arc.

© The Author(s), under exclusive license to Springer Nature Switzerland AG 2021
J. Gray, *Change and Variations*, Springer Undergraduate Mathematics Series,
https://doi.org/10.1007/978-3-030-70575-6_29

It is quite otherwise when the characteristics are imaginary. To see this, it suffices to take the simple example of the equation

$$\frac{\partial^2 z}{\partial x^2} + \frac{\partial^2 z}{\partial y^2} = 0.$$

In general one cannot have a solution of this equation that is continuous in the rectangle $ABA'B'$ along with its first-order partial derivatives, and for which $\frac{\partial z}{\partial x}$ and $\frac{\partial z}{\partial y}$ take on the arc AB the succession of values denoted above by $\varphi(x)$ and $\psi(y)$, these functions being subject to no other condition than being continuous. In the contrary case one could, in effect, form an analytic function $z + iz_1$ that will be holomorphic in the rectangle under consideration, the real part of this function being arbitrary on the curve AB, which is impossible because a holomorphic function determined on an arc of a curve however small can only be extended in a unique way.

29.3   Hadamard and Mathematical Physics

Jacques Hadamard (Fig. 29.1) was a remarkable analyst, the first (with de la Vallée Poussin, independently) to prove the prime number theorem, which says that the number of prime numbers less than a real number x is well approximated by $x/\ln x$, but his particular fields were integral equations, the calculus of variations, and partial differential equations.

But before we proceed, I cannot resist this anecdote, which dates from a scientific Jubilee in honour of Hadamard in 1937. Picard had wondered if Hadamard would still remember his lectures on rational mechanics and Hadamard replied[1]:

> It is perfectly true that you had accepted the task – should I say the burden – of involving us in that artificial and lamentably monotonous exercise that is the problem of mechanics for the degree. You had been able to render it almost interesting; I always asked myself how you were able to do that, because I was never able to when it was my turn.

In his paper [132] he made some important remarks about the two types of partial differential equation that typically arise in physics: the Dirichlet problem and the Cauchy problem. In the Dirichlet problem in which the unknown function (say, a function of two variables) is required to satisfy a given condition at each point on the boundary of its domain. In the Cauchy problem, the boundary information is the value of a function and one of its first derivatives at each point on some boundary.

"These problems", he said, "are presented in every sort of question in mathematical physics. However, there is an extensive list of cases in which one or the other is presented as if it is well posed, I want to say as *possible* and *determined*". What he wished to point out, he said, was that "these two circumstances are intimately related, the one to the other, and this in a sufficiently close way that of the two problems, entirely analogous in appearance, one can be possible and the other impossible according to how they correspond or not to a physical given".

---

[1] See Cartwright ([33], 77).

Fig. 29.1 Jacques
Hadamard (1865–1963)

Hadamard's language needs a little unpicking. By a possible problem, he meant one admitting a solution, and by a determined problem one that has a unique solution. His discovery was that the two boundary conditions work very differently, even though the equations look very similar, and that one problem may have a solution when the other does not depend on whether the boundary conditions make physical sense or not.

He illustrated his point with two examples. Laplace's equation (A) in three dimensions leads to Dirichlet's problem, which is a possible and determined problem. On the other hand, the Cauchy problem for equation (A) asks for the determination for $x \geq 0$ of a solution such that, for $x = 0$,

$$u = u_0, \frac{\partial u}{\partial x} = u_0',$$

where u_0 and u_0' are given functions of y, z. "This problem" he said (p. 214), "which has no physical significance, can always be solved when u_0 and u'_0 are analytic, but we know today that it is quite otherwise in the general case", and he gave a brief explanation of why this was so.

Suppose, for example, that u_0 has been defined on a circle C in the plane $x = 0$ and the hemisphere S in the region $x > 0$ bounded by that circle and within which the Dirichlet problem is known to be solved. The unknown function u will be determined by the values it takes on C and S, because they define a region for which the Dirichlet problem is known to be solved. The part corresponding to the values on S defines an analytic function of x, y, z inside C, so one can say that W is the potential of a double layer distributed in the (y, z) plane, the thickness (density?) of this double layer being represented at each point by u_0. However, the Cauchy problem is only possible if

$$u'_0 - \frac{1}{2\pi} \frac{\partial W}{\partial x}$$

is an analytic function.

Then he considered the wave equation (B). The Cauchy problem in this case asks, find the solution for $t > 0$ such that when $t = 0$

$$u = U \quad \text{and} \quad \frac{\partial u}{\partial t} = U',$$

U and U' being given functions of x, y, z. Such a problem is, in general, possible and determined, he said. The solution is given by Poisson's formula

$$u(x, y, z, t) = \frac{\partial [U](t)}{\partial t} + [U'](t),$$

where $[U]$ and $[U']$ are the mean values of U and U' on the sphere centre (x, y, z) and radius t.

But, said Hadamard, one should not infer that the Cauchy problem for the equation is always possible and determined. That is the case when t is taken as the principal variable, but it is false for the same equation not when the principal variable is x, y, or z. For example, taking it with respect to x and with functions u_0 and u'_0 independent of t problem (B) now reduces to problem (A) and is therefore impossible in general.

For example, taking x as the principal variable, the Cauchy problem asks for the solution for $x > 0$ such that when $x = 0$

$$u = U \quad \text{and} \quad \frac{\partial u}{\partial x} = U',$$

U and U' being given functions of y, z, and t. Suppose, he said, that the functions U and U' are independent of t. In this case, if the solution is unique then u will certainly be independent of t, but this reduces problem (B) to problem (A), which we have just seen is impossible.

Could there, instead, be infinitely many solutions u that take the same value on $x = 0$ and for which also the values of $\frac{\partial u}{\partial x}$ are the same? If so, then there is a solution u of problem (B) that is not identically zero but vanishes on $x = 0$ along with $\frac{\partial u}{\partial x}$. Any such solution can be defined in the region $x < 0$ by the simple formula $u(-x, y, z, t) = u(x, y, z, t)$. A consideration of the implications of Poisson's formula allowed Hadamard to prove that the only solution of (B) that on $x = 0$ satisfies $u_0 = 0 = u'_0$ is the function that vanishes everywhere. Therefore the Cauchy problem in this case cannot be indeterminate and it follows that it is in general impossible.

Hadamard concluded that the Cauchy problems for $t = 0$ and $x = 0$ are very different, and the problem in the second case is much closer to the theory of equations with imaginary characteristics.

29.4 The Cauchy Problem

Hadamard began his book by introducing Cauchy's use of boundary conditions when solving a second-order, linear partial differential equation, and he concluded the first chapter by writing (§14):

> The result of Cauchy's and Sophie Kowalewsky's analysis would therefore be that Cauchy's problem has one (and only one) solution every time the surface which bears the *data is not characteristic, nor tangent anywhere to a characteristic.* (emphasis Hadamard's)

But then he immediately went on in Chapter Two to say that in fact the true situation was not so simple, and indeed was almost paradoxical.

> The reasonings of Cauchy, S. Kowalewsky and Darboux, the equivalent of which has been given above, are perfectly rigorous; nevertheless, their conclusion must not be considered as an entirely general one. The reason for this lies in the hypothesis, made above, that Cauchy's data, as well as the coefficients of the equations, are expressed by *analytic* functions; and the theorem is very often likely to be false when this hypothesis is not satisfied. [...] Indeed, one of the most curious facts in this theory is that apparently very slightly different equations behave in quite opposite ways in this matter.

To defend his position, he compared the way Cauchy data and Dirichlet data work. There are occasions when Cauchy's approach to a second-order partial differential equation, which involves specifying initial data in the form of values of the solution function and its first derivatives on a hypersurface is valid without any requirement of analyticity. (Hadamard defined an analytic function on an interval as one admitting a power series expansion.)

In contrast, the Dirichlet problem for a region requires only that the boundary values of the solution function be specified. For a region V bounded by a surface S

> It is a known fact that this problem is correctly set: i.e. it has one (and only one) solution. This fact immediately appears as contradictory to Cauchy–Kowalewsky's theorem: for, if the knowledge of numerical values of u, at the points of S (together with the partial differential equation) is by itself sufficient to determine the unknown function within V, we evidently have no right to impose upon u any additional condition, and we cannot therefore, besides values of u, choose arbitrarily those of $\frac{\partial u}{\partial n}$.

To understand the deep reason for this, Hadamard noted but set aside the fact that the data in a Cauchy problem and a Dirichlet problem are specified on topologically distinct regions. He thought it more important that Cauchy data can only supply a solution in a neighbourhood of the hypersurface, whereas the Dirichlet data leads to a solution valid throughout the enclosed region.

He then argued that in fact if data on even a small part S of a hypersurface is non-analytic there will be no solution of Laplace's equation valid in a neighbourhood of S. For if there were there would be harmonic functions defined on either side of S with the same normal derivatives at each point of S. But this means that the two harmonic functions are analytic extensions of each other, and so their values on S must be analytic, contrary to assumption.

The way out of this paradox would come, he suggested (§15), by following Poincaré's advice, for

> No question offers a more striking illustration of the ideas which Poincaré developed at the first International Mathematical Congress at Zurich, 1897 (see also *La Valeur de la Science*, pp. 137–155), viz. that it is physical applications which show us the important problems we have to set, and that again Physics foreshadows the solutions.

Hadamard then gave a more detailed and technical examination of the nature of boundary data, before returning to his broad theme. This was that theorems proved for analytic functions may not be true when more general types of function are considered and that the physical interpretation of the problem is a sure guide to whether Cauchy data or Dirichlet boundary data are appropriate. He strongly suggested (§18) that

> This remarkable agreement between the two points of view appears to me as an evidence that the attitude which we adopted above – that is, making a rule not to assume analyticity of data – agrees better with the true and inner nature of things than Cauchy's and his successors' previous conception.

There then followed one of Hadamard's more famous observations, that is worth savouring for its own sake. He had in mind a theorem of Weierstrass's that ensured that any continuous function may be approximated arbitrarily well by an analytic function. This being the case, why not replace a non-analytic partial differential equation and non-analytic data with very good analytic approximations? Surely this will produce arbitrarily good approximations to the solution of the original non-analytic problem?

Hadamard remarked (§18):

> I have often maintained, against different geometers, the importance of this distinction. Some of them indeed argued that you may always consider any functions as analytic, as, in the contrary case, they could be approximated with any required precision by analytic ones. But, in my opinion, this objection would not apply, the question not being whether such an approximation would alter the data very little, but whether it would alter the solution very little. It is easy to see that, in the case we are dealing with, the two are not at all equivalent.
>
> Let us take the classic equation of two-dimensional potentials

$$\frac{\partial^2 u}{\partial x^2} + \frac{\partial^2 u}{\partial y^2} = 0,$$

> with the following data of Cauchy's

$$(15) \quad u(0, y) = 0, \quad \frac{\partial u}{\partial x}(0, y) = u_1(y) = A_n \sin(ny),$$

> n being a very large number, but A_n a function of n assumed to be very small as n grows very large (for instance $A_n = n^{-p}$). These data differ from zero as little as can be wished.

The Dirichlet problem with the boundary data

$$u(0, y) = 0, \quad \frac{\partial u}{\partial x}(0, y) = 0$$

Fig. 29.2 The graph of $\frac{1}{n^2}\sin(ny)\sinh(nx)$, $-\pi < y < \pi$, $0 < x < 2$, $n = 10$

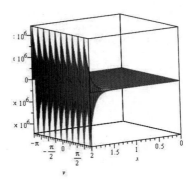

has the unique solution $u(x, y) = 0$. As for the new problem, with boundary data differing from zero by an arbitrarily small amount, Hadamard continued (see Fig. 29.2):

> Nevertheless, a Cauchy problem has for its solution
>
> $$u = \frac{A_n}{n}\sin(ny)\sinh(nx),$$
>
> which, if $A_n = \frac{1}{n}$, $\frac{1}{n^p}$, $e^{-\sqrt{n}}$ is very large for any determinate value of x different from zero on account of the mode of growth of e^{nx} and consequently $\sinh(nx)$.
>
> In this case, the presence of the factor $\sin ny$ produces a "fluting" of the surface, and we see that this fluting, however imperceptible in the immediate neighbourhood of the y-axis, becomes enormous at any given distance of it however small, provided the fluting be taken sufficiently thin by taking n sufficiently great.

After some more technical matters Hadamard then observed

> 21. Another paradoxical consequence furthermore appears if we consider things from the concrete point of view.
>
> Strictly, mathematically speaking, we have seen (this is Holmgren's theorem) that one set of Cauchy's data u_0, u_1 corresponds (at most) to one solution of [Laplace's equation], so that, if these quantities u_0, u_1 were "known," u would be determined without any possible ambiguity.[2]
>
> But, in any concrete application, "known," of course, signifies "known with a certain approximation," all kinds of errors being possible, provided their magnitude remains smaller than a certain quantity; and, on the other hand, we have seen that the mere replacing of the value zero for u_1, by the (however small) value (15) changes the solution not by very small but by very great quantities. Everything takes place, physically speaking, as if the knowledge of Cauchy's data would *not* determine the unknown function.
>
> This shows how very differently things behave in this case and in those which correspond to physical questions. If a physical phenomenon were to be dependent on such an analytical problem as Cauchy's for $\nabla^2 u = 0$, it would appear to us as being governed by pure chance (which, since Poincaré, has been known to consist precisely in such a discontinuity in determinism) and not obeying any law whatever.

---

[2]Holmgren published this theorem in [146].

After having been led by physical interpretation to the need of the above distinctions, we must now try to formulate them analytically. This is subordinate to the classification of linear partial differential equations of the second order into different types.

These are the hyperbolic, parabolic, and elliptic types, but, because Hadamard always emphasised the importance of working in any number m of variables, he distinguished among the hyperbolic types between those in which all but one of the m squares have the same sign—which he called the normal hyperbolic type—and the others.

He then observed that the normal hyperbolic type is the only one known in which Cauchy's problem can be correctly set, and the non-normal hyperbolic types are not known to be connected to any physical problem and do not lead to any problem known to comply with Cauchy's condition. Finally, for reasons Hadamard explained later in the book, elliptic equations never lead to correctly set Cauchy problems.

29.4.1 *Commentary and Concluding Remarks*

Hadamard's work, which we have done little more than sample here, established three things. First, that any theory of partial differential equations deals not only with a differential equation but with some boundary or initial conditions. Second, that elliptic and hyperbolic equations are very different in this respect (and that parabolic equations exhibit some features of each type). Third, that there is a class of partial differential equations that are what he called well posed: they have solutions, these solutions are unique, and they depend continuously on the initial data and any parameters that enter the problem.

The first point makes clear what Kovalevskaya seems to have suspected, and Riemann quite likely understood, that the solution of a partial differential equation is not some general expression that is made precise when some extra information is supplied (as in the theory of ordinary differential equations). This is a natural view, it was held by Euler and Lagrange, and it is ultimately shallow. The theory of partial differential equations is instead a dialogue between the equation and its boundaries.

The second point was surely understood by Riemann, but Hadamard's insight can be amplified here. Solutions to a hyperbolic partial differential equation propagate at a given speed that reflects some aspect of the situation being described; solutions to elliptic equations propagate instantaneously. The other side of that coin is that the solution to an elliptic equation at a point depends on all the boundary values, but the solution to a hyperbolic equation at a point depends only on the nearby boundary values.

The third point is Hadamard's most original. He believed that the partial differential equations that arise in science are well posed and this is why they can be profitably studied, and that problems that are not well posed (ill-posed, as they are called) are likely to be both difficult and artificial. Although it is true that today some naturally occurring ill-posed problems are studied, Hadamard's observation is deeper than it looks and may yet have useful things to say.

29.5 Exercises

Questions

1. Hadamard's remarks brought finally into light a fundamental failure of the first generations of people who studied partial differential equations, in that he showed that these equations cannot be studied without the accompanying boundary conditions. To what extent does this course suggest that boundary conditions were initially almost ignored (they were to be fitted in after the equation was solved), then incorporated, then appreciated (and given equal status with the equation)?

Chapter 30
Revision

30.1 Revision and Assessment 3

This chapter is given over to revision and discussion of the final assignment, see H.4.

However, I would like to repeat my recommendation that students read the essay [156], which restates and reinvigorates many of the concerns that surfaced towards the end of the nineteenth century in partial differential equation theory. In particular, there is a stimulating return to the concerns that animated Hadamard and Poincaré.

© The Author(s), under exclusive license to Springer Nature Switzerland AG 2021
J. Gray, *Change and Variations*, Springer Undergraduate Mathematics Series,
https://doi.org/10.1007/978-3-030-70575-6_30

Chapter 31
Translations

31.1 Cauchy: Note on the Integration of First-Order Partial Differential Equations in Any Number of Variables

This is [34], in *Oeuvres* (2) 2, 238).

Until now there has been no treatise on the differential and integral calculus where one is given the means to integrate completely partial differential equations of the first order in any number of independent variables. Having been occupied for several months with this object, I was happy to have obtained a general method appropriate to fulfilling this desire. But, having finished my work, I learned that M. Pfaff, a German geometer, had been led on his side to the solution of the equations mentioned above. As this concerns one of the most important questions in the integral calculus, and M. Pfaff's method is different from mine I believe that geometers will not be without interest in a short analysis of one and the other. I will first expound the method that I have used, profiting, in order to simplify the exposition, from some remarks made by M. Coriolis, an engineer at Ponts et Chaussées, and some others that have since occurred to me.

Suppose in the first place that we are to integrate a first-order partial differential equation with two independent variables. One already has several methods for integrating an equation of this kind, of which one (due to M. Ampère) is based on the change of a single independent variable. The method that I propose, based on the same principle as in the admitted hypotheses, reduces to this:

Let

$$f(x, y, u, p, q) = 0 \tag{31.1}$$

be the given equation, in which x and y denote the two independent variables, u an unknown function of these two variables, and p, q the partial derivatives of u relative

© The Author(s), under exclusive license to Springer Nature Switzerland AG 2021
J. Gray, *Change and Variations*, Springer Undergraduate Mathematics Series,
https://doi.org/10.1007/978-3-030-70575-6_31

to the variables x and y. In order to completely determine the sought-for function u it is not enough to know that it must satisfy Eq. (31.1); it is also necessary that it satisfies another condition, for example, that it yields a certain particular value for a function y for a given value of the variable x. Let us suppose in consequence that the function u must receive, for $x = x_0$, the particular value $\varphi(y)$: the function q or the partial derivative of u relative to y, will on this hypothesis receive the particular value $\varphi'(y)$. On the same hypothesis the general value of u is, as one knows, completely determined. It now remains to calculate this value: one can proceed in the following manner.

Let us replace y by a function of x and a new independent variable y_0. The quantities u, p, q, being functions of x and y, become themselves functions of x and y_0; and on differentiating on this supposition,[1]

$$\frac{\partial u}{\partial x} = p + q \frac{\partial y}{\partial x},$$ (31.2)

$$\frac{\partial u}{\partial y_0} = q \frac{\partial y}{\partial y_0}.$$ (31.3)

If one takes one of these two equations from the other, after differentiating the first with respect to y_0 and the second with respect to x, one finds that

$$\frac{\partial p}{\partial y_0} = \frac{\partial q}{\partial x} \frac{\partial y}{\partial y_0} - \frac{\partial y}{\partial x} \frac{\partial q}{\partial y_0}.$$ (31.4)

If one also writes the total differential of the first member of Eq. (31.1) as

$$X dx + Y dy + U du + P dp + Q dq = 0.$$

one finds, on differentiating this equation with respect to y_0, that

$$Y \frac{\partial y}{\partial y_0} + U \frac{\partial u}{\partial y_0} + P \frac{\partial p}{\partial y_0} + Q \frac{\partial q}{\partial y_0} = 0.$$ (31.5)

and consequently, in view of Eqs. (31.3) and (31.4) that

$$\left(Y + qU + P \frac{\partial q}{\partial x} \right) \frac{\partial y}{\partial y_0} + \left(Q - P \frac{\partial y}{\partial x} \right) \frac{\partial q}{\partial y_0} = 0.$$ (31.6)

Let us now observe that the value of y as a function of x and y_0 being entirely arbitrary one can dispose of it in such a way that it satisfies the differential equation

$$Q - P \frac{\partial y}{\partial x} = 0,$$ (31.7)

---

[1] See the comment at the end of the translation.

and reduces to y_0 on the particular supposition $x = x_0$. The value of y at x and y_0. being chosen in the way just described, the particular values of u and q corresponding to $x = x_0$, that is to say $\varphi'(y)$ and $\varphi'(y)$, become, respectively $\varphi(y_0)$. and $\varphi'(y_0)$. Representing these values by u_0, q_0 one will have

$$u_0 = \varphi(y_0), \quad q_0 = \varphi'(y_0). \tag{31.8}$$

As for formula (31.6), it is reduced by Eq. (31.7) to

$$\left(Y + qU + P \frac{\partial q}{\partial x} \right) \frac{\partial y}{\partial y_0} = 0$$

and as, y depending on y_0 by hypothesis, $\frac{\partial y}{\partial y_0}$ cannot be constantly zero, the same formula becomes

$$Y + qU + P \frac{\partial q}{\partial x} = 0. \tag{31.9}$$

This done, the integration of Eq. (31.1) is reduced to the following question: Find for y, u, p, q four functions of x and y_0, which satisfy the Eqs. (31.1), (31.2), (31.3), (31.7), (31.9), and of which three, namely, y, u, q, reduce, respectively, to y_0, u_0, q_0 on the supposition $x = x_0$.

We do not speak of Eq. (31.4), because it is a necessary consequence of Eqs. (31.2) and (31.3). As for the particular value of p corresponding to $x = x_0$, it will not enter into the general values of y, u, p, q determined by the preceding conditions. If one denotes it by p_0 it will be deduced from the formula[2]

$$f(x_0, y_0, u_0, p_0, q_0) = 0. \tag{31.10}$$

It is essential to remark that the general values of y, u, p, q as functions of x and y_0, remain completely determined if, among the conditions that they must satisfy one fails to take account of the verification of Eq. (31.3). This last condition must therefore be an immediate consequence of all the others. To show this, let us suppose for a moment that the other conditions having been verified, the two members of Eq. (31.3) are unequal. The difference between these two members can only be a function of x and y_0. Let α be this function and α_0 be what it becomes when $x = x_0$. One will have

$$\alpha = \frac{\partial u}{\partial y_0} - q \frac{\partial y}{\partial y_0}, \quad \alpha_0 = \frac{\partial u_0}{\partial y_0} - q_0 \frac{\partial y_0}{\partial y_0} = \varphi'(y_0) - \varphi'(y_0) = 0. \tag{31.11}$$

Consequently, instead of Eqs. (31.3) and (31.4), one finds

---

[2] Cauchy used the same subscript notation for the initial conditions that he had used for the new variable, but the confusion this causes is slight.

$$\frac{\partial u}{\partial y_0} = q\frac{\partial y}{\partial y_0} + \alpha, \quad \frac{\partial p}{\partial y_0} = \frac{\partial q}{\partial x}\frac{\partial y}{\partial y_0} - \frac{\partial y}{\partial x}\frac{\partial q}{\partial y_0} + \frac{\partial \alpha}{\partial x}, \tag{31.12}$$

then, instead of (31.6) the following:

$$\left(Y + qU + P\frac{\partial p}{\partial x}\right)\frac{\partial y}{\partial y_0} + \left(Q - P\frac{\partial y}{\partial x}\right)\frac{\partial q}{\partial y_0} + U\alpha + P\frac{\partial \alpha}{\partial x} = 0. \tag{31.13}$$

This last equation will reduce, by Eqs. (31.7) and (31.9), which one supposes verified, to

$$U\alpha + P\frac{\partial \alpha}{\partial x} = 0. \tag{31.14}$$

On integrating it, and treating $\frac{U}{P}$ as a function of x and y_0, one will find

$$\alpha = \alpha_0 e^{-\int \frac{U}{P}\left(\frac{x_0}{x}\right)dx}; \tag{31.15}$$

and consequently, taking account of the second of the Eqs. (31.11), one will generally have

$$\alpha = 0. \tag{31.16}$$

The two members of Eq. (31.3) cannot, therefore, be unequal on the admitted hypothesis. One must conclude from this that the quantities y, u, p, q satisfy all the conditions required if these quantities, considered as functions of x satisfy Eqs. (31.1), (31.2), (31.7), (31.9), and if in addition y, u, q reduce, respectively, to y_0, $u_0 = \varphi(y_0)$, and $q_0 = \varphi'(y_0)$ for $x = x_0$. It is useless to add that on the same supposition p must obtain the particular value p_0; in fact this value will not be contained in the integrals of Eqs. (31.1), (31.2), (31.7), (31.9), because none of these equations contain $\frac{\partial p}{\partial x}$.

If, in Eq. (31.2), one substitutes the value of $\frac{\partial y}{\partial x}$ drawn from Eq. (31.7), one will find

$$\frac{\partial u}{\partial x} = p + \frac{Qq}{P} = \frac{Pp + Qq}{P}. \tag{31.17}$$

Furthermore, if one differentiates Eq. (31.1) with respect to x, one obtains the following:

$$X + Y\frac{\partial y}{\partial x} + U\frac{\partial u}{\partial x} + P\frac{\partial p}{\partial x} + Q\frac{\partial q}{\partial x} = 0, \tag{31.18}$$

which the values of $\frac{\partial y}{\partial x}, \frac{\partial u}{\partial x}, \frac{\partial q}{\partial x}$ drawn from Eqs. (31.7), (31.17), and (31.9) reduce to

$$X + pU + P\frac{\partial p}{\partial x} = 0. \tag{31.19}$$

This done, one can substitute Eq. (31.17) in Eq. (31.2), and Eq. (31.19) in one of Eqs. (31.1), (31.17), (31.7), (31.9). If besides one observes that, in the case where one considers y, u, p, q as functions of x alone, one can include Eqs. (31.7), (31.9), (31.17), (31.19) in the algebraic formula[3]

$$\frac{dx}{P} = \frac{dy}{Q} = \frac{du}{Pp + Qq} = -\frac{dp}{X + pU} = -\frac{dq}{Y + qU}; \qquad (31.20)$$

one definitively concludes that to determine the sought-for values of the quantities y, u, p, q, it is enough to work with four of the five equations contained in the two formulae

$$f(x, y, u, p, q) = 0, \quad \frac{dx}{P} = \frac{dy}{Q} = \frac{du}{Pp + Qq} = -\frac{dp}{X + pU} = -\frac{dq}{Y + qU};$$
$$(31.21)$$

and to know, for $x = x_0$ the particulars y_0, u_0, p_0, q_0, for the three latter ones are determined as a function of the first by Eqs. (31.8) and (31.10).

Suppose, to fix ideas, that by means of the equation

$$f(x, y, u, p, q) = 0$$

one eliminates p from three equations in the formula

$$\frac{dx}{P} = \frac{dy}{Q} = \frac{du}{Pp + Qq} = -\frac{dq}{Y + qU}. \qquad (31.22)$$

On integrating the last three, on will obtain three finite equations that involve, with the quantities

$$x, y, u, q$$

the particular values represented by

$$x_0, y_0, \varphi(y_0), \varphi'(y_0).$$

If after the integration one eliminates q, the remaining two equations involve, with the quantities x, y, u, and the constant quantity x_0, only the new variable y_0, the elimination of which can only be carried out when one has assigned a particular form to the arbitrary function denoted by φ. Whatever it may be, the system of two equations with which we are concerned can always be considered as equivalent to the general integral of Eq. (31.1).

As, in all that has been done so far, one can substitute the variable x for the variable y, and reciprocally, it follows that the integrals of Eqs. (31.21) again furnish a solution of the question proposed, if one in the integrals one considers y_0 as constant, x_0 as a new variable that one must eliminate, and u_0, p_0, q_0 as functions of this new variable that are determined by equations of the form

---

[3]Cauchy mistakenly wrote $X + PU$ for $X + pU$.

$$u_0 = \varphi(x_0), \quad p_0 = \varphi'(x_0) \tag{31.23}$$

$$f(x_0, y_0, u_0, p_0, q_0) = 0. \tag{31.24}$$

Let us apply the principles we have just established to the solution of the partial differential equation

$$pq - xy = 0. \tag{31.25}$$

[The extract from Cauchy [34] ends here.]

Cauchy showed that in this case that Eqs. (31.21) become

$$pdx = qdy = \frac{1}{2}du = xdp = ydq. \tag{31.26}$$

These imply that

$$\frac{dp}{p} = \frac{dx}{x}, \ \frac{dq}{q} = \frac{dy}{y}, \ du = \frac{p}{x}2xdx = \frac{q}{y}2ydy; \tag{31.27}$$

then on integrating and taking note of the condition $p_0 q_0 = x_0 y_0$

$$\frac{p}{p_0} = \frac{x}{x_0}, \frac{q}{q_0} = \frac{y}{y_0}, \tag{31.28}$$

$$u - u_0 = \frac{p_0}{x_0}(x^2 - x_0^2) = \frac{q_0}{y_0}(y^2 - y_0^2) = \frac{y_0}{q_0}(x^2 - x_0^2) = \frac{x_0}{p_0}(y^2 - y_0^2). \tag{31.29}$$

In these equations, he said, x_0 is an arbitrary constant and y_0 a new variable that one can only eliminate after fixing a value for the arbitrary function φ. Finally Cauchy deduced that the general integral is represented by the equations

$$(u - \varphi(x_0))^2 = (x^2 - x_0^2)(y^2 - y_0^2), \ (u - \varphi(x_0)) \, \varphi'(x_0) = x_0(y^2 - y_0^2),$$

and he noted that the second of these equations is the derivative of the first with respect to x_0.

Cauchy concluded his paper with the observation that the method worked without essential change when there were more than two independent variables, and illustrated his point by going over the method in the case of three independent variables. He also worked through the example

$$pqr = xyz.$$

Comment The change of variables argument at the start of Cauchy's paper may be easier to follow on introducing new variables s and t, where

$$s = x, \quad t = t(x, y), \quad \text{so} \quad x = s \quad \text{and} \quad y = y(s, t).$$

This means that $x_s = 1$ and $x_t = 0$. At the end, restore $x = s$, $y_0 = t$.

31.2 Riemann's Lectures on Partial Differential Equations and Physics

Riemann lectured three times on physics at Göttingen: in 1854/55, in 1860/61, and in 1862. After Riemann's death his former pupil Karl Hattendorff edited Riemann's notes (mostly from the course of 1860/61) and published them as a book [237].[4] In the preface, he noted that while mathematicians drawn to the theory of partial differential equations took their lead from Dirichlet, as indeed Riemann had done, what was here was not restricted to potential theory but included a slew of applications.

The book became the principal introduction to mathematical physics for over a generation, because it was taken up and re-edited by Heinrich Weber, and "Riemann–Weber", as it came to be known, grew to two volumes.

31.2.1 Riemann, Introduction to Partial Differential Equations

The object of these lectures is the treatment of partial differential equations and their application to physical questions. Therefore it is convenient to make some introductory remarks on the relationship of the theory of partial differential equations to physics.

It is well known that a scientific physics first began with the discovery of the differential calculus. Since one first learned how to follow the course of natural events continuously, research into the connection of appearances to abstract consequences has succeeded. This involves two things: first the simple basic ideas with which we construct, and second a method with which, from the simple basic laws of this construction that concern points of time and space, laws can be derived for finite intervals of time and space that alone are accessible to observations (and can be compared to experience).

Galileo took the first step in respect of the basic ideas, when he constructed the laws of motion for freely falling bodies from the operation of weight at every moment of time; he found the law of accelerating force, the idea of a simple cause of motion. To this step Newton added a second: he found the idea of an attracting centre, the idea of a simple cause of force. With these two basic ideas, the idea of accelerating force and of an attracting or repelling centre,

---

[4]I have used the third edition, 1882, which Hattendorff said is a careful revision of the first edition.

physics is still constructed to this day. The present-day speculations of Laplace, Poisson, Cauchy, where the thread of observations stops, are attributable only to the struggles with the appearance of these two laws. In respect of the ideas that one places at the basis of the physical explanation of nature, we therefore take today the standpoint of Newton. No new step has been taken since Newton; all research into basic ideas that penetrate into the heart of nature have up to now failed; the influence of later philosophical systems that have been applied in the physical literature have only had the success of disfiguring Newton's original perception with inconsistencies.

But the method, by which the simple basic laws for moments of time and space are obtained— differential equations—are turned into laws for finite intervals and extended bodies, is essentially perfected. At first, after the discovery of the differential calculus, one could handle certain abstract cases: in the study of free fall one connected the mass of a body with its centre of gravity, one treated the heavenly bodies as mathematical points, in the study of pendulums one first treated only the mathematical pendulum i.e. a rigid movable line connected to a heavy point; so that one only had to take one step from the infinitely small to the finite in only one dimension with respect to one variable, the time. But in general, in order to derive the experiences from the elementary laws, one must take the step from the infinitely small to the finite in more than one dimension. For the elementary laws involve space and time points, experiences involve extended bodies. Such problems lead, to speak generally—in special cases the problem can be simplified—to partial differential equations.

Sixty years after the appearance of Newton's *Principia* the first physical problem was solved that led to a partial differential equation. It was the one that d'Alembert showed determined the oscillations of a stretched string. It was then a long time until the general method was found by which the physical problems that lead to partial differential equations can be solved. For this we thank Fourier, who first applied such methods in his study of the diffusion of heat in solid bodies. This took almost as long from the origin of partial differential equations as that had from the creation of the differential calculus. Newton's *Principia* appeared in 1687, d'Alembert's solution of the problem of the vibrating string in 1747, again 60 years later, on 21 December 1807, Fourier presented the first part of his work on heat to the Paris Academy.

After these selective and not entirely accurate historical pages Riemann turned to list the many areas in physics where partial differential equations provided the appropriate foundations. These included oscillations in gases, liquids, and solid bodies, elasticity of bodies, and optics. He noted that most of this work involved making assumptions about molecules that make up these bodies, and so the determination of the constants that enter the partial differential equations depended on assumptions about the molecular composition of bodies that, he said, we were far from having the key to being able to do. The same was true, he went on, for gravitation, electricity, and magnetism: the fundamental laws involve partial differential equations. He then concluded:

What then emerges as a fact by means of induction arises also a priori: the proper foundations for mathematical physics are partial differential equations. True elementary laws can only occur in the infinitely small, for space and time points. In general, such laws will be partial differential equations, and the derivation of laws for extended bodies and times requires their integration. So methods are necessary by which the finite laws can be derived from the laws of the infinitely small, and indeed derived with complete rigour neglecting nothing. Only then can they be tested against experience.

The book itself opens with just under a hundred pages of mathematical methods and twenty on the basics of ordinary and partial differential equations. Then it turns to a more detailed investigation of heat diffusion in solid bodies, oscillations of solid bodies, fluid motion, oscillations in compressible media, and finally the motion of a solid body in an unbounded incompressible fluid. Nothing, one notes, on magnetism and electricity—an omission that became the main reason for Weber's new editions— but one that Riemann had addressed in another series of lectures, later published as *Schwere, Elektricität und Magnetismus* (Gravity, Electricity, and Magnetism). That said, as Weber noted, Riemann's book on partial differential equations was not a physics textbook but a mathematical book devoted to the solution of various mathematical problems.

31.3 Extracts from Schwarz, "Ueber eine Abbildungsaufgaben", 1869

[At this stage in his paper Schwarz has shown that the most general map of an angular sector of angle $\alpha\pi$ to the upper half-plane that is given by a map that it analytic everywhere except at the origin, and maps the origin to itself, is one of the forms

$$v = u^{1/\alpha}, \quad t = Cv(1 + a_1 v + a_2 v^2 + \cdots),$$

where C is a non-zero constant and the coefficients a_j are all real. The inverse function is

$$u = v^\alpha, \quad U = \frac{1}{C}t^\alpha(1 + c_1 t + c_2 t^2 + \cdots),$$

where C the coefficients c_j are all real.
He then continued:]
In a problem about conformal maps the position and absolute size of the figure in the u-plane on which a figure in the t-plane is to be represented conformally is usually unimportant. So the general solution of the representation problem introduces two arbitrary constants that determine the position and absolute size, for, if $u = f(t)$ is a function that maps a figure T in the t-plane onto a figure U in the u-plane then $u' = C_1 u + C_2$ is another such function, only it places the corresponding figure U' in another position, is of another proportion, and can be dragged to the position of the figure U. So if we have to obtain the characteristic properties of a figure T on a figure U we must look at the dependence between the quantities u and t to determine which are independent of the particular position and absolute size of the figure U in the u-plane; that is, to determine the differential equation in whose general solution the constants C_1 and C_2 enter as constants of integration.
This leads to

$$\frac{du'}{dt} = C_1 \frac{du}{dt},$$

$$\frac{d}{dt} \log \frac{du'}{dt} = \frac{d}{dt} \log \frac{du}{dt}.$$

This function is then independent of the particular position and absolute size of the figure U in the u-plane.

The passage from u to $\frac{du}{dt}$ and $\frac{d}{dt} \log \frac{du}{dt}$ is all the more important a step because all the values of the argument t, for which the quantity $\frac{du}{dt}$ becomes infinitely large or infinitely small, and $\frac{d}{dt} \log \frac{du}{dt}$ infinitely large, are singular points for the representation problem, in that conformal representation in the strict sense cannot hold for them.

In the case already considered of the conformal representation of an angle π on an angle $\alpha\pi$

$$\frac{d}{dt} \log \frac{du}{dt} = \frac{\alpha - 1}{t} + d_1 + d_2 t + \cdots.$$

This function has the character of a rational function in the neighbourhood of the value $t = 0$. The coefficients $d_1, d_2, \ldots$ all have real values and therefore the values of the function $\frac{d}{dt} \log \frac{du}{dt}$, for those real values of the argument t for which the series converges, are likewise real.

Therefore, when the problem is to conformally map the surface of a figure T in the t-plane onto another bounded by a simple curve (i.e. one that goes through no point more than once) lying entirely in the finite part U of the u-plane, then it is immediately assumed that the quantity $\frac{du}{dt}$ can never become infinitely small or infinitely large at any point in the interior of T, and therefore that the function $\frac{d}{dt} \log \frac{du}{dt}$ has the character of an entire function for all values of the argument t.

In the present case the singular values of t lying in the finite part of the plane are $t = -1$, $t = 0$, $t = +1$; α is equal to $\frac{1}{2}$. The function

$$\frac{d}{dt} \log \frac{du}{dt} + \frac{1}{2} \left(\frac{1}{t+1} + \frac{1}{t} + \frac{1}{t-1} \right),$$

which for all real values of the argument likewise has real finite values, has the character of an entire function for all finite values of t with positive imaginary part, and so for all finite values of t it has the character of an entire function.[5] For the infinite value of t there is the development

$$u - u_0 = -\frac{Ci}{\sqrt{t}} \left(1 + c_1' t + c_2' t^2 + \cdots \right),$$

from which

$$\frac{d}{dt} \log \frac{du}{dt} = -\frac{3}{2} \frac{1}{t} + d_1' \frac{1}{t^2} + d_2' \frac{1}{t^3} + \cdots,$$

---

[5] This follows from the Schwarz reflection principle that Schwarz had introduced earlier in his paper.

so the function $\frac{d}{dt} \log \frac{du}{dt}$ is infinitely small for all infinite values of t, and therefore $\frac{d}{dt} \log \frac{du}{dt}$ is a rational function of t and indeed equal to $-\frac{1}{2} \left(\frac{1}{t+1} + \frac{1}{t} + \frac{1}{t-1} \right)$. From this it follows by integration that

$$\log \frac{du}{dt} = -\frac{1}{2} \log 4t \, (1 - t^2) + \log C_1,$$

$$u = C_1 \int_0^t \frac{dt}{\sqrt{4t \, (1 - t^2)}} + C_2.$$

One can easily recognise that in this case the lemniscatic integral

$$u' = \int_0^t \frac{dt}{\sqrt{4t \, (1 - t^2)}}$$

represents the interior of each of the two half-planes in which the plane is divided by the real axis conformally on the interior of a square with sides $\int_0^1 \frac{dt}{\sqrt{4t(1-t^2)}}$. Through the substitution $s = \frac{t-i}{t+1}$ one goes from the half-plane lying on the positive side of the real axis [i.e. the upper half-plane, JJG] to the surface of a circle of radius 1 drawn in the s-plane around the point $s = 0$.

[Schwarz then considered a number of other cases, of which I only translate these two.]

If the problem is to represent the interior of a half-plane T conformally on the interior of a straight-sided triangle with angles $\alpha\pi$, $\beta\pi$, $\gamma\pi$ then by an analogous argument one deduces that the representation is provided by a function of the form

$$C_1 u + C_2 = \int_{t_0}^t (t - a)^{\alpha - 1} (t - b)^{\beta - 1} (t - c)^{\gamma - 1} dt.$$

In this case the three real values that correspond to the three vertices of the straight-sided triangle can be chosen arbitrarily provided they follow the same order on the boundary of the half-plane T the angles $\alpha\pi$, $\beta\pi$, $\gamma\pi$ of the triangle are encountered on a circuit around the interior of the triangle.

With this result the form of the function is found that conformally represents the surface of a half-plane onto the simply connected surface of any straight-sided polygon. In the general case of an n-gon, only three of the n real quantities $a, b, c, \ldots$ that correspond to the vertices of the straight-sided polygon van be chosen arbitrarily, the remaining $n - 3$ are determined by the given ratios of the lengths of the individual sides of the polygon under consideration.

[A little later in the paper, Schwarz remarked:]

Concerning the problem of representing the surface of a straight-sided polygon on the surface of a circle I had the pleasure of seeing the researches of Herr Christoffel on the subject: Sul problema delle temperature stazionare e la rappresentazione di una data superficie, *Annali di matematica*, IIe serie, tomo $1°$, 1867, that seem to agree with mine.

31.3.1   The Schwarz–Christoffel Transformation

It will be helpful to give a more modern account. Consider the problem of finding a map $z \mapsto f(z)$ from the upper half-plane onto a triangle with vertices at the points b_1, b_2, b_3 and angles at these points of $\alpha_1\pi, \alpha_2\pi, \alpha_3\pi$, respectively. We require that the map sends the points a_1, a_2, a_3 on the real axis to the points b_1, b_2, b_3, and we look for a map that is holomorphic everywhere except at the points a_1, a_2, a_3.

The map will fail to be holomorphic precisely at the points where its derivative vanishes, which suggests that the map must be such that

$$f'(z) = (z - a_1)_1^\mu (z - a_2)_2^\mu (z - a_3)_3^\mu.$$

Now, the map $z \mapsto z^\alpha$ maps the origin to the origin and an interval around the origin to two line segments meeting at an angle of $\alpha\pi$, and its derivative is

$$f'(z) = \alpha z^{\alpha - 1},$$

so this suggests that when mapping the half-plane to a triangle we try maps for which

$$f'(z) = (z - a_1)^{\alpha_1 - 1}(z - a_2)^{\alpha_2 - 1}(z - a_3)^{\alpha_3 - 1}.$$

Notice that away from the pre-images of the vertices the map is locally one-to-one, because f' does not vanish. So we can find the image of a domain that does not have a pre-image of a vertex in its interior by finding the image of the boundary of the domain.

Integration produces the map

$$f(z) = C_0 \int_0^z (z - a_1)^{\alpha_1 - 1}(z - a_2)^{\alpha_2 - 1}(z - a_3)^{\alpha_3 - 1}dz + C_1.$$

Certainly, this map maps angular segments of π around each a_j to angular segments of $\alpha_j\pi$ around each of three points. But are they the points b_j, and are the sides—the images of the segments joining each a_j to the next one (via ∞, if need be)—straight?

The first question is easy, and the answer is Yes. Consider the interval (a_1, a_2). The integral can be written as

$$\int_{a_1}^{a_2} (x-a_1)^{\alpha_1-1}(x-a_2)^{\alpha_2-1}(x-a_3)^{\alpha_3-1}dx,$$

which is real, and so the image is part of the real axis, and therefore straight. We can adjust the arbitrary constants to map a_1 to b_1 and rotate the line segments around b_1 to point in the right directions, and because maps of the form $z \mapsto C_0z + C_1$ maps line segments to line segments, the images of the sides through b_1 will be straight. We can do this for any vertex, to the image of the upper half-plane has straight sides.

That brings us to the first question, because everything now depends on the lengths of the sides. In the case of a triangle this is easy, because a triangle is determined up to size by its angles, and the constant C_0 takes care of any potential problem.

But what happens if we want to map the upper half-plane onto a quadrilateral, or more generally an n-gon with prescribed vertices and sides? Can it be done? It turns out that the answer to this question is also Yes, but the computation of the lengths of the sides of the polygon as functions of choice of positions of the pre-images of the vertices and the angles is difficult and cannot be discussed here.[6] For maps of the upper half-plane onto n-gons with given angles ($n > 3$) the positions, a_j, of the pre-images of the vertices must also be specified and the proof that a solution can always be found is delicate. In particular, the length of the side with vertices $f(a)$ and $f(b)$ is given by the integral

$$\int_a^b |f'(z)dz|.$$

Notice that the function $f(z)$ involves all the pre-images of the vertices. It can be done, and the formula that does it is the natural generalisation of the above integral to any number of vertices is called the Schwarz–Christoffel formula.

31.4 An Extract from Schwarz, On the Alternating Method

(1870) [This comes from Schwarz's paper "On a passage to a limit by an alternating method" [244].]

The rigour of the well-known inference that goes under the name of the Dirichlet principle, and that in a certain sense must be seen as the foundation of the branch of the theory of analytic functions developed by Riemann, is subject, as is now quite generally admitted, to very well-founded objections whose complete resolution to my knowledge the efforts of mathematicians have not yet achieved.

By developing some enquiries, which involve a certain kind of representation, and part of which I have published in vol. 70 of Borchardt's *Journal* and in the paper "On the theory of representation" in the programme of the polytechnic school for the

---

[6]For a complete discussion, see Nehari ([204], 189–198).

Winter semester 1869–70, I have been led to a method of proof by means of which
I am convinced that all the theorems that Riemann used the Dirichlet principle to
prove in his published works can be proved rigorously.

The following report is essentially a summary of a work on the integration of the
partial differential equation $\Delta u = 0$, that I reported on to Herr Kronecker and some
other mathematicians in November last year.

It is concerned essentially only with the proof of the existence of a function u
that on a given domain T of the independent variables x and y satisfies the partial
differential equation

$$\Delta u = \frac{\partial^2 u}{\partial x^2} + \frac{\partial^2 u}{\partial y^2} = 0$$

and also satisfies certain prescribed boundary and discontinuity conditions.

For brevity, I restrict myself here to the case in which the auxiliary conditions are
only boundary conditions and therefore imply that the function u is always continuous
and takes prescribed finite values on the boundary of the domain T, which consists
of one or more continuous parts. The general case can be reduced to this case by a
method to be described.

For the applicability of this method of proof it is in no way necessary to assume
that the boundary curve of T has only finitely many corners, nor that in general at
every point it has a finite radius of curvature, an assumption that Herr Weber and Herr
Carl Neumann made for this purpose in their researches (see Borchardt's *Journal*,
vol. 71, p. 29 and the *Berichte der mathematisch-physischen Classe der Königlich
Sächsichen Geselleschaft der Wissenschaften*, 21 April 1870). At no point will the
tangent to the boundary curve be assumed to vary continuously; rather, it is enough to
know that the boundary curve can be divided into a finite number of pieces such that
in the interior of each piece the change in the direction of the tangent is always in the
same sense even though it may also have infinitely many jumps, and so, therefore,
the boundary curve can have infinitely many corners.

Cusps on the boundary curve are also not excluded. I have carried out the analysis
of such cusps, which arise from the contact of two analytic curves that have the
character of algebraic curves in a neighbourhood of the point of contact; but to avoid
unnecessary complications here no reference is made in what follows to the presence
of cusps.

The success of the proof whose basic idea is reported here rests in the last analysis
on the following lemma:

The boundary line of the domain T for which it is possible to integrate the partial
differential equation $\Delta u = 0$ with arbitrary boundary conditions, will be divided
into a finite number of segments (parts). These can be arranged in two groups in such
a way that each group contains at least one segment. One can give the individual
segments, according as they belong to the first or second group, an odd or even
number and denote the points that separate the segments with an even number from
those with an odd number by P. In the interior of T one considers a finite number of
analytic curves L that have either no point or only an end point P in common with
the odd-numbered segments and are not tangents at these points.

In this way we determine a function u for the domain T that satisfies the partial differential equation $\Delta u = 0$ and at all points of the boundary of T has the value 0 or 1 according as the number of the segment in the interior of which the given point lies is even or odd. Then the upper bound, respectively, the maximum of all values that the function u takes on the curves L, is a positive number q that is less than 1.

We now determine a function u_1 that satisfies the partial differential equation $\Delta u_1 = 0$ on the same domain T, with the same division of the boundary into odd and even numbered segments and the same curves L that takes the value 0 on the even numbered segments of the boundary, and on the odd numbered segments takes arbitrarily prescribed values that do not exceed a quantity g in absolute value; so the absolute values of the values that the function u_1 can take on the curves L never exceeds the value gq, where q has the previously ascribed significance and so is less than 1.

For the surface of the circle, and for all simply connected surfaces that are known to be conformal images of the circle, the integration of the partial differential equation $\Delta u = 0$ with prescribed boundary conditions presents no difficulty. In this regard the task may be treated as in a paper in this journal (pp. 113–128 of the current year); there, breaks in the continuity for the function u in the prescribed series of values for the function u are exceptionally excluded; in this way the inference there developed can, mutatis mutandis, also be given if a finite number of boundary points in the series of boundary points are subjected to a break in continuity.

After it is shown that for a number of simpler domains the differential equation $\Delta u = 0$ can be integrated for arbitrary boundary conditions, the proof has to be found to show that for a less simple domain that is composed of these in a certain way the differential equation is also possible with arbitrary boundary conditions. For the proof of this theorem a limiting argument can serve that has a great analogy with a two-chamber air pump used to produce an evacuated space. The periods of the operation consist indeed in that in one and the other case involve two alternately operating single operations, which indeed have the same purpose, but are not identical in respect of the way and manner in which they work, but are rather in a certain sense symmetric (Fig. 31.1).

Such a limiting argument may be called a limiting argument by an alternating method.

Let two domains T_1 and T_2 be given which have one or more domains T^* in common, and whose boundary lines are not tangent. (In the schematic figure 10 T_1 is the surface of a circle, T_2 the surface of a square.)

The totality of all parts of the boundary of T_1 that lie outside T_2 will be denoted L_0, the totality of all remaining parts of the boundary of T_1 that lie inside T_2 will be denoted L_2.

Likewise the boundary of T_2 divides into the two parts L_1 and L_3, if indeed the totality of all pieces of the boundary that lie inside T_1 will be denoted by L_1, and the totality of all parts of the boundary that lie outside T_1 will be denoted L_3.

It will be assumed that equally for the domains T_1 and T_2 it is possible to integrate the partial differential equation $\Delta u = 0$ with arbitrary boundary conditions; it then remains to show that this is also possible for the domain $T_1 + T_2 - T^* = T$ that has

Fig. 31.1 Schwarz's Figure
10. Schwarz, *Gesammelte
mathematische
Abhandlungen*, vol. II, p. 136

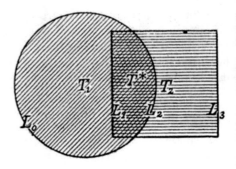

the domains T_1 and T_2 as parts and in which their common domain T^* is counted only once.

The conditions of the previous lemma are satisfied as well for the domain T_1 and the curve L_1 as for the domain T_2 and the curve L_2; in the first case the curve L_0 and in the second the curve L_3 may be taken as the location of the group of segments of even order. It is therefore possible to determine two numbers q_1 and q_2 that play the role of q in the lemma and are therefore both less than 1.

To the recipients of the air pump corresponds—maintaining the above analogy— the domain T^*, to the interior of two pump cylinders correspond the domains the domains $T_1 - T^*$, $T_2 - T^*$, the vents to the curves L_1 and L_2.

On the boundary of T, thus along L_0 and L_3, let the values of the function be given arbitrarily; let g be the upper bound and k the lower bound of these values; the difference $g - k$ will be denoted G.

Now one takes along L_2 an arbitrary sequence of values, for example, the value k at every point of L_2, and determines for the domain T_1 a function u_1 that takes the prescribed values along L_0, has the value k along L_2 and satisfies the differential equation $\Delta u_1 = 0$ in the interior of T_1. By the assumption about the domain T_1 there is such a function. (First push of the first piston.)

The values that the function u_1 has along L_1 one thinks of as fixed, and determines a function u_2 for the domain T_2 that along L_3 has the prescribed values and agrees with the previously determined function u_1 along L_1, and for which $\Delta u_2 = 0$. By the assumption about the domain T_2 there is such a function. (First push of the second piston.)

The value of $u_2 - u_1$ or $u_2 - k$ along L_2 is smaller than $g - k = G$.

One determines for the domain T_1 a function u_3 that takes the prescribed values along L_0, has the value u_2 along L_2 and for which $\Delta u_3 = 0$ in the interior of T_1. (Second push of the first piston.)

At no point in the interior of T_1 is the difference $u_3 - u_1$ negative; in absolute value the difference $u_3 - u_1$ is less than G along L_1 but by the earlier lemma less that Gq_1, because $u_3 - u_1$ had the value 0 along L_0 and along L_2 it is smaller than G.

The values that the function u_3 has along L_1 one thinks of as fixed, and determines a function u_4 for the domain T_2 that along L_1 agrees with u_3 and along L_3 has the prescribed values and for which $\Delta u_4 = 0$. By the assumption about the domain T_2 there is such a function. (Second push of the second piston.)

The difference $u_4 - u_2$ along L_3 has the value 0, and along L_1, where it agrees with $u_3 - u_1$ it is positive and smaller than Gq_1; therefore $u_4 - u_2$ is never negative in the interior of T_2 and is always less than Gq_1 but along L_2 smaller than Gq_1q_2.

By continuing this alternating method one obtains a sequence of infinitely many functions with odd and even index. The ones for the domain T_1 and the others for the domain T_2 are so explained that, respectively, along L_0 and L_2 they have the prescribed values and in the interior of the domains on which they are explained they satisfy the partial differential equation $\Delta u = 0$.

For the domain T^* the functions with odd and even index are explained and indeed they agree with each other alternately along L_1 and L_2. Indeed along L_1 $u_{2n-1} = u_{2n}$ and along L_2 $u_{2n+1} = u_{2n}$.

It is now not difficult to prove that the functions with odd and even index approach definite limiting functions u' and u'' as the index increases, as is shown by the equations

$$u' = u_1 + (u_3 - u_1) + (u_5 - u_3) + \cdots + (u_{2n+1} - u_{2n-1}) + \cdots ,$$

$$u'' = u_2 + (u_4 - u_2) + (u_6 - u_4) + \cdots + (u_{2n+2} - u_{2n}) + \cdots .$$

The series on the right-hand side converge unconditionally and uniformly ("in gleichem Grade") for all pairs of values of , y under consideration, indeed

$$(u_{2n+1} - u_{2n-1}) < G(q_1q_2)^{n-1} \text{ and}$$

$$(u_{2n+2} - u_{2n}) < G(q_1q_2)^{n-1}q_1.$$

Along L_1 as well as along L_2 $u' = u''$. In the interior of T_1 $\Delta u' = 0$, in the interior of T_2 $\Delta u'' = 0$, therefore at every point of T^* $u' = u''$, because along the entire boundary of T^* both functions agree with each other.

Therefore both functions u' and u'' are values of the same function u, and it is explained that for the interior of the whole domain $T = T_1 + T_2 - T^*$ the partial differential equation $\Delta u = 0$ is satisfied and takes the prescribed values along the boundary $L_0 + L_3$.

Thus the proof of the correctness of the above considerations established: Under the given conditions it is possible for the domain T to integrate the partial differential equation with arbitrary prescribed boundary conditions.

By repeated application and suitable modification of the explained limiting process of the alternating method the existence of a function u on a given domain can be established also for boundary conditions with discontinuities, or the prescribed discontinuities such as Abelian integrals possess, for which Riemann required the existence in his work and sought to prove using the Dirichlet principle.

The outlined method of proof extends not only to the case in which the domain T is represented geometrically as a simply or multiply connected Riemann surface in its entire extension in the plane or the surface of a sphere, but is essentially unaltered also in the case in which this surface is spread over one or many plane or spherical surfaces and polyhedral surfaces.

By means of this extension can the proof be given, among other things, that a simply connected domain spread over a polyhedral surface can be conformally represented on the surface of a circle if this surface has a closed boundary curve and on the surface of a sphere if it is a simply connected and closed domain.

In this way the question is answered of the possibility of determining the constants to which the conformal representation of a simply connected covering surface of polyhedron bounded by plane figures on the surface of a sphere can be reduced (see Borchardt's *Journal*, vol. 70, p. 119).

A special case of the just-mentioned problem occurs when it is required to map a simply connected surface in the form of a plane figure with polygonal boundary conformally onto the surface of a circle, where the surface of the polygon may lie entirely in the finite or contain the infinitely distant point once or several times in its interior; branch points in the interior are also not excluded. For this problem the sole difficulty consists in the proof of the possibility that on a certain number of parts real and on some parts complex conjugate constants can be determined, upon which the function providing the conformal representation depends, so that all the conditions of the problem are satisfied.

This difficulty can be overcome by the method that Herr Weierstrass has developed. The application of the above limiting process offers a new way to overcome it.

Similarly, the proof is provided of the possibility of determining the constants to which the problem of the conformal representation of a simply connected figure bounded by circular arcs upon the surface of a circle is reduced.

[The later *Nachtrag*, a dispute with Christoffel on conformal representation, is not translated.]

31.5 Schwarz on the Hypergeometric Equation (1873)—A Summary

Report on those cases in which the Gaussian hypergeometric equation $F(\alpha, \beta, \gamma, x)$ is an algebraic function of its fourth element. *Schweizerischen Naturforschenden Gesellschaft*, 1871, 74–77, (session of 22 August 1871), in *Gesammelte Mathematische Abhandlungen* 2, 172–174.

The problem of studying when a given (ordinary) algebraic differential equation has a particular algebraic solution, and, if this is the case, of finding all its particular algebraic solutions, still belongs today to the most difficult problems in analysis. It seems, in the present state of knowledge, that the problem must be tackled in isolated cases with the help of such methods as are appropriate to the special cases under consideration.

For the second-order linear ordinary differential equation that the Gaussian hypergeometric equation $F(\alpha, \beta, \gamma, x)$ satisfies, considered as a function of its fourth element, the following train of thought leads to a complete solution of the given problem.

The general solution of the given differential equation can, as it is easy to see, only be an algebraic function of the independent variable x if the first three elements α, β, γ are real and indeed rational numbers. If on the assumption that these conditions are met, one considers in addition to the general solution of the differential equation the quotient of two linearly independent particular solutions, then this latter is related to the general solution in such a way that either both are algebraic functions of the argument x or neither of the two functions depend algebraically on the quantity x.

The independent variable x is an unrestricted variable quantity that can take all real and complex values. If one now thinks of the plane, whose points represent the values of the complex quantity x geometrically, as divided by the real axis into two half-planes, and considers the conformal representation that is provided by a branch s of the above-mentioned quotient as a function of the complex variable x, then there corresponds to each of the two half-planes a figure whose points represent geometrically the values of the complex quantity s; one, generally, a domain bounded by three circular arcs and which can therefore be called a circular-arc triangle.

By analytic continuation of the branch of s under consideration there arise in general infinitely many circular-arc triangles in the plane of the complex variable s, and indeed every neighbouring two of them have a side in common. If, in a special case, this side is straight, then both triangles correspond one to the other in the usual way, i.e. they are symmetrical figures with respect to the line. If however as a consequence of the development the common side is a circular arc—and this is the general case—then in place of the usual symmetry a Möbius circular transformation occurs, and indeed the circle which the common arc belongs, is the directrix of this transformation. This relationship can rather be called symmetry with respect to an arc.

Through these considerations the problem that is to be solved, that was originally function-theoretic, reduces to the following geometric one: Find all circular-arc triangles that on being multiplied by this symmetry law occupy only a finite number of positions and have the form of various different circular-arc triangles.

Through geometric arguments one now finds that the number of different symmetric repetitions of a circular-arc triangle can only be finite when it is possible to map this triangle conformally onto the surface of a sphere so that it corresponds to a spherical triangle. Since now for a spherical triangle all corresponding repetitions are either symmetric figures in the strict sense or congruent figures, in this way the question is reduced to the following purely geometric problem:

"A body has only a finite number of symmetry planes: find all of the different positions these can have".

This problem, already solved by Steiner, leads either to a family of n planes with a common axis, in which each plane meets the next at an angle of π/n, and with a plane that cuts each plane of this family at right angles, or to the symmetry planes of a regular polyhedron.

This is connection of the question "When is the general solution of the differential equation of the hypergeometric series $F(\alpha, \beta, \gamma, x)$ an algebraic function of the argument x?" with the theory of regular polyhedra.

The case in which not the general but only one particular integral of this differential equation is an algebraic function of the argument x—a case that is easy to analyse— can be set aside here.

31.6   Darboux on the Solution of Riemann's Equation (1887)

We now follow the account given in Darboux ([58], vol. 2, Sects. 358–360).[7]

Section 358 The adjoint equation to a given linear equation was presented for the first time in a memoir by Riemann on the propagation of sound. […] In what follows, we discuss only the equation, already studied in the previous chapter,

$$\frac{\partial^2 z}{\partial x \partial y} + a\frac{\partial z}{\partial x} + b\frac{\partial z}{\partial y} + cz = 0,$$

[where a, b, c are functions of x and y]. We shall give Riemann's results and indicate the consequences that one can deduce. One has[8]

$$F(z) = \frac{\partial^2 z}{\partial x \partial y} + a\frac{\partial z}{\partial x} + b\frac{\partial z}{\partial y} + cz,$$

$$G(u) = \frac{\partial^2 u}{\partial x \partial y} - a\frac{\partial u}{\partial x} - b\frac{\partial u}{\partial y} + (c - \frac{\partial a}{\partial x} - \frac{\partial b}{\partial y})z,$$

$$M = auz + \frac{1}{2}\left(u\frac{\partial z}{\partial y} - z\frac{\partial u}{\partial y}\right),$$

$$N = buz + \frac{1}{2}\left(u\frac{\partial z}{\partial x} - z\frac{\partial u}{\partial x}\right).$$

Riemann showed that if S is a region bounded by a simple closed curve σ then

---

[7] See also (Courant–Hilbert Vol. 2, Ch. 5, Sect. 5).

[8] If we regard F as an operator, then we regard G as the adjoint operator.

$$\iint_S (uF(z) - zG(u))dxdy = \int_\sigma (Mdy - Ndx). \tag{31.30}$$

Suppose that z and u are, respectively, some solutions of the given equation and its adjoint, so

$$F(z) = 0, \quad G(u) = 0.$$

Then the integral over S in Eq. (31.30) vanishes, and we therefore have

$$\int_\sigma (Mdy - Ndx) = 0.$$

Let A be an arbitrary point in the plane and $B'C'$ a curve placed arbitrarily in the plane. Draw through A the lines AB and AC parallel to the coordinate axes, and suppose that the solutions z and u and the coefficients of the differential equations and their first derivatives are continuous in ABC. The above equation gives

$$\int_A^C Mdy + \int_C^B (Mdy - Ndx) - \int_B^A Ndx = 0.$$

Insert the above values of M and N into this equation, and one gets

$$\int_A^C Mdy = \int_A^C \left(\frac{1}{2}\frac{\partial uz}{\partial y}dy - z\left(\frac{\partial u}{\partial y} - au\right)dy \right),$$

$$\int_A^B Ndx = \int_A^B \left(\frac{1}{2}\frac{\partial uz}{\partial x}dx - z\left(\frac{\partial u}{\partial x} - bu\right)dx \right).$$

If quite generally one denotes by φ_P the value of a function φ at a point P, one has

$$\int_A^C Mdy = \frac{1}{2}((uz)_C - (uz)_A) - \int_A^C z\left(\frac{\partial u}{\partial y} - au\right)dy,$$

$$\int_A^B Ndy = \frac{1}{2}((uz)_B - (uz)_A) - \int_A^B z\left(\frac{\partial u}{\partial x} - bu\right)dx,$$

so

$$(uz)_A = \frac{1}{2}((uz)_B + (uz)_C) - \int_B^C (Mdy - Ndx) - \int_A^B z\left(\frac{\partial u}{\partial x} - bu\right)dx - \int_A^C z\left(\frac{\partial u}{\partial y} - au\right)dy. \tag{31.31}$$

We examine each term on the RHS.

We imagine that, with Riemann, we have to find the solution z of the given partial differential equation that takes given values, along with one of its derivatives, at all points of the curve $B'C'$. The equation

$$dz = \frac{\partial z}{\partial x}dx + \frac{\partial z}{\partial y}dy$$

applied to a displacement along the curve evidently determines whichever one of the partial derivatives that was not given a priori, so we can consider that they are both known at each point of the curve $B'C'$. It follows that if one has chosen a solution u of the adjoint equation, the three terms

$$(uz)_B, \ (uz)_C, \ \int_B^C (Mdy - Ndx)$$

that enter the RHS of Eq. (31.31) are known and depend only on the bounding conditions on z. If one could calculate the latter two integrals on the RHS one would know z_A, that is to say the value of z at an arbitrary point of the plane. Now, in general these integrals depend on the entirely unknown values that the sought-for solution z takes on the line segments AB and AC. For these values not to intervene it is necessary that the solution u has been chosen so that

$$\frac{\partial u}{\partial x} - bu = 0 \quad \text{everywhere on } AB$$

$$\frac{\partial u}{\partial y} - au = 0 \quad \text{everywhere on } AC.$$

If these two equations can be satisfied, then the fundamental Eq. (31.31) reduces to the following:

$$(uz)_A = \frac{1}{2}((uz)_B + (uz)_C) - \int_C^B (Ndx - Mdy), \tag{31.32}$$

which determines the value of z at an arbitrary point of the plane as a function of the boundary conditions only.

Thus, *to obtain the general solution of the equation in the form most appropriate for problems in mathematical physics, it is enough to find a solution of the adjoint equation that satisfies the two equations stated above.*

These conditions can be transformed as follows. One must have $\frac{\partial u}{\partial x} - bu = 0$ everywhere on AB. Because only x varies on this segment, one can integrate this equation, which gives

$$u_M = u_A \exp\left(\int_A^M bdx\right)$$

for all points M between A and B. Likewise, the second condition can be replaced by

$$u_N = u_A \exp\left(\int_A^N a\,dy\right)$$

for all points N between A and C.

One can always reduce the constant u_A to unity, so, if the coordinates of A are (x_0, y_0) the question is reduced to finding a solution $u(x, y, x_0, y_0)$ of the adjoint equation depending on two parameters x_0, y_0, that reduces to unity for $x = x_0$, $y = y_0$, taking the value $\exp(\int_{x_0}^y b\,dx)$ for $y = y_0$ and the value $\exp(\int_{y_0}^y a\,dy)$ for $x = x_0$.

This is the fundamental result established by Riemann. The great mathematician had been able to determine a function u for the equation that he had discussed and which is no other than equation $E(\beta, \beta)$. We shall see that the determination of this function can also be carried out for the more general equation $E(\beta, \beta')$, but first staying with the general theory we shall add an essential remark to the result we have just given.

Section 359 Suppose that the primitive curve BC reduces to two straight lines $C'D$ and DB parallel to $B'D$ and DC' [the y and x axes, respectively], and let (x_1, y_1) be the coordinates of the point D. One will have

$$\int_C^B (N\,dx - M\,dy) = \int_C^D N\,dx - \int_D^B M\,dy.$$

Moreover, one can write

$$\int_C^D N\,dx = \int_C^D \left(\frac{1}{2}\left(u\frac{\partial z}{\partial x} - z\frac{\partial u}{\partial x}\right) + buz\right)dx =$$

$$\int_C^D \left(-\frac{1}{2}\frac{\partial (uz)}{\partial x} + u\left(\frac{\partial z}{\partial x} + bz\right)\right)dx,$$

and so

$$\int_C^D N\,dx = \frac{1}{2}\left((uz)_C - (uz)_D\right) + \int_C^D u\left(\frac{\partial z}{\partial x} + bz\right)dx.$$

Likewise one has

$$\int_D^B M\,dy = \frac{1}{2}\left((uz)_B - (uz)_D\right) + \int_B^D u\left(\frac{\partial z}{\partial y} + az\right)dy.$$

Therefore, substituting these values in the above equations, one has

$$(uz)_D = \int_C^D u\left(\frac{\partial z}{\partial x} + bz\right)dx - \int_D^B u\left(\frac{\partial z}{\partial y} + az\right)dy. \qquad (31.33)$$

This formula applies to every solution z of the given equation. It offers the greatest analogy with the general Eq. (31.32), but it is distinguished by an essential property.

Indeed, one recognises that it is not now necessary to prescribe one of the derivatives of z on the contour $C'DB'$. Knowing only the values of the sought-for solution on the lines $C'D$ and DB' allows one to calculate the two integrals that the preceding formula contains and to obtain the value of this solution. It is necessary to look for the origin of this very interesting result in the circumstance that the new contour is formed with the *characteristics* of the given linear equation.

Suppose now that one takes for z this particular solution $z(x, y, ; x_1, y_1)$ of the given equation that is given by entirely similar conditions to those indicated for $z(x, y; x_0, y_0)$ considered as a solution of the adjoint equation. As it is necessary to change the sign of the coefficients a and b when one passes from one equation to the other, one sees that this solution must reduce

$$\text{for } y = y_1 \text{ to } \exp\left(-\int_{x_1}^{x} bdx\right)$$

$$\text{for } x = x_1 \text{ to } \exp\left(-\int_{y_1}^{y} ady\right)$$

and consequently to 1 for $x = x_1$, $y = y_1$. Therefore one has

$$\frac{\partial z}{\partial x} + bz = 0 \text{ at all points of } CD$$
$$\frac{\partial z}{\partial y} + az = 0 \text{ at all points of } BD$$
$$z = 1 \text{ at } D.$$

Consequently, Eq. (31.33) reduces here to $z_A = u_D$, that is to say

$$z(x_0, y_0; x_1, y_1) = u(x_1, y_1; x_0, y_0).$$

This equality implies the following proposition: The solution $u(x, y; x_0, y_0)$ of the adjoint equation that we defined before can be considered as a function of the parameters x_0, y_0; it is then a solution of the primitive equation (where one will have replaced x, y by x_0, y_0) and possesses, with respect to that equation and the variables x_0, y_0, the properties by which is has been defined as a function of the variables x, y and is a solution of the adjoint equation. In other words, *the definition of u does not change if one switches the linear equation and its adjoint, on condition that one switches the two systems of variables x, y and x_0, y_0.*

It follows that the determination of this function $u(x, y; x_0, y_0)$ also allows one to integrate the adjoint equation by a formula analogous to what has been given above. *The integration of two linear equations, the given one and its adjoint, therefore lead to one and the same problem, the determination of the function $u(x, y; x_0, y_0)$. This function can be completely defined, either as the solution of the given equation, or as the solution of the adjoint equation, via the boundary conditions to which they are subjected.*

31.7 Picard and Elliptic Partial Differential Equations (1890)

Picard showed in his [211] that these equations have regular solutions, just like the Laplace equation. The paper opens as follows:

Introduction

Let us consider a second-order partial differential equation of the form

$$A\frac{\partial^2 u}{\partial x^2} + 2B\frac{\partial^2 u}{\partial x \partial y} + C\frac{\partial^2 u}{\partial y^2} = F\left(u, \frac{\partial u}{\partial x}, \frac{\partial u}{\partial y}, x, y\right), \qquad (31.34)$$

A, B, C depending only on the two independent variables x and y. In order to solve this equation under certain specified boundary conditions, one can proceed by successive approximations in the following manner. In the second member we insert an arbitrary function u_1 of x and y, and form the equation

$$\Delta u_2 = F\left(u_1, \frac{\partial u_1}{\partial x}, \frac{\partial u_1}{\partial y}, x, y\right)$$

(here putting, for brevity, $\Delta u = A\frac{\partial^2 u}{\partial x^2} + B\frac{\partial^2 u}{\partial x \partial y} + C\frac{\partial^2 u}{\partial y^2}$). Let us suppose that we have solved this equation for u_2 and provided certain boundary conditions that, we suppose, completely determine an integral that we denote by u_2. One then forms the equation

$$\Delta u_3 = F\left(u_2, \frac{\partial u_2}{\partial x}, \frac{\partial u_2}{\partial y}, x, y\right),$$

and solves it for u_3 under the same boundary conditions as above, and continue in this fashion indefinitely. If the solution u_n tends to a definite limit u as n increases indefinitely on then obtains the solution u of Eq. (31.34) that satisfies the given conditions.

These generalities only have interest when one can make the boundary conditions precise and put in place conditions that allow us to establish rigorously the convergence of u_n to the limit u; this is the point of this memoir. We make the essential supposition that in the region of the plane containing the point (x, y) the discriminant $B^2 - AC$ does not change its sign. Consequently, we can reduce our equation to one of the two following types:

$$\frac{\partial^2 u}{\partial x^2} + \frac{\partial^2 u}{\partial y^2} = F\left(u, \frac{\partial u}{\partial x}, \frac{\partial u}{\partial y}, x, y\right), \qquad (31.35)$$

$$\frac{\partial^2 u}{\partial x \partial y} = F\left(u, \frac{\partial u}{\partial x}, \frac{\partial u}{\partial y}, x, y\right), \qquad (31.36)$$

for which the problems posed are entirely different.

For equations of the first form we provide as boundary conditions the values of the functions u_n along a closed contour C, and we require them to remain continuous along with their first two differentials inside the contour. The study of u_n shows that it converges to a limit provided that C satisfies certain conditions that, in particular, are satisfied *when the contour bounds a sufficiently small area*. In this case one obtains *a solution of Eq. (31.35) that takes a given continuous succession of values on the contour*. This solution, as we shall show, is moreover *unique* if the equation is linear, when the contour is sufficiently small.

One cannot affirm in general that the solution is unique when F is not linear in u, $\frac{\partial u}{\partial x}$, and $\frac{\partial u}{\partial y}$.

In the case of Eq. (31.36), the boundary conditions must be taken in an entirely different manner. Here we take an arc of a curve C, and along C we prescribe the values of $\frac{\partial u_n}{\partial x}$ and $\frac{\partial u_n}{\partial y}$ as well as the value of u_n at a point A of C. Let B be a second point of C, such that the coordinates of a point M of the curve constantly vary in the same sense as M goes from A to B; consider the rectangle parallel to the axes of which A and B are opposite vertices. If B is sufficiently close to A, u_n will tend to a limit u for all points of this rectangle, and one will have the solution u of Eq. (31.36) that takes a given value at A and for which $\frac{\partial u}{\partial x}$ and $\frac{\partial u}{\partial y}$ take a given continuous succession of values on the arc AB; u and its two first partial derivatives are continuous functions of x and y as one traverses the arc AB.

The theorems indicated above for Eq. (31.35) are only correct if the contour C encloses a sufficiently small area. It is very interesting to find equations where *without restriction* a solution that is continuous together with its partial derivatives will always be determined by its values on an arbitrary closed contour.

One can give some detailed examples. This will happen for the Eq. (31.35)

$$\frac{\partial^2 u}{\partial x^2} + \frac{\partial^2 u}{\partial y^2} = F\left(u, \frac{\partial u}{\partial x}, \frac{\partial u}{\partial y}, x, y\right),$$

if, on replacing $\frac{\partial u}{\partial x}$ by v and $\frac{\partial u}{\partial y}$ by w one has the inequality

$$4F'_u > (F'_v)^2 + (F'_w)^2,$$

whatever u, v, w may be. In particular, if F depends on neither $\frac{\partial u}{\partial x}$ nor $\frac{\partial u}{\partial y}$, this condition will be satisfied.

We shall make a special study of the case where the equation can be written as

$$\frac{\partial^2 u}{\partial x^2} + \frac{\partial^2 u}{\partial y^2} = F(u, x, y),$$

and F increases continually with u.

Supposing first of all that F is always positive, we shall see what our method of successive approximations can give here. Its use leads to a very curious result. *This method leads not to one limit, but to two limits u and v.* These functions take the given values on the contour and satisfy the two equations

$$\Delta u = F(v, x, y), \quad \Delta v = F(u, x, y).$$

In order that the problem of finding a solution to the given equation and taking the given values on the contour can be solved, it is necessary that $u = v$; this will not be the case for an arbitrary contour, but this identity is verified if the contour is sufficiently small.

In this particular case we shall show in what follows that one can pass to an arbitrary contour. In fact, *the problem being treated for two contours having a part in common can be solved for the bounding contour exterior to the two areas.* The alternating process, that M. Schwarz and M. Neumann used in their memorable works on the Laplace equation $\Delta u = 0$, can, with modifications that are in any case quite obvious, be extended to our general equation, and, as a result, prove completely effective in the study of the integral, which moreover is unique, of the equation

$$\Delta u = F(u, x, y)$$

that takes a given continuous succession of values on an arbitrarily closed contour.

I also consider an interesting case in which the function F, which always increases with u, vanishes for $u = 0$.

The solutions considered up to now are continuous inside the area. Taking in particular the equation

$$\frac{\partial^2 u}{\partial x^2} + \frac{\partial^2 u}{\partial y^2} = A(x, y)e^u,$$

I examine the case where the integral has logarithmic singular points, and I particularly direct my attention to the following equation which is of great interest, in geometry as much as in analysis,

$$\frac{\partial^2 u}{\partial x^2} + \frac{\partial^2 u}{\partial y^2} = ke^u,$$

where k denotes a positive constant, and which one can call Liouville's equation. I deepen the study of the solutions of this equation by considering them in the whole plane and by studying especially those which are continuous in the whole plane with the exception of a certain number of logarithmic singular points for which one regards the corresponding coefficients as given (only satisfying certain inequalities); I draw attention here to the following result: *These solutions depend only on an arbitrary constant, and a solution of this kind is determined when its value is given at a point of the plane distinct from the singular points.*

Having studied these solutions in the ordinary plane, I extend this to the multiple plane, that is to say to the plane covered by a certain number of leaves forming a Riemann surface.

I remark, in ending this chapter, that the above results relative to the equation

$$\Delta u = ke^u$$

are not without interest for the theory of Fuchsian functions. I shall come back to this special application of the general theory that I have tried to develop in this memoir.

In a final chapter I apply to ordinary differential equations the approximation methods that I have used. It is particularly interesting to consider a system of differential equations of the form

$$\frac{d^2 y_1}{dx^2} = f_1(x, y_1, y_2, \ldots, y_m),$$

$$\frac{d^2 y_2}{dx^2} = f_2(x, y_1, y_2, \ldots, y_m),$$

$$\ldots$$

$$\frac{d^2 y_m}{dx^2} = f_m(x, y_1, y_2, \ldots, y_m),$$

and to obtain system of solutions for these equations continuous between $x = a$ and $x = b$ and taking given values at these extremes.

The same considerations apply to a system of partial differential equations of the form

$$\frac{\partial^2 u_1}{\partial x^2} + \frac{\partial^2 u_1}{\partial y^2} = f_1(x, y, u_1, u_2, \ldots, u_m).$$

$$\frac{\partial^2 u_2}{\partial x^2} + \frac{\partial^2 u_2}{\partial y^2} = f_2(x, y, u_1, u_2, \ldots, u_m).$$

$$\ldots$$

$$\frac{\partial^2 u_m}{\partial x^2} + \frac{\partial^2 u_m}{\partial y^2} = f_m(x, y, u_1, u_2, \ldots, u_m).$$

By imposing certain very general hypotheses on the f one can determine a system of solutions $u_1, u_2, \ldots, u_m$ taking given values on a contour. [Extract ends.]

31.8   Picard and Hyperbolic Partial Differential Equations (1890)

Picard showed that these equations can be solved with initial conditions on suitable arcs.

Among the equations of the type to be considered, it will be enough to consider the equation

$$\frac{\partial^2 z}{\partial x \partial y} = a \frac{\partial z}{\partial x} + b \frac{\partial z}{\partial y} + cz;$$

because once this equation has been dealt with it will be enough to repeat the same arguments as above to deal with the more general equation

$$\frac{\partial^2 z}{\partial x \partial y} = F\left(z, \frac{\partial z}{\partial x}, \frac{\partial z}{\partial y}, x, y\right).$$

There can be no question here of finding the solution in terms of its values along a closed contour; we have to study the method of successive approximations with a view to determining the general integral.

Let us consider in the (x, y)-plane the arc of an arbitrary curve C for which we suppose only that either of the coordinates is a function of the other, and always increases in the same sense. We want to obtain a solution of this equation for which the partial derivatives $\frac{\partial z}{\partial x}$ and $\frac{\partial z}{\partial y}$ take a given succession of values on C and which itself takes a specified value at A on C.

One first tackles the problem for the equation

$$\frac{\partial^2 z_1}{\partial x \partial y} = 0.$$

Let z_1 be a solution, one must then consider the equation

$$\frac{\partial^2 z_2}{\partial x \partial y} = a \frac{\partial z_1}{\partial x} + b \frac{\partial z_1}{\partial y} + cz_1.$$

One looks for a solution on which $\frac{\partial z_2}{\partial x}$ and $\frac{\partial z_2}{\partial y}$ vanish on C and z_2 itself vanishes at A.

One then considers the equation in z_3

$$\frac{\partial^2 z_3}{\partial x \partial y} = a \frac{\partial z_2}{\partial x} + b \frac{\partial z_2}{\partial y} + cz_2$$

which one solves under the same conditions, and continues in this way indefinitely.

It is now necessary to study the series

$$z_1 + z_2 + \cdots + z_n + \cdots \tag{31.37}$$

and to see if it gives a solution of the problem stated. [...]

Let us remark right away that the solution of the equation

Fig. 31.2 The curve C and
the point P

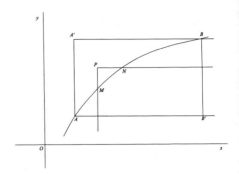

$$\frac{\partial^2 z_1}{\partial x \partial y} = 0$$

is immediate if *on* C one gives $\frac{\partial z}{\partial x}$ as a function of x and $\frac{\partial z}{\partial y}$ as a function of y. Let $\varphi(x)$ and $\psi(y)$ be these two functions. One will evidently have

$$z_1 = z_0 + \int_{x_0}^{x} \varphi(x)dx + \int_{y_0}^{y} \psi(y)dy,$$

where A has coordinates (x_0, y_0).

On the other hand, let the equation be

$$\frac{\partial^2 z}{\partial x \partial y} = F(x, y),$$

where F is a continuous function of x and y. The solution of this equation that vanishes at A and for which $\frac{\partial z}{\partial x}$ and $\frac{\partial z}{\partial y}$ vanish on C can be represented in the following way: let P be a point with coordinates (x, y), and draw through this point parallels to the x and y axes meeting the curve C at points M and N: the required solution is given by the double integral

$$-\iint F(\xi, \eta)d\xi d\eta$$

taken over the curvilinear triangle PMN.

One assumes, as I have already said, that from A to B on the arc C either of the coordinates is a continuous function of the other and always varies in the same sense (increasing in the case of the figure).

This done, suppose that the point P lies in the rectangle $ABA'B'$ [see Fig. 31.2], and let $AB' = \alpha$ and $BB' = \beta$; we are going to look for upper bounds on the different terms of the series (31.37).

Let us denote by M the maximum value of $a\frac{\partial z_1}{\partial x} + b\frac{\partial z_1}{\partial y} + cz_1$ in the rectangle $ABA'B'$. One then has

$$|z_2| < M\alpha\beta, \quad \left|\frac{\partial z_2}{\partial x}\right| < M\beta, \quad \left|\frac{\partial z_2}{\partial y}\right| < M\alpha.$$

If moreover the maximum absolute values of a, b, and c in the rectangle are A, B, and C, respectively, then in the rectangle

$$\left|a\frac{\partial z}{\partial x} + b\frac{\partial z}{\partial y} + cz\right| < M(A\alpha + b\beta + C\alpha\beta).$$

Consequently,

$$|z_3| < M(A\alpha + b\beta + C\alpha\beta)\alpha\beta, \quad \left|\frac{\partial z_3}{\partial x}\right| < M(A\alpha + b\beta + C\alpha\beta)\beta$$

Continuing thus, one arrives in general at

$$|z_n| < M(A\alpha + b\beta + C\alpha\beta)^{n-1}\alpha\beta.$$

It follows that the terms of the series (31.37) can be compared with a geometric progression; if, therefore,

$$A\alpha + b\beta + C\alpha\beta < 1 \tag{31.38}$$

then the series (31.37) and the two series

$$\frac{\partial z_1}{\partial x} + \frac{\partial z_2}{\partial x} + \frac{\partial z_3}{\partial x} \cdots,$$

$$\frac{\partial z_1}{\partial y} + \frac{\partial z_2}{\partial y} + \frac{\partial z_3}{\partial y} + \cdots.$$

will converge. As for condition (31.38), it will evidently hold if the point B is sufficiently close to the point A. The series converges inside the rectangle $ABA'B'$. The function z, the limit of the series $z_1 + z_2 + \cdots + z_n + \cdots$ will obviously have first partial derivatives and the second derivative $\frac{\partial^2 z}{\partial x \partial y}$; furthermore it satisfies the equation

$$\frac{\partial^2 z}{\partial x \partial y} = a\frac{\partial z}{\partial x} + b\frac{\partial z}{\partial y} + cz.$$

Thus, under the above hypotheses, we have for the given partial differential equation a solution z that takes a given value at the point A on the curve, and for which the partial derivatives $\frac{\partial z}{\partial x}$ and $\frac{\partial z}{\partial y}$ take, respectively, on C a prescribed continuous succession of values. These functions $\varphi(x)$ and $\psi(y)$ in our analysis are subject only to the single condition of being continuous. *Let us remark that z, $\frac{\partial z}{\partial x}$, and $\frac{\partial z}{\partial y}$ are continuous functions of x and y even when one crosses the arc C*; here there is an interesting point in the theory of partial differential equations that it good to insist upon.

A solution z of a linear second-order partial differential equation is, one says in a general way, determined when one prescribes the values of z and $\frac{\partial z}{\partial x}$ on a curve C, or, which comes to the same thing, the values of $\frac{\partial z}{\partial x}$ and $\frac{\partial z}{\partial y}$ on this curve and the value of z at a particular point of C. But this general conception is only valuable for a curve C traced in a region of the plane where the characteristics are real, that is to say, only in this case, when one is certain to have a solution satisfying the given conditions that is continuous along with its first-order partial derivatives when one crosses the arc C; our preceding analysis shows very neatly that z, $\frac{\partial z}{\partial x}$ and $\frac{\partial z}{\partial y}$ are continuous in the passage over C.

It is quite otherwise when the characteristics are imaginary. To see this, it suffices to take the simple example of the equation

$$\frac{\partial^2 z}{\partial x^2} + \frac{\partial^2 z}{\partial y^2} = 0.$$

In general one cannot have a solution of this equation that is continuous in the rectangle $ABA'B'$ along with its first-order partial derivatives, and for which $\frac{\partial z}{\partial x}$ and $\frac{\partial z}{\partial y}$ take on the arc AB the succession of values denoted above by $\varphi(x)$ and $\psi(y)$, these functions being subject to no other condition than being continuous. In the contrary case one could, in effect, form an analytic function $z + iz_1$ that will be holomorphic in the rectangle under consideration, the real part of this function being arbitrary on the curve AB, which is impossible because a holomorphic function determined on an arc of a curve however small can only be extended in a unique way.

Thus the proof that we have given of the existence of a solution of the equation

$$\frac{\partial^2 z}{\partial x \partial y} = a\frac{\partial z}{\partial x} + b\frac{\partial z}{\partial y} + cz,$$

and its development in series allows us to raise a question that we must necessarily put on one side, when one supposes that a, b, c are analytic functions and that the conditions on the bounds are expressed by means of analytic functions.

The linear equation

$$\frac{\partial^2 z}{\partial x \partial y} = a\frac{\partial z}{\partial x} + b\frac{\partial z}{\partial y} + cz$$

has been the object of a remarkable chapter in Darboux's *Leçons sur la théorie des surfaces* (Vol. II, Chap. IV). Following an idea of Riemann's, Darboux reduced the solution of this equation to the study of a particular solution z; this solution z is determined by the condition that it reduces for $x = x_0$ to a given function $\varphi(x)$. Darboux established the existence of such a solution in supposing that a, b, c are *analytic* functions of x and y, and he uses as an intermediary the celebrated equation considered by Euler and Poisson. Staying with our point of view of successive approximations, the proof of the existence of such a solution z is quite easy, without making any other assumption about a, b, c, φ, and ψ other than that of continuity. [Evidently one has $\varphi(x_0) = \psi(y_0)$, and one assumes that φ and ψ have first derivatives.] [Here the extract ends.]

Picard then sketched the slight modifications of his earlier proof needed to adapt it to the new situation.

Appendix A
Newton's *Principia Mathematica*

Newton's theory of celestial mechanics, set out in his *Principia Mathematica* (Fig. A.1) [206], is based on his analysis of the concept of motion and its causes, a thorough-going mathematical analysis of the motion of bodies under the action of forces, and a meticulous study of the observed motion of the Moon, the planets, and their satellites. This led him to proclaim his highly novel theory of gravity, and to refute the earlier and widely accepted ideas of Descartes's. A notable success on the way was his integration of Kepler's laws, and his study of the motion of a body under a central force. He came up with a remarkably accurate description of the motion of the planets and their satellites based on his inverse-square law of gravity, but it nonetheless failed to account for the motion of the moon, and this was a cause of great controversy the generation after his death.

A.1   Newton's Laws of Motion in His *Principia*

Newton's *Principia* was published in late 1687. It is a book of 547 pages, written in scholarly Latin, and after some introductory remarks and a few definitions it opens with these three laws of motion.[1]

> Law 1 *Every body perseveres in its state of being at rest or of moving uniformly straight forward except insofar as it is compelled to change its state by forces impressed.* Projectiles persevere in their motions, except insofar as they are retarded by the resistance of the air and are impelled downward by the force of gravity. A spinning hoop, which has parts that by their cohesion continually draw one another back from rectilinear motions, does not cease to rotate, except insofar as it is retarded by the air. And larger bodies – planets and comets – preserve for a longer time both their progressive and their circular motions, which take place in spaces having less resistance.

---

[1] See *Principia* Axioms, or the laws of motion in the Cohen and Whitman translation of 1999, pp. 416–417, and (in the Motte–Cajori translation) pp. xvii–xix and F&G 12. B2.

© The Author(s), under exclusive license to Springer Nature Switzerland AG 2021
J. Gray, *Change and Variations*, Springer Undergraduate Mathematics Series,
https://doi.org/10.1007/978-3-030-70575-6

Fig. A.1 Title page of
Newton's *Principia*

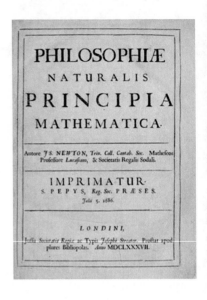

Law 2 *A change in motion is proportional to the motive force impressed and takes place along the straight line in which that force is impressed.* If some force generates any motion, twice the force will generate twice the motion, and three times the force will generate three times the motion, whether the force is impressed all at once or successively by degrees. And if the body was previously moving, the new motion (since motion is always in the same direction as the generative force) is added to the original motion if that motion was in the same direction or is subtracted from the original motion if it was in the opposite direction or, if it was in an oblique direction, is combined obliquely and compounded with it according to the directions of both motions.

Law 3 *To any action there is always an opposite and equal reaction; in other words, the actions of two bodies upon each other are always equal and always opposite in direction.* Whatever presses or draws something else is pressed or drawn just as much by it. If anyone presses a stone with a finger, the finger is also pressed by the stone. If a horse draws a stone tied to a rope, the horse will (so to speak) also be drawn back equally toward the stone, for the rope, stretched out at both ends, will urge the horse toward the stone and the stone toward the horse by one and the same endeavor to go slack and will impede the forward motion of the one as much as it promotes the forward motion of the other. If some body impinging upon another body changes the motion of that body in any way by its own force, then, by the force of the other body (because of the equality of their mutual pressure), it also will in turn undergo the same change in its own motion in the opposite direction. By means of these actions, equal changes occur in the motions, not in the velocities – that is, of course, if the bodies are not impeded by anything else. For the changes in velocities that likewise occur in opposite directions are inversely proportional to the bodies because the motions are changed equally. This law is valid also for attractions, as will be proved in the next scholium.

Newton's laws of motion are stated as axioms, and accordingly neither derived from other statements nor based on experiments. Newton gave an explanation and elucidation of each law, but not a justification. And, as befits axioms, the laws are the basis for subsequent deductions concerning the behaviour of moving bodies and bodies acted upon by forces. Newton's laws state presumed properties of matter in motion; they are not specifically mathematical, being neither geometric nor algebraic. They are not stated in the form of equations involving symbols and their manipulation.

In a book devoted to the study of motion in general and planetary motions in particular Newton had to decide what were the crucial astronomical ideas that he needed. He chose to rely on all three of Kepler's laws, and Newton's vast generalisation of Kepler's somewhat controversial second (equi-area) law, prominently placed near the front of the book, was to play a vital role in the theory he presented (see Sect. A.1.1).

The Newton scholar I.B. Cohen has observed that[2]:

> It was an unusual and a very daring step to erect an astronomical system encompassing Kepler's three laws, as Newton did. Following the imaginative leap forward that Newton made, in showing the physical meaning and conditions of mathematical generality or applicability of each of Kepler's laws, this whole set of three laws gained a real status in exact science.

A.1.1   The Content of the **Principia**

Before Book I begins, the *Principia* has an introduction in which Newton spelled out the mathematically precise concepts used in his three laws of motion. He defined the quantity of matter and the quantity of force, and he discussed forces of various kinds. He then gave a complicated distinction between relative and absolute motion and relative and absolute time: in many ways Newton treated all motion as relative, but he also regarded the centre of the universe (which he regarded as the centre of the solar system) as being absolutely at rest. Only then come the three axioms or laws of motion we looked at above, and the first of their elementary consequences.

This book then gives a long, careful, cumulative discussion of "the method of first and last ratios of quantities": a geometrical study of curves and their tangents in the spirit in which Newton conducted his investigations of the calculus.

Then Newton turned to a study of the motion of a point under a centripetal force. He showed that the line joining a fixed point to a moving one sweeps out equal areas in equal times if and only if the force on the moving point is directed towards the fixed point. Remarkably, the size of the force can depend in any way on the length of the radial line; the orbit can be any shape determined by the law, not just a circle or an ellipse.

Among the special cases that are then worked out is this one: if the moving body traverses a conic section under a centripetal force directed towards one focus, then the magnitude of the force is inversely proportional to the square of the distance. Newton knew well, as his acceptance of Kepler's laws indicates, that this is the relevant case in astronomy. In the first edition of the *Principia* he also stated the converse (it is

---

[2]See Cohen ([46], 229).

Cor. 1 to Prop. 13): under an inverse-square law, bodies move in curves which are conic sections having the centre of force as one focus. Controversies surrounding this statement led Newton to enrich it with a skeleton proof in the second edition (1713).

To find where in its orbit a planet can be found at any particular time Newton used Kepler's equi-area law. He also showed that if planets traverse ellipses under the action of a force that obeys an inverse-square law, then they necessarily obey Kepler's third law (the 3/2 power law, which says that if r is the radius of the orbit of a planet, and t is the time to complete one orbit, then $t^2 \propto r^3$).

Newton then investigated the attraction between solid bodies under an inverse-square law. He established that a spherical shell exerts no force on a point inside it and attracts a point outside it in the same way as a point mass concentrated at the centre. This remarkable result, which surprised him as much as his contemporaries, enabled him to reduce the study of large spherical objects like planets to the study of points and centripetal forces, which he had already described. In Newton's theory of gravity, large solid spheres may be replaced by points (of the same mass)—a considerable simplification in the theory. For much work in astronomy, the assumption that planets and the sun are spherical in shape is entirely reasonable.

Newton discussed many topics in Book II, subtitled "The Motion of Bodies (in resisting media", but we need to note only that at the end of the book Newton demolished Descartes's theory of motion in vortices and concluded: "Hence it is manifest that the planets are not carried round in corporeal vortices".[3]

In Book III, "The System of the World (in mathematical treatment)" Newton demonstrated that the theory of an inverse-square law for gravity acting as a force between bodies can explain the motion of the planets and of their satellites, the motion of comets, and enable the shape of the Earth to be determined. But the Moon gave him trouble and could deduce only that its motion obeys the equi-area law.

A.2   The Motion of the Moon

The motion of the Moon is far from simple, and it has been intensively studied not only because it is our nearest neighbour in space but because, if it could be understood accurately, it would provide an excellent clock and so be an aid to navigation.

The problem is that the Moon is part of a system of three bodies: the Earth, the Moon, and the Sun, and while Newton could deal well with two bodies acting on each other by gravity, the three-body problem, as it became known, is (strictly speaking) unsolved to this day. No one can yet answer the question: given three arbitrary bodies acting on each other by gravity and released initially with such-and-such velocities, what will be their orbits for all future times? Will the Moon always orbit the Earth, move away, or eventually collide with it? We do not know.

---

[3] See Newton, *Principia*, Book II, 790.

But if the mathematical problem is too difficult to solve exactly, exhaustive computer simulations enable scientists to navigate satellites around the solar system with astonishing accuracy and to make a variety of predictions about the long-term fate of the solar system (and, equally interesting, about its origins). These predictions say that the Moon will still be orbiting the Earth 50 million years from now.

In the seventeenth and eighteenth centuries conclusions could only be reached if some simplifying assumptions are made. Newton assumed, for example, that the only effect of the Sun is to perturb slightly an otherwise elliptical orbit of the Moon around the Earth, and then tried to calculate that perturbation exactly (as measured by the motion of the apse of the ellipse). His success was less than complete, and because the Moon is an easy object to observe the mismatch between his, and indeed any, prediction and reality was apparent.

Newton's calculations in *Principia* I, 45 showed the elliptical orbit rotates slowly around the centre of gravity of the Earth and the Moon, returning to its original place every 18 years.[4] But, as Newton conceded in later editions of the *Principia* in a single crisp sentence: "The [advance of the] apsis of the Moon is about twice as swift".[5] He was unable to come up with significantly more convincing and accurate theory, although unpublished papers show that he was able to get a better approximation to the Moon's motion, and it says a lot about his high standards that he was displeased that his approximate theory was out by a factor of 2.

But this point is not merely technical; its implications were profound. This failure of Newton's, it was thought, might be the loose thread that would unravel his theory. And his theory of universal gravitation was unpopular in Cartesian circles, and initially too difficult to understand in all circles. As a result, Newtonian gravity (more precisely, the inverse-square law) now came to stand or fall by its ability to describe sufficiently accurately the motion of the Moon.

Significant progress on the question had to wait until 1747. In that year, Clairaut, wrote to Euler that it was "a proven fact that Newtonian gravitation is inadequate to account for the [lunar] phenomena".[6] He therefore proposed to add a small inverse fourth-power term to the inverse-square law (making a law of the form $f(r) = ar^{-2} + br^{-4}$.) D'Alembert had come independently to the same opinion that Newton's theory was incorrect, as did Euler, on the basis of his study of a different three-body system (the Sun, Jupiter, and Saturn), although each man had a different remedy in mind.[7]

Euler had already submitted his essay on the motion of Saturn to the Académie des Sciences in Paris for consideration in their prize competition. In it, he expressed his doubts about the inverse-square law, particularly Newton's failure with the motion of the apse of the Moon, and he hinted that he wished to re-introduce vortices (thus he framed what Newton would have called "an hypothesis", an ad hoc mechanism). Clairaut, one of the judges, read the essay in September 1747, recognised Euler's

---

[4]See Newton *Principia* I, Sect. 9, 534–545.

[5]See the Cohen and Whitman translation, p. 545.

[6]For this exchange of letters, see Euler *Opera Omnia* (4A) 5, 173–175 and F&G 14.B2 and 14.B4.

[7]See Wilson [274] for a good account, and for references to the primary literature.

354 Appendix A: Newton's *Principia Mathematica*

handwriting, and wrote to Euler on 11 September to say that he was delighted that Euler had thought about Newtonian attraction. Clairaut went on

It is true that on adding some other term one feels that the theory will better accord with the phenomena. But it seems to me that this term must be such that at the distances of Mercury, Venus, the Earth and Mars it must be almost insensible, in view of the extreme smallness of the motion of the apsides. And if, as it seems initially from your work, the law of squares is palpably in error at the distance of Saturn and Jupiter it would still be necessary to add terms which were significant only at that distance. I confess that the whole of gravitation seems to me to be only a speculative hypothesis.

He then remarked that

It seems to me, and I am not a candidate for the prize, much more important to know if Newtonian attraction holds or not than to treat simply of Saturn. And in seeing if the square law of attraction must suffer some correction which can only be for small distances it seems to me to be necessary to begin by finishing the theory of the moon.

However, Clairaut soon withdrew the suggestion that the modifying term should be an inverse-fourth power, because it predicted that objects near the surface of the Earth should be heavier than they are. He also rejected Euler's vortices, which he thought Euler himself had shown to be no help at all.[8]

In Clairaut's view, part of the problem was that Newton's *Principia* was difficult to understand.[9] He praised it, which was still a controversial thing to do in France, by saying

The famous book *The Mathematical Principles of Natural Philosophy* has been the occasion of a great revolution in Physics. The method which Mr Newton, its illustrious author, has followed to derive facts from their causes, has shed the light of mathematics on a science which up till then had been in the shadows of conjectures and hypotheses.

and then he turned to say what had to be done next. The problem was not that Newton concealed his fluxional calculus that was easy to supply. Rather,

is it not right to reproach him for another wrong which without doubt has struck all those who have studied his book with a true desire to understand it? Namely, that in most of the difficult places he employed too few words to explain his principles [...].

That said, Clairaut reflected, so much else was right—"Kepler's laws [...], the movement of the nodes of the moon [...], the tides, [...] and finally several other questions equally favourable to attraction [that] it appeared to me as difficult to reject as to accept".

Clairaut began to work intensively on the law of gravitational attraction, and on 17 May 1749 he announced his surprising conclusion that, by taking a new point of view, he had found that the problem disappeared, and the inverse-square law could give the correct prediction for the apse line of the Moon.

Euler was not immediately convinced, and in 1749 he persuaded the St Petersburg Academy to have a prize competition, and suggested several propitious topics, all

---

[8] See Euler *Opera Omnia* (4) 5, letter 421 and F&G 14.B2(b).
[9] See Clairaut [45] and the extract in F&G 14.B3.

of an astronomical nature. They chose this one "To demonstrate whether all the inequalities observed in lunar motion are in accordance with Newtonian theory— and if they are not, to demonstrate the true theory behind all these inequalities, such that the exact position of the Moon at any time can be computed by means of it", and Euler became one of the judges, indeed the decisive one.[10]

Clairaut hesitated over whether to enter the competition. He published his paper, which d'Alembert criticised, and only submitted his entry in December 1750. The Academy had by then decided to extend the competition to June 1, 1751, but when they sent Euler Clairaut's entry he replied that it "is superb, and it is hardly likely that anything better will be received prior to June 1". He repeated his endorsement, and admitted that he had changed his own opinion, in the official statement he wrote on 5 June, and the result was announced on 6 September 1751.[11]

In his paper of 1749, and again in 1753, Clairaut argued that the error lay in the poor way in which exact, unsolvable equations for the motion of the Moon had been reduced to inexact, approximate, but solvable equations.

Clairaut formulated the problem of the motion of the Moon in terms of differential equations, and after integrating twice found this expression for the solution, in which Ω is an unknown function of r, the radial distance of the Moon, and of the perturbative force of the Sun:

$$\frac{f^2}{Mr} = 1 - g \sin v - q \cos v + \sin v \int \Omega \cos v dv - cos v \int \Omega \sin v dv. \quad (A.1)$$

Here, f, g, and q are constants of integration, M is the sum of the masses of the Earth and the Moon, v is an astronomical quantity called the true anomaly (which may be taken to represent the velocity of the Moon).

To find Ω, Clairaut employed a process of successive approximations. The apse of the Moon was understood to move rather as if the Moon precesses on an ellipse. So Clairaut, following Newton, first wrote the equation of an ellipse in polar coordinates

$$\frac{k}{r} = 1 - e \cos mv. \quad (A.2)$$

Here, k, e, and m are constants that are either to be determined from the constants f, g, and q or otherwise found from observation. In particular, e was already known empirically to be about 0.05. This means that as $\cos mv$ varies between at most $+1$ and -1, r varies between $\pm k$.

Clairaut substituted this approximation into his original equation, and obtained this better approximation to r:

$$\frac{k}{r} = 1 - e \cos mv + \beta \cos \frac{2v}{n} + \gamma \cos \left(\frac{2}{n} - m\right)v + \delta \left(\frac{2}{n} + m\right)v. \quad (A.3)$$

---

[10]See Kopelevich [163].

[11]Clairaut then published his own theory of the motion of the Moon in 1753. D'Alembert now also withdrew his criticisms of the inverse-square law.

Here n is another quantity determined by observations (and therefore known) and β, γ, and δ are constants determined from the other constants so far; the new terms describe the way the ellipse slowly changes.

Clairaut now explained his original mistake, which had led him to deny the inverse-square law. He evaluated β, γ, and δ and found that to nine decimal places

$$\beta = 0.007090988, \gamma = -0.00949705, \text{ and } \delta = 0.00018361 .$$

These numbers are much smaller than e, and he felt that they were already too small to allow his method to double the value of m, so he did not seek a better, second approximation. Accordingly, he had been inclined to believe that the error must lie in the inverse-square law, which consequently needed amending. But in the spring of 1749 he calculated the next approximation and found that his hunch had been wrong. It turned out that the contributions coming from the γ term were not only quite large, they were proportional to the transverse perturbing force, whereas the initial contribution to m related only to the radial perturbing force. It was only by going to the second approximation that Clairaut could pick up the effect that was making the Moon's ellipse precess. Now, on calculating these numbers, Clairaut found that the monthly apsidal motion was $3°2'6''$, which was just $2'$ less than the empirical value that he accepted.

It must be said that even Euler found Clairaut's method difficult to follow in detail.[12] But in the end Newtonianism became accepted very much as Newton had presented it, in that

1. it was a highly mathematical theory of the solar system;
2. the predictions that it made rested on a highly theoretical analysis;
3. if its conclusions were accepted then its theoretical presuppositions seemed inevitable, provided the mysterious force of gravity was accepted as really existing.

This vindication of Newton's theoretical approach gave mathematicians the confidence to deal for the first time with many more of the most interesting aspects of the physical world.

---

[12]See Euler *Opera Omnia* (4) V, 195–196 and F&G 14.B4.

Appendix B
Characteristics

B.1   First-Order Linear Partial Differential Equations

The simplest partial differential equation in two variables x and y that one could hope to solve is $u_x = 0$—which is essentially an ordinary differential equation— and its solution is $u(x, y) = f(y)$, where f is any function of y. Notice, crucially, that any solution is constant along the curves $y = const.$, and also that it can vary arbitrarily—not necessarily even continuously, from curve to curve.

A modern approach to linear partial differential equations in two variables aims to reduce them to this form, and so reduce the problem to one in ordinary differential equations.[13]

Consider the equation

$$au_x + bu_y = 0,$$

where a and b are constants, not both zero. One solution method notices that the equation says that the directional derivative of the function u vanishes in the direction (a, b) and so any solution u is constant along lines with equations of the form $bx - ay = c$, and so the solution to the partial differential equation is

$$u(x, y) = f(bx - ay),$$

which is constant along the lines $bx - ay = c$.

A second solution method changes variables to

$$\xi = ax + by, \quad \eta = bx - ay,$$

observes that as a result

---

[13] See Grigoryan's account: http://www.math.ucsb.edu/~grigoryan/124A.pdf or google Grigoryan partial differential equations.

© The Author(s), under exclusive license to Springer Nature Switzerland AG 2021
J. Gray, *Change and Variations*, Springer Undergraduate Mathematics Series,
https://doi.org/10.1007/978-3-030-70575-6

$$u_x = au_\xi + bu_\eta, \quad u_y = bu_\xi - au_\eta$$

and so the partial differential equation becomes

$$(a^2 + b^2)u_\xi = 0.$$

This gives the solution

$$u(\xi, \eta) = f(\eta) = f(bx - ay),$$

as before.

Once again the solution is constant along a family of curves—called the characteristics of the partial differential equation—and is therefore determined by the values specified on any curves transversal to the characteristics.

The method of characteristics applies to linear first-order partial differential equations with variable coefficients:

$$a(x, y)u_x + b(x, y)u_y = 0.$$

The characteristics are now curves along which the directional derivative of u vanishes.

Grigoryan gives this example

$$u_x + yu_y = 0.$$

We want the directional derivative in the direction $(1, y)$ to vanish, that is, along curves for which

$$\frac{dy}{dx} = \frac{y}{1},$$

and these are the curves with equations $y = Ce^x$, where $C = ye^{-x}$ is a parameter that varies from curve to curve. As before, the solution to the partial differential equation is constant along these curves, so it is given by

$$u(x, y) = f(ye^{-x}).$$

Or we could argue that if $u(x, y) = const.$ then we always have

$$u_x dx + u_y dy = 0,$$

and so

$$-\frac{u_x}{u_y} = \frac{dy}{dx} = \frac{y}{1},$$

which leads to the same conclusion.

As before, we can also solve the equation

$$a(x, y)u_x + b(x, y)u_y = 0,$$

by changing variables to

$$\xi = \xi(x, y), \quad \eta = \eta(x, y).$$

We have

$$u_x = u_\xi \xi_x + u_\eta \eta_x, \quad u_y = u_\xi \xi_y + u_\eta \eta_y,$$

and so the partial differential equation becomes

$$(a\xi_x + b\xi_y)u_\xi + (a\eta_x + b\eta_y)u_\eta = 0.$$

If we set $a\eta_x + b\eta_y = 0$ then the partial differential equation becomes the ordinary differential equation $u_\xi = 0$, which we solve. The solutions are of the form $u(\xi, \eta) = f(\eta)$, and once again they are constant along the curves $\eta = const$.

So we now have to solve the equation $a\eta_x + b\eta_y = 0$, and (mimicking an earlier argument) we deduce from $u(x, y) = 0$ and $u_x dx + u_y dy = 0$ that

$$-\frac{u_x}{u_y} = \frac{dy}{dx} = \frac{b}{a},$$

as before.

Indeed, the solution to the ordinary differential equation

$$\frac{dy}{dx} = \frac{b(x, y)}{a(x, y)}$$

is given by an equation of the form $f(x, y, c) = 0$. This can be written locally, if we allow ourselves to share the confidence of the eighteenth century authors, as $y = f(x) + c$.

Define $\eta = y - f(x)$ and $\xi = x$, so

$$u_x = u_\xi + u_\eta \left(-\frac{df}{dx} \right), \quad u_y = u_\eta \eta_y,$$

and in the new variables the partial differential equation becomes

$$a \left(u_\xi + \left(-\frac{b}{a} \right) a_\eta \right) + bu_\eta = 0,$$

or

$$au_\xi = 0,$$

the solutions of which are

$$u(\xi, \eta) = g(\eta),$$

and these values can be prescribed arbitrarily along the curve $\eta = constant$.

To sum up the story so far, the first step in solving a linear first-order partial differential equation is to change variables, so almost automatically we use the formulae

$$u_x = u_\xi \xi_x + u_\eta \eta_x; \, u_y = u_\xi \xi_y + u_\eta \eta_y.$$

The second step is to change variables in such a way that one term in the partial differential equation vanishes. This requires solving an ordinary differential equation for the characteristic curves. The third step observes that the solutions are constant along the characteristics, and so (fourth and final step) these values are picked up from a given set of initial conditions that specify the values of a solution along a curve transversal to all the characteristics.

B.2 Burgers' Equation, a Non-linear Equation

Burgers' equation,

$$u_t + u u_x = 0,$$

is a non-linear equation, and it displays interesting phenomena.[14] The characteristics are straight lines of varying slopes, and the solution is constant along the characteristics. But now suppose, for example, that the slopes of all the lines through the x-axis for negative x are steep and positive, and that the slopes of all the lines through the x-axis for positive x are shallow and positive. Then none of the first kind of characteristic lines meet any of the second kind, and as t increases a gap opens up between the first kind and the second kind. This is called rarefaction.

It is more interesting if, instead, the slopes of all the lines through the x-axis for negative x are shallow and positive, and that the slopes of all the lines through the x-axis for positive x are steep and positive. Then all of the first kind of characteristic lines meet characteristic lines of the second kind, and as t increases it becomes impossible to say what happens when the characteristics cross (because the solution cannot have two different values). This is the phenomenon of a shockwave.

We are given the quasi-linear equation

$$u_t + a(u)u_x = 0$$

with initial values on the line $t = 0$.

The characteristics in this case are the solutions of the ordinary differential equation

---

[14] See, for example, Grigoryan's account.

$$\frac{dx}{dt} = a(u).$$

The theory of characteristics says that every solution u remains constant on each characteristic, and so the slope of each characteristic is constant and the characteristics are straight lines.

There is a characteristic for each point $(x_1, 0)$ on the line $t = 0$ and its slope is given by the value of u at that point. So if the values of u at $(x_1, 0)$ and $(x_2, 0)$, with $x_1 < x_2$, are u_1 and u_2 and $0 < a(u_2) < a(u_1)$ then the characteristic through $(x_1, 0)$ has a lesser slope than the characteristic through $(x_2, 0)$ and will eventually cross it. This shows that the solution cannot be continued beyond that point.

Courant and Hilbert (Vol. 2, Appendix 2 to Chap. II) also pointed out that the partial differential equation with initial values given by $u(x, 0) = \varphi(x)$ has a solution in the form

$$u - \varphi(x - ta(u)) = 0,$$

and the implicit function theorem implies that u is a differentiable function of x and t as long as the u derivative of $-\varphi(x + ta(u))$ does not vanish, a condition that holds whenever $ta'(u)\varphi \neq 1$. Whenever this condition is violated one can expect u to become singular.

Another good discussion of Burgers' equation

$$u_t + uu_x = 0$$

is given in Evans ([100], 140–144). He takes initial data on the axis $t = 0$ that is given by a function $u = g(x)$ that is defined as

$$g(x) = \begin{cases} 1 & \text{if } x \leq 0 \\ 1 - x & \text{if } 0 \leq x \leq 1 \\ 0 & \text{if } x \geq 1. \end{cases}$$

The characteristic through x_0 is $x(t) = g(x_0)t + x_0, t \geq 0$ and along it any smooth solution takes the constant value $z_0 = g(x_0)$. It is instructive to plot some of these before proceeding. Accordingly, the solution function is

$$g(x) = \begin{cases} 1 & \text{if } x \leq t \quad 0 \leq t \leq 1 \\ \dfrac{1 - x}{1 - t} & \text{if } t \leq x \leq 1 \, 0 \leq t \leq 1 \\ 0 & \text{if } x \geq 1 \quad 0 \leq t \leq 1. \end{cases}$$

The method breaks down when $t \geq 1$—note that u is apparently infinite when $t = 1$—and in this case the characteristics cross. The most visible case is the way the characteristics through points $x_0 \leq 0$ meet the ones through points $x_0 \geq 1$.

Appendix C
The First-Order Non-linear Partial Differential Equation

For reference, here is a statement of the existence and uniqueness theorem for the first-order partial differential equation

$$F(x, y, z, p, q) = 0$$

(from John ([153], 29)).

The function F has continuous second derivatives;

Along an initial curve $(x_0(s), y_0(s))$ initial values $z_0(s)$ are assigned, and x_0, y_0, z_0 have continuous second derivatives;

There are two continuously differentiable functions $p_0(s)$ and $q_0(s)$ such that

$$F(x_0(s), y_0(s), z_0(s), p_0(s), q_0(s)) = 0$$

and

$$\frac{dz_0}{ds} = p_0 \frac{dx_0}{ds} + q_0 \frac{dy_0}{ds}.$$

The transversality condition:

$$\frac{dx_0}{ds} F_q(x_0, y_0, z_0, p_0, q_0) - \frac{dy_0}{ds} F_p(x_0, y_0, z_0, p_0, q_0) \neq 0.$$

Then in some neighbourhood of the initial curve there exists a unique solution that contains the initial strip, i.e.

$$z(x_0(s), y_0(s)) = z_0(s), z_x(x_0(s), y_0(s)) = p_0(s), z_y(x_0(s), y_0(s)) = q_0(s).$$

We shall now see that the equation has a unique solution.

Solutions of the general first-order partial differential equation

$$F(x, y, z, p, q) = 0,$$

© The Author(s), under exclusive license to Springer Nature Switzerland AG 2021
J. Gray, *Change and Variations*, Springer Undergraduate Mathematics Series,
https://doi.org/10.1007/978-3-030-70575-6

with whatever conditions on F that seem necessary will form a two-parameter family of surfaces together with whatever envelopes one-parameter families may form and any envelopes the two-parameter family has (there may be none).

We now make a standard move in the subject and attempt to accumulate so many necessary conditions on any candidate for a solution to the partial differential equation that together they form a set of sufficient conditions that enable the problem to be solved.

Fix attention on one of these surfaces and suppose it passes through the point $P_0 = (x_0, y_0, z_0)$. The equation

$$F(x_0, y_0, z_0, p, q) = 0 \tag{C.1}$$

is an equation relating p and q at P_0, and we can think of it as defining q as a function of p: $q = q(x_0, y_0, z_0, p)$. So we have a one-parameter family of planes through P_0 with equations:

$$z - z_0 = (x - x_0)p + (y - y_0)q. \tag{C.2}$$

These planes envelope a cone through P_0 that is tangent to the surface. To find this envelope, we differentiate the above equation with respect to p and obtain

$$0 = x - x_0 + (y - y_0)\frac{dq}{dp}. \tag{C.3}$$

From the first equation we obtain

$$\frac{dF}{dp} = F_p + F_q\frac{dq}{dp} = 0, \tag{C.4}$$

and so we may write

$$\frac{x - x_0}{F_p} = \frac{y - y_0}{F_q}, \tag{C.5}$$

and therefore

$$\frac{x - x_0}{F_p} = \frac{y - y_0}{F_q} = \frac{z - z_0}{pF_p + qF_q}. \tag{C.6}$$

From Eqs. (C.2) and (C.5) we deduce that these values of p and q define a tangent line in the cone that is tangent to the surface and is given by Eq. (C.6).

The tangent line is a tangent to the characteristic curve in the surface through P_0, which therefore satisfies the equations

$$\frac{dx}{F_p} = \frac{dy}{F_q} = \frac{dz}{pF_p + qF_q}, \tag{C.7}$$

which we can regard as

$$\frac{dx}{dt} = F_p, \quad \frac{dy}{dt} = F_q, \quad \frac{dz}{dt} = pF_p + qF_q. \tag{C.8}$$

This is not enough information to determine the surface. But we also know that any characteristic curve in the surface (regarded as a curve parameterised by t) must satisfy the equations

$$\frac{dp}{dt} = p_x \frac{dx}{dt} + p_y \frac{dy}{dt} = p_x F_p + p_y F_q$$

and

$$\frac{dq}{dt} = q_x \frac{dx}{dt} + q_y \frac{dy}{dt} = q_x F_p + q_y F_q.$$

We can also differentiate the equation $F(x, y, z, p, q) = 0$ with respect x and with respect to y and obtain

$$F_x + F_z p + F_p p_x + F_q q_x = 0 \quad \text{and} \quad F_y + F_z q + F_p p_y + F_q q_y = 0.$$

Therefore

$$\frac{dp}{dt} = -F_x - F_z p, \quad \frac{dq}{dt} = -F_y - F_z q.$$

Now we have five equations for the functions that define the characteristic curves and fill out the solution surface, $x(t), y(t), z(t), p(t), q(t)$. This is too many, but the condition $F(x, y, z, p, q) = 0$ implies that $\frac{dF}{dt} = 0$, so it merely says that F is constant along any of these curves, which is as it should be.

It remains to check that these characteristic curves define a surface that is a solution to the partial differential equation, and that a surface can be found passing through any initial curve $(x_0(s), y_0(s), z_0(s))$, $0 \le s \le 1$ that is not a characteristic curve. The five equations for $\frac{dx}{dt}, \frac{dy}{dt}, \frac{dz}{dt}, \frac{dp}{dt}, \frac{dq}{dt}$ are all ordinary differential equations. We suppose that when $t = 0$ $x = x_0(s), y = y_0(s), z = z_0(s), p = p_0(s), q = q_0(s)$—the first three of these equations place (x_0, y_0, z_0) on the initial curve. We appeal to the existence of solutions to an ordinary differential equation to deduce that in the neighbourhood of the initial curve there are functions such that

$$x = X(s, t), y = Y(s, t), z = Z(s, t), p = P(s, t), q = Q(s, t)$$

and

$$X(s, 0) = x_0(s), Y(s, 0) = y_0(s), Z(s, 0) = z_0(s), P(s, 0) = p_0(s), Q(s, 0) = q_0(s).$$

It remains to check that these curves define a surface $z = z(x, y)$ and that this surface is a solution of the partial differential equation. To eliminate s and t so that we can write

$$z = Z(s, t) = Z(s(x, y), t(x, y)) = z(x, y)$$

(if you forgive the ambiguous notation for a moment), requires that the Jacobian $\begin{pmatrix} X_s & X_t \\ Y_s & Y_t \end{pmatrix}$ be invertible when $t = 0$, which it is because its determinant $\frac{dx_0}{ds} F_q - \frac{dy_0}{ds} F_p$ is non-zero.

Finally, to see that $z = z(x, y)$ is a solution of the partial differential equation we check that

$$F(x, y, z(x, y), z_x(x, y), z_y(x, y)) = 0,$$

which involves checking only that $p = x_x$ and $q = z_y$.

However, the argument to establish this conclusion is somewhat roundabout. One defines

$$U = z_t - p x_t - q y_t,$$

$$V = z_s - p x_s - q y_s.$$

If we see these equations as equations for p and q, we can deduce that

$$\begin{pmatrix} x_t & y_t \\ x_s & y_s \end{pmatrix} \begin{pmatrix} p \\ q \end{pmatrix} = \begin{pmatrix} z_t - U \\ z_s - V \end{pmatrix}.$$

We can solve these equations provided that $x_t y_s - x_s y_t \neq 0$, which we have already observed above.

Then we treat the equations

$$0 = z_t - z_x x_t - z_y y_t,$$

$$0 = z_s - z_x x_s - z_y y_s,$$

in the same way, and deduce that

$$\begin{pmatrix} x_t & y_t \\ x_s & y_s \end{pmatrix} \begin{pmatrix} u_x \\ u_y \end{pmatrix} = \begin{pmatrix} z_t \\ z_s \end{pmatrix}.$$

Now we can see that our remaining problem is solved if $U = 0$ and $V = 0$. Happily for us, $V = 0$ is a consequence of the characteristic equations. The requirement on U is more difficult. We regard U and V both as functions of s and t. Then we calculate $\dfrac{\partial U}{\partial s} - \dfrac{\partial V}{\partial t}$ and use the fact that $V = 0$ to deduce that $\dfrac{\partial V}{\partial t} = 0$ to obtain an expression for $\dfrac{\partial U}{\partial s}$. Then, from the equation $F = 0$ we deduce that

$$\frac{\partial U}{\partial s} = -F_u U.$$

We treat this as an ordinary differential equation for U as a function of u for each fixed t and solve it to obtain

$$U(s) = U(0)e^{\int_0^s F_u \, ds}.$$

But $U(0) = 0$ and so $U(s) = 0$ for all s. This concludes the proof.

Note also that nothing more has been required of these functions than that they are at least two times continuously differentiable with respect to the obvious variables.

Appendix D
Green's Theorem and Heat Conduction

Green's theorem proved endlessly useful in the study of the partial differential equations of mathematical physics, as the following examples illustrate.

D.1   Explicit Representations

To proceed further, we come down to two dimensions. The same process applies as it did in three dimensions, except that v is to be infinite like $\frac{1}{2\pi} \ln r$ at a point of D. If the Dirichlet problem is posed for a two-dimensional region D with boundary C then the relation between the Dirichlet problem and Green's problem is given by

$$u(P) = \int_C f(\mathbf{n}.\nabla v)$$

for a suitable function v.

The Dirichlet problem for the unit disc asks for a harmonic function u on the disc that agrees with a given function f on the unit circle. If we take the approach of finding a Green's function then we have to find a function $v(x, y)$.

It helps to be clear about domains and coordinates. We define v on $D \times D$ and write $v(x, y, \xi, \eta)$. We then define

$$v(x, y, \xi, \eta) = \frac{1}{2\pi} \ln \left(\left((x - \xi)^2 + (x - \eta)^2 \right)^{1/2} \right) + h(x, y, \xi, \eta),$$

for some suitable function $h(x, y, \xi, \eta)$ that we have to find, where we require that

$$\nabla^2 h = 0 \quad \text{on } D,$$

and, for v to vanish when $\xi, \eta \in C$ that

© The Author(s), under exclusive license to Springer Nature Switzerland AG 2021
J. Gray, *Change and Variations*, Springer Undergraduate Mathematics Series,
https://doi.org/10.1007/978-3-030-70575-6

$$h(x, y, \xi, \eta) = -\frac{1}{2\pi} \ln \left(\left((x - \xi)^2 + (x - \eta)^2 \right)^{1/2} \right) \quad (\xi, \eta) \in C.$$

So we have set $v(x, y) = \frac{1}{2\pi} \ln r + h(x, y)$, where h is to be harmonic everywhere in the disc and has the prescribed behaviour on the boundary that $h(x, y, \xi, \eta) = -\frac{1}{2\pi} \ln r$, where r is the distance from (x, y) to (ξ, η), when $(\xi, \eta) \in C$.

If we let D be the upper half-plane, so the boundary C is the real axis, then we define $h(x, y, \xi, \eta)$ by looking at the mirror image of the point (x, y) in the real axis, which is the point $(x, -y)$. We now define

$$h(x, y, \xi, \eta) = -\frac{1}{2\pi} \ln r',$$

where r' is the distance from $(x, -y)$ to (ξ, η). This function fails to be harmonic only at $(x, -y)$ which is not in D, and it is equal to $-\frac{1}{2\pi} \ln r$ on the boundary, as required.

If we let D be the unit disc and take $v = \ln r$. The unit normal vector at a point (x, y) on the unit circle is (x, y). The above argument says we have

$$u(0) = \int_{|z|=1} f \mathbf{n}.\nabla v.$$

Here

$$\nabla v = \frac{1}{2\pi} (\partial_x \log r, \partial_y \log r).$$

We find

$$\partial_x \log r = \frac{\partial_x r}{r} = \frac{\partial_x (x^2 + y^2)^{1/2}}{r} = \frac{x}{r^2},$$

and similarly $\partial_y \log r = \frac{y}{r^2}$, so

$$\mathbf{n}.\nabla v = (x, y).\left(\frac{x}{r^2}, \frac{y}{r^2}\right) = \frac{x^2 + y^2}{r^2} = 1,$$

and therefore

$$u(0) = \int_{|z|=1} f \, dz.$$

This expresses the averaging property of harmonic function on a disc: its value at the centre is the average of its values on the boundary circle.

What happens if we move the point where Green's function becomes infinite to somewhere off-centre but still in the disc, say $z = \alpha$, $|\alpha| < 1$? There is a Möbius transformation that maps the unit circle to itself and the point α to the origin,

Fig. D.1 Finding Green's function at an off-centre point

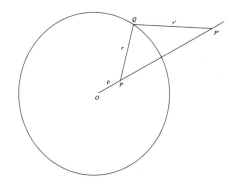

$$\gamma(z) = \frac{az - b}{-\bar{b}z + \bar{a}}$$

where $\alpha = b/a$. The corresponding Green's function is

$$\frac{1}{2\pi} \ln \left| \frac{az - b}{-\bar{b}z + \bar{a}} \right|.$$

There is a classical argument called the method of images that finds the harmonic function explicitly from its boundary values when D is the unit disc. We move the infinite point to $P = (x, y)$ or (ρ, θ) in polar coordinates, define $P' = (\rho', \theta)$ where $\rho\rho' = 1$—the point P' is called the *image* of P because it is obtained by inverting P in the unit circle (Fig. D.1).

Now we look at the point Q with (Cartesian) coordinates (ξ, η) and polar coordinates $(\tilde{\rho}, \tilde{\theta})$. We let O denote the origin, and let r denote the distance PQ and r' denote the distance $P'Q$. Then by applying the cosine rule first to triangle OPQ and then to triangle $OP'Q$ we find that

$$r^2 = \tilde{\rho}^2 + \rho^2 - 2\rho\tilde{\rho}\cos(\tilde{\theta} - \theta),$$

$$r'^2 = \tilde{\rho}^2 + \rho'^2 - 2\rho'\tilde{\rho}\cos(\tilde{\theta} - \theta).$$

When Q is on the boundary of the circle, so $|OQ| = \tilde{\rho} = 1$, we find that

$$\frac{r^2}{r'^2} = \frac{1 + \rho^2 - 2\rho\cos(\tilde{\theta} - \theta)}{1 + 1/\rho^2 - 2/\rho\cos(\tilde{\theta} - \theta)} = \rho^2.$$

So the Green's function for the Laplacian is now

$$v(x, y) = \frac{1}{4\pi} \ln \frac{r}{r'\rho} = \frac{\tilde{\rho}^2 + \rho^2 - 2\rho\tilde{\rho}\cos(\tilde{\theta} - \theta)}{\rho^2\tilde{\rho}^2 + 1 - 2\rho\tilde{\rho}\cos(\tilde{\theta} - \theta)}.$$

In this case,

$$\mathbf{n}.\nabla v = \frac{\partial v}{\partial \tilde{\rho}}$$

evaluated at $\tilde{\rho} = 1$, which is

$$\frac{1}{2\pi} \frac{1 - \rho^2}{1 + \rho^2 - 2\rho\cos(\tilde{\theta} - \theta)}.$$

This gives the Poisson integral formula for the disc:

$$u(\rho, \theta) = \frac{1}{2\pi} \int_0^{2\pi} \frac{1 - \rho^2}{1 + \rho^2 - 2\rho\cos(\tilde{\theta} - \theta)} f(\tilde{\theta}) d\tilde{\theta}.$$

Note that if $\rho = 0$ this collapses to the averaging result that we had before, as it should.

D.1.1 Adjoint Equations

A later generalisation of Green's argument became known as the method of adjoint equations. The idea of introducing the adjoint equation of a given ordinary or partial differential equation is to make the original equation easier to solve, and as was briefly mentioned in Sect. 1.5 it was introduced by Lagrange in the context of ordinary differential equations.

We write, following Sommerfeld ([248], Sect. 10)

$$L(u) = A\frac{\partial^2 u}{\partial x^2} + 2B\frac{\partial^2 u}{\partial x \partial y} + C\frac{\partial^2 u}{\partial y^2} + D\frac{\partial u}{\partial x} + E\frac{\partial u}{\partial y} + Fu = 0,$$

where A, B, C, D, E, F are sufficiently differentiable functions of x and y.

The adjoint equation to $L(u) = 0$ will be another second-order partial differential equation, $M(v) = 0$ such that

$$vL(u) - uM(v) = \frac{\partial X}{\partial x} + \frac{\partial Y}{\partial y},$$

for two functions $X(x, y)$ and $Y(x, y)$ that have also to be found.

It turns out that

$$M(v) = \frac{\partial^2 Av}{\partial x^2} + 2\frac{\partial^2 Bv}{\partial x \partial y} + \frac{\partial^2 Cv}{\partial y^2} - D\frac{\partial v}{\partial x} - E\frac{\partial v}{\partial y} + Fv = 0,$$

and

$$X = A\left(v\frac{\partial u}{\partial x} - u\frac{\partial v}{\partial x}\right) + B\left(v\frac{\partial u}{\partial y} - u\frac{\partial v}{\partial y}\right) + \left(D - \frac{\partial A}{\partial x} - \frac{\partial B}{\partial y}\right),$$

$$Y = B\left(v\frac{\partial u}{\partial x} - u\frac{\partial v}{\partial x}\right) + C\left(v\frac{\partial u}{\partial y} - u\frac{\partial v}{\partial y}\right) + \left(E - \frac{\partial B}{\partial x} - \frac{\partial C}{\partial y}\right).$$

In the last two equations, we can replace (x, A, D) by (y, C, E).

You can check that $L(v)$ is the adjoint of $M(u)$, and that $L(u) = M(u)$ – when the equation $L(u) = 0$ is said to be self-adjoint—if and only if

$$\frac{\partial A}{\partial x} + \frac{\partial B}{\partial y} = D, \text{ and } \frac{\partial B}{\partial x} + \frac{\partial C}{\partial y} = E.$$

In particular, if the coefficients are constant and $D = E = 0$ the equation is self-adjoint.

It is true, but will not be proved here, that if a second-order linear partial differential equation arises as an Euler–Lagrange equation then it is self-adjoint. It is also true that a second-order linear partial differential equation with constant coefficients can be made self-adjoint by multiplying it by a factor of the form $\exp(\sum \lambda x + \mu y)$ unless it is the heat equation.

Let us now integrate $\frac{\partial X}{\partial x} + \frac{\partial Y}{\partial y}$ over a region S with area element $d\sigma$ bounded by a simple closed curve C with line element ds. The theorem of Gauss and Green says that

$$\int_S (vL(u) - uM(v))d\sigma = \int_S \left(\frac{\partial X}{\partial x} + \frac{\partial Y}{\partial y}\right)d\sigma = \int_C (X, Y).\mathbf{n}ds.$$

In the elliptic case in normal form, where $A = C = 1$, $B = 0$, the RHS becomes

$$\int_C (D, E).\mathbf{n}.$$

This generalises the usual Green's theorem of potential theory (the case where also $D = E = 0$), which says that

$$\int_S (v\Delta(u) - u\Delta(v))D\sigma = \int_S \left(\frac{\partial X}{\partial x} + \frac{\partial Y}{\partial y}\right)d\sigma,$$

where

$$\Delta(u) = \frac{\partial^2 u}{\partial x^2} + \frac{\partial^2 u}{\partial y^2}.$$

The importance of the adjoint equation arises from the fact that if u and v are such that

$$L(u) = 0, \quad M(v) = 0$$

then the LHS's above vanish, and the corresponding equations become

$$\int_C (X, Y).\mathbf{n}ds = 0,$$

and

$$\int_S \left(\frac{\partial X}{\partial x} + \frac{\partial Y}{\partial y}\right) d\sigma = \int_C (X, Y).\mathbf{n}ds = 0.$$

This holds provided that u, v, and their derivatives are continuous throughout the region S. If v, say, is discontinuous at a point $Q = (\xi, \eta)$ in S then it is excluded by drawing an arbitrarily small contour K around it, and taking the integral over both s in the positive direction and K in the negative direction.

The most important single case is where the discontinuity of v at Q represents a point source of unit strength, one for which the yield q is the gradient of v. This means that at a radial distance ρ from Q

$$q = \int_K \frac{\partial v}{\partial \rho} ds.$$

If, very close to Q, v depends only on ρ then

$$q = \int_{-\pi}^{\pi} \frac{\partial v}{\partial \rho} \rho d\rho = 2\pi \rho v.$$

So

$$\frac{\partial v}{\partial \rho} = \frac{1}{2\pi\rho}, \quad v = \frac{1}{2\pi} \log \rho + const.$$

as $\rho \to 0$.

So we have

$$v = U \log \rho + V, \quad \rho = \sqrt{(x - \xi)^2 + (y - \eta)^2},$$

where U and V are analytic functions of (x, y) and (ξ, η) such that $U \to \frac{1}{2\pi}$ as $(x, y) \to (\xi, \eta)$.

This function v is called a Green's function or a principal solution of the differential equation $M(v) = 0$. Similarly, the function u is called a Green's function or a principal solution of the differential equation $L(u) = 0$. The functions U and V will be analytic if D, E, and F are analytic.

Next, an account of how this comes about in the context of heat conduction, following (Sommerfeld [248], Sect. 12).

The partial differential equation for heat conduction is (with $y = kt$)

$$L(u) = \frac{\partial^2 u}{\partial x^2} - \frac{\partial u}{\partial y} = 0.$$

It is not self-adjoint; the adjoint equation is

$$M(v) = \frac{\partial^2 v}{\partial x^2} + \frac{\partial v}{\partial y} = 0.$$

Furthermore,

$$X = v\frac{\partial u}{\partial x} - u\frac{\partial v}{\partial x}, \text{ and } Y = -uv.$$

Because x is a space variable but y is a time variable, we choose to consider this only for simple regions bounded by sides parallel to the x or y axes. For such regions, along a side AB that is parallel to the x-axis

$$ds = dx \text{ and } dn = -dy; \quad \cos(n, x) = 0, \cos(n, y) = -1,$$

and so

$$\int_A^B (X\cos(n, x) + Y\cos(n, y))ds = -\int_A^B Ydx.$$

Similarly, along a side CD parallel to the y-axis we obtain

$$\int_C^D (X\cos(n, x) + Y\cos(n, y))ds = \int_B^C Xdy.$$

The general form of Green's theorem then says

$$\int (vL(u) - uM(v))dxdy = \int uvdx + \int \left(v\frac{\partial u}{\partial x} - u\frac{\partial v}{\partial x}\right)dy.$$

We apply this to a heat conductor infinite in both directions and for which the temperature at time $t = 0$ is given as $u = f(x)$. The Fourier integral representation of the function $f(x)$ is

$$f(x) = \frac{1}{2\pi}\int_{-\infty}^{\infty}\left(\int_{-\infty}^{\infty} f(\xi)e^{iw(x-\xi)}d\xi\right)dw.$$

To obtain a solution to the heat equation, we multiply the exponential term by a function $\varphi(y)$ and plug this expression into the equation. We find that we require

$$-w^2\varphi(y) = \frac{d\varphi}{dy},$$

so

$$\varphi(y) = Ce^{-w^2 y}.$$

As we require $\varphi(0) = 1$, we obtain the solution to the heat equation in the form

$$u(x, t) = \frac{1}{2\pi} \int_{-\infty}^{\infty} \left(\int_{-\infty}^{\infty} f(\xi) e^{iw(x-\xi) - w^2 y} d\xi \right) dw.$$

Curiously, the resulting integral converges for a wider class of functions than does the original representation of the function f, and moreover the order of integration is now reversible.

We now write $y = kt$ and obtain the solution in the form

$$u(x, t) = \frac{1}{2\pi} \int_{-\infty}^{\infty} \left(\int_{-\infty}^{\infty} f(\xi) e^{iw(x-\xi) - w^2 kt} d\xi \right) dw.$$

The exponent is of the form $-\alpha w^2 + \beta w$, so we complete the square

$$-\alpha w^2 + \beta w = -\alpha \left(w - \frac{\beta}{2\alpha} \right)^2 + \frac{\beta^2}{4\alpha}.$$

Set $p = w - \beta/2\alpha$, and then

$$\frac{1}{2\pi} \int_{-\infty}^{\infty} e^{-w^2 ky + iw(x-\xi)} dw = \frac{1}{2\pi} e^{-\frac{(x-\xi)^2}{4kt}} \int_{-\infty}^{\infty} e^{-\alpha p^2} dp.$$

Write the RHS as U for the moment.

There is a Laplace transform that was well known in the nineteenth century (and doubtless still is) that says

$$\int_{-\infty}^{\infty} e^{-p^2} dp = \sqrt{\pi},$$

so

$$\int_{-\infty}^{\infty} e^{-\alpha p^2} dp = \sqrt{\frac{\pi}{\alpha}}$$

and

$$U = \frac{1}{\sqrt{4\pi kt}} \exp \left(\frac{-(x-\xi)^2}{4kt} \right).$$

As $t \to 0$ we have $u(x, t) \to f(x)$, so

$$f(x) = \int_{-\infty}^{\infty} f(\xi) U d\xi,$$

and

$$\int_{x-\varepsilon}^{x+\varepsilon} U d\xi = 1.$$

For a source of heat concentrated at the origin we have

$$U = \frac{1}{\sqrt{4\pi kt}} \exp\left(\frac{-x^2}{4kt}\right). \tag{D.1}$$

D.1.2 Boundary Value Problems

We shall suppose that the heat conductor is infinite in both directions and can be represented by the real line.[15] We write

$$u(x, t) = \int_{-\infty}^{\infty} f(\xi)U d\xi, \quad \text{where } U = \frac{1}{\sqrt{4\pi kt}} \exp\left(\frac{-(x - \xi)^2}{4kt}\right).$$

We shall consider two separate boundary conditions. The isothermal one for a temperature distribution $u(0, t)$, where we impose $u = 0$; and the adiabatic one for a given heat flow $G(0, t)$, where we impose $\partial u/\partial x = 0$.

Both these conditions are satisfied if the function f, which is given only for $0 < x < \infty$, is extended to the negative real function either as an even or an odd function, and so by a pure sine or a pure cosine integral. This yields

$$u(x, t) = \int_{0}^{\infty} f(\xi)U(\xi)d\xi \mp \int_{0}^{\infty} f(\xi)U(-\xi)bd\xi.$$

The principal solution $U(\xi)$ becomes

$$U(-\xi) = \frac{1}{\sqrt{4\pi kt}} \exp\left(\frac{-(x + \xi)^2}{4kt}\right)$$

and describes a point source of heat at $x = -\xi$, $t = 0$.
    So we have

$$u(x, t) = \int_{0}^{\infty} f(\xi)G(\xi)d\xi, \quad \text{where } G(\xi) = U(\xi) \mp U(-\xi).$$

The function G is called a Green's function. It has only one pole (or heat source) in the interval $0 < x < \infty$, and it satisfies the adjoint equation as a function of ξ and τ because it is independent of τ and so the change of sign with respect to the τ variable is irrelevant.

    The initial conditions are now to be represented by not a single point source but a continuum of them—first sum over a finite number of point sources of heat and then pass to an integral. They are placed at every $\eta < -\xi$.
    The corresponding function G is given by

---

[15]See Sommerfeld ([248], 66).

$$G = U(\xi) + AU(-\xi) + \int_{-\infty}^{-\xi} a(\eta)U(\eta)d\eta,$$

where the point source of heat contributes an amount A and the continuous source is represented by $a(\eta)d\eta$.

The boundary conditions can the be used to determine A and $a(\eta)$.

Appendix E
Complex Analysis

It is impossible to explain Riemann's contribution to the study of differential equations without describing his approach to the theory of complex analysis, of which he is one of the three creators, alongside Cauchy and Weierstrass. But obviously there is no room here for a proper historical account of the emergence of key ideas on complex function theory. This chapter is therefore a series of glimpses into some of these ideas.

E.1 Harmonic Functions

We have seen that it is an elementary consequence of the Cauchy–Riemann equationsthat the real and imaginary parts of a complex analytic function are harmonic functions. We shall now see that given the real part (say) of a complex function on a simply connected domain the imaginary part is determined up to a constant (we assume the necessary partial derivatives exist and are continuous). Suppose that we are given $u(x, y)$ and required to find $v(x, y)$ such that $u(x, y) + iv(x, y)$ is a complex analytic function. Then we may solve the Cauchy–Riemann equations

$$u_x = v_y \text{ and } u_y = -v_x.$$

For these equations to have a solution v it is necessary that $u_{xx} = -u_{yy}$, because the equations we are trying to solve say that these expressions are each equal to v_{xy}, but this condition is met when the given function u is harmonic.

We now solve the equation $v_x = -u_y$ as follows. We can integrate both sides with respect to x to obtain

$$v = -\int u_y dx,$$

which is single-valued because u is defined on a simply connected domain, so this determines v up to a constant. Then we check that $v_y = u_x$ by differentiating under

© The Author(s), under exclusive license to Springer Nature Switzerland AG 2021   379
J. Gray, *Change and Variations*, Springer Undergraduate Mathematics Series,
https://doi.org/10.1007/978-3-030-70575-6

the integral sign:

$$v_y = -\int u_{yy} dx = \int u_{xx} dx = u_x,$$

as required. The function v is said to be the "harmonic conjugate" of the given function u.

By the time Riemann was writing his Ph.D. thesis in 1850, the basic properties of harmonic functions were well known. For example, the value of a harmonic function $h(x, y)$ at a point (x_0, y_0) in its domain is the average of its values on a disc containing that point (this is little more than a statement of the Cauchy integral theorem, but it was known independently as a theorem in harmonic functions due to Gauss). Because an average value can never also be a largest (or a smallest) value unless all the values are the same, this means that a harmonic function takes its maximal value(s) on the boundary of its domain. This means that if a harmonic function is defined everywhere in the plane it tends to infinity as either x or y tends to infinity. In the context of complex analytic functions, this is Liouville's theorem. Moreover, if two harmonic functions defined on the same domain have the same values on the boundary of the domain, they are equal. For, consider their difference. It is a harmonic function whose boundary values are zero, and by the earlier remark, this means that the difference of the two functions is zero everywhere in the domain, and so the two functions are equal.

E.2   Branch Points and Many-Valued "Functions"

Throughout the nineteenth century many-valued "functions" such as the nth root function, the infinitely many-valued log function, the arcsine function, and others were considered as legitimate functions. Let us consider the square root "function", which assigns to a non-zero complex number its two square roots

$$z \mapsto z^{1/2}, \text{ or } re^{i\theta} \mapsto r^{1/2} e^{i\theta/2} \text{ and } r^{1/2} e^{i(\theta/2+\pi)} = -r^{1/2} e^{i\theta/2}$$

and investigate what happens as the domain variable is moved along a circle around the origin.

For a fixed value of r, as θ goes from 0 to 2π and z goes once the origin on a circle of radius r, the square root that was initially positive goes on a semicircle from $r^{1/2}$ to $-r^{1/2}$, the negative square root, and the root that was initially negative goes on a semicircle from $-r^{1/2}$ to $+r^{1/2}$, the positive square root. The two roots are interchanged in this process.

Mathematicians of nineteenth century would say that the square root function is two-valued, and that it is branched at the point $z = 0$. Each single-valued determination of the function, either of the square roots, is called a branch of the function, and it is necessary to define the domain of these functions carefully, as we discuss below.

The behaviour of the many-valued function $z \mapsto z^\alpha$ is not very different from that of $z \mapsto z^{1/2}$. We set $z = re^{i\theta}$ and observe that

$$re^{i\theta} = re^{i(\theta+2k\pi)},$$

so the values of z^α are all of the form

$$z^\alpha = re^{i(\theta+2k\pi)\alpha} = re^{i(\theta\alpha+2k\pi\alpha)} = re^{i\theta\alpha}e^{2ki\pi\alpha}$$

and are obtained from each other by using a suitable multiple of $e^{2ki\pi\alpha}$. If $\alpha = p/q$ then the function takes q distinct values, and if α is irrational then the function is infinitely many-valued.

The same account holds for the function

$$z \mapsto (z-a)^\alpha,$$

which is branched at the point $z = a$.

Now let $h(z)$ be an analytic function, and suppose that its power series expansion in a neighbourhood of the origin is

$$h(z) = h_0 + h_1 z + h_2 z^2 + \cdots .$$

Because each power of z in this expansion is integral we deduce that the function $h(z)$ is single-valued in a neighbourhood of the origin. It follows that the function

$$z \mapsto z^\alpha h(z)$$

takes as many values as does the function $z \mapsto z^\alpha$. The function $(z-a)^\alpha h(z)$ is said to be branched at the point $z = a$; Riemann said more informally that it behaves like $(z-a)^\alpha$ near the point a.

Conversely, if the function $z \mapsto f(z)$ is not single-valued near the origin one can look for a value of α such that the function

$$z \mapsto z^\alpha f(z)$$

is single-valued in a neighbourhood of the origin.

For future reference, we note that if we differentiate the function $(z-a)^\alpha h(z)$ we find

$$\frac{d}{dz}\left((z-a)^\alpha h(z)\right) = \alpha(z-a)^{\alpha-1}h(z) + (z-a)^\alpha h'(z) = (z-a)^{\alpha-1}(\alpha h(z) + (z-a)h'(z)),$$

so the derived function is branched like $(z-a)^{\alpha-1}$ at $z = a$.

As we shall see, we are free to consider a complex many-valued function that is branched at several points $z = a, b, c, \ldots$.

On the other hand, if we stay away from the branch points, and let the domain variable vary only along curves that do not go round a branch point, we can recover a single-valued function, albeit on a restricted domain. This function will be the restriction of a branch of the many-valued function to the restricted domain. In the main, this is what Cauchy did in developing his theory of complex functions.

For example, if we pick a value of the square root function at the point $z = 4$, say the value $+2$. Then near to $z = 4$ the value of the square root function that we shall choose is, of course, the value near to 2. We can proceed in this way by varying z and assigning the unique value to the square root function that makes our new function continuous, as long as we do not take z on a loop around the origin. One way to do this is to decide not to consider values of the square root function on the negative real axis, the points z for which $z \leq 0$. The domain of the square root function has been restricted to the plane of complex numbers with the negative real axis removed—this is commonly called a cut—and on that domain it is possible to choose a single-valued branch of the function. More complicated many-valued functions could similarly be restricted to other simply connected subsets of the plane of complex numbers and in this way studied, but at the price of introducing an element of arbitrariness into the theory.

More precisely, on the domain $\{re^{i\theta} : -\pi < \theta < \pi\}$ we define the function $re^{i\theta} \mapsto re^{i\theta/2}$. The image of this function is the right-hand half-plane $\{\rho e^{i\varphi} : -\pi/2 < \varphi < \pi/2\}$.

If, however, we cross the negative real axis, the value of the square root function would be multiplied by -1. More precisely, on the domain $\{re^{i\theta} : -\pi < \theta < \pi\}$ we also define the function $re^{i\theta} \mapsto -re^{i\theta/2}$. The image of this function is the left-hand half-plane $\{\rho e^{i\varphi} : \pi/2 < \varphi < 3\pi/2\}$.

We are free to choose either branch in order to assign a value to the square root function on the cut, and it is natural to assign points $re^{i\pi}$ the value $r^{1/2}e^{i\pi/2}$.

We shall refer to this later, so let us say in general that a many-valued function f is locally single valued if it is possible to choose a domain D, such as a disc, that contains no branch points of f, a value of f at a point z_0 in the domain D, and a single-valued function F on the domain D that agrees with the branch of f on D that takes the specified value at the point z_0.

E.3   Analytic Continuation

We have seen that Gauss was clear about the distinction between a complex-valued function of a complex variable and a power series representation of that function. The rule of thumb for a power series representation is that it is valid on an open disc centred at a point P and of radius $r > 0$ that is determined by the fact that there is a point on the boundary of the disc where the function becomes infinite (more precisely, ceases to be defined and analytic).

Thus the power series representation of the function $(1 - z)^{-1}$, which is

$$1 + z + z^2 + \cdots + z^n + \cdots$$

converges on the open disc centre the origin and radius 1, because the function is not defined at the point $z = 1$. The same is true of the function $(1 + z^2)^{-1}$, because it is not defined at the points $z = \pm i$ that lie on the boundary of the unit disc.

Very often one obtains a function by first finding a power series representation of it.[16] If, for example, one has the series

$$1 + z + z^2 + \cdots + z^n + \cdots$$

then one has a function defined on the open unit disc. But it is possible to obtain a power series representation for this function on a disc centred at the point $z = -\frac{1}{2}$ by introducing the new variable $z_1 = z + \frac{1}{2}$, and the radius of convergence of this power series is $3/2$—the distance from $z = -\frac{1}{2}$ to $z = 1$. The disc on which the new power series converges partly lies outside the unit disc, so the function defined by the power series can be extended to a larger domain. In this way, the function can be extended to its maximal domain, which will be the plane of complex numbers with the point $z = 1$ removed, because, of course, the power series are has been obtained from the function $(1 - z)^{-1}$. This process of extending the domain of definition of a function is called analytic continuation. It can be applied to any function defined initially on some open disc by a convergent power series, and in the present context the examples we shall consider will produce many-valued functions defined everywhere except at a finite set of points.[17]

The basic facts about analytic continuation concern what happens if a function is continued in the way just described along a chain of discs, each one overlapping with the one before. For as long as this can be done, one says that the function is obtained by analytic continuation from the original disc. If you wish, you may suppose that the domain variable moves along a closed curve and at each point is the centre of a disc of possibly varying radius which is the disc of convergence of a power series representation of that function. But a question arises when a disc overlaps one much earlier in the chain of discs. In this situation, it can happen that the values of the function on the first and last discs are the same, or that they differ. They differ only if the chain goes round a branch point (which must not lie in any of the discs). The square root function is a case where something can go wrong.

If, on the other hand, a function is extended analytically from the same initial disc along one chain to one disc, D_1, and along another chain to another disc D_2 and the extended functions agree in the intersection of D_1 and D_2, then nothing can be said. It might be, for example, that the function is the square root function but one of the chains goes twice round the branch point at the origin. In this case both chains extend the function to the same value.

---

[16]This is because a good way of solving many a problem involving analytic functions is the method of undetermined coefficients.

[17]Many other things can happen, but that is not our subject.

However, if we add the condition that the first path can be deformed until it agrees with the second path and at no stage does the path as it is deformed pass over a branch point, then the result of analytic continuation along the chains will necessarily be the same. For the purposes of this chapter, let us call this result the deformation principle.

The basic principles of analytic continuation were certainly known to Riemann, but it was Weierstrass who preferred to build a theory of analytic functions this way.

E.4   Liouville's Theorem

There are many ways of distinguishing between a complex analytic function and a map from $\mathbb{R}^2$ to $\mathbb{R}^2$. We have already remarked that a complex analytic map is infinitely complex differentiable, and (as a result of theorems due to Cauchy) it is expressible locally as a convergent power series. As noted above, Riemann was the first to draw attention to a point known earlier to Gauss: at every point where a complex analytic function has a non-zero derivative it is conformal or angle-preserving. But one of the most surprising properties of a complex analytic function was discovered by Joseph Liouville; it was also known to Riemann, but it is not clear on what grounds he believed it.

The theorem says that a complex analytic function that is bounded everywhere, even at infinity, is a constant. One proof uses the Cauchy integral theorem to estimate the coefficients in a power series expansion of the function and to show that all terms but the constant term are arbitrarily small and must therefore be zero. Another proof uses the fact, also proved by Riemann, that the real and imaginary parts of a complex function are harmonic, together with the fact that at any point the value of a harmonic function is the average of the values the function takes on a neighbourhood of the point. This means that a harmonic function can only take a maximum or minimum value on the boundary of its domain, so if that domain is the entire plane including the point at infinity a bounded harmonic function must be constant.

The use of Liouville's theorem is to show that if the quotient of two complex analytic functions is bounded then the functions are complex multiples of each other. So for example, if a polynomial $p(z)$ of degree n and the function 1 are such that the quotient $1/p(z)$ is bounded everywhere then $p(z)$ is a constant. Consider now a non-constant polynomial with no zeros. It is therefore bounded away from zero, and so its quotient is bounded everywhere, and is therefore constant, which is a contradiction. Therefore the polynomial $p(z)$ must have a zero—the fundamental theorem of algebra. In the same way a function that has zeros at the points $a_1, a_2, \ldots, a_m$ and becomes infinite like $1/z$ (simple poles) at the points $b_1, b_2, \ldots, b_m$ and is otherwise neither zero nor infinite (including at infinity) must be a constant multiple of the rational function

$$\frac{(z - a_1)(z - a_2) \cdots (z - a_m)}{(z - b_1)(z - b_2) \cdots (z - b_m)}.$$

Appendix F
Möbius Transformations

Möbius transformations are needed for a look at the work of Schwarz and Poincaré on the hypergeometric equation, which requires a modest amount of complex analysis.

F.1 Möbius Transformations

A Möbius transformation is a map of the complex plane to itself of the form

$$z \mapsto \frac{a\bar{z} + b}{c\bar{z} + d},$$

where a, b, c, d are complex numbers and $ad - bc \neq 0$. We say that the point $z = -\bar{d}/\bar{c}$ goes to ∞ and that ∞ goes to a/c, and strictly speaking we should say that the map is of the extended complex plane to itself.

A proper Möbius transformation is a map of the complex plane of the form

$$z \mapsto \frac{az + b}{cz + d}.$$

A proper Möbius transformation is obtained by following the map $z \mapsto \bar{z}$ with a Möbius transformation, and for most purposes it is enough to work with proper Möbius transformations. I shall drop the work "proper" when it can be inferred from the context.

Möbius transformations have the convenient property that the transformations

$$z \mapsto \frac{az + b}{cz + d} \quad \text{and} \quad z \mapsto \frac{kaz + kb}{kcz + kd}, \quad k \neq 0,$$

are the same, and so the inverse of the Möbius transformation

© The Author(s), under exclusive license to Springer Nature Switzerland AG 2021
J. Gray, *Change and Variations*, Springer Undergraduate Mathematics Series,
https://doi.org/10.1007/978-3-030-70575-6

$$z \mapsto \frac{az+b}{cz+d},$$

is easy to write down: it is, as you should now check,

$$z \mapsto \frac{dz+-b}{-cz+a}.$$

Show that if μ is the Möbius transformation

$$z \mapsto \frac{az+b}{cz+d},$$

and μ' is the Möbius transformation

$$z \mapsto \frac{a'z+b'}{c'z+d'},$$

then the Möbius transformation $\mu\mu'$ (performing μ' first) is

$$z \mapsto \frac{a''z+b''}{c''z+d''},$$

where a'', b'', c'', d'' are given by the matrix product

$$\begin{pmatrix} a'' & b'' \\ c'' & d'' \end{pmatrix} = \begin{pmatrix} a & b \\ c & d \end{pmatrix} \begin{pmatrix} a' & b' \\ c' & d' \end{pmatrix}.$$

This allows us to use the convenient notation for the Möbius transformation

$$z \mapsto \frac{a\bar{z}+b}{c\bar{z}+d},$$

$$z \mapsto A(\bar{z}),$$

where $A = \begin{pmatrix} a & b \\ c & d \end{pmatrix}$. So, writing A' the matrix for μ' in the obvious way, the matrix for $\mu\mu'$ is AA'.

The derivative of the Möbius transformation

$$f(z) = \frac{az+b}{cz+d},$$

is

$$f'(z) = \frac{a(cz+d)-c(az+b)}{(cz+d)^2} = \frac{ad-bc}{(cz+d)^2}.$$

This means that the Möbius transformation is angle-preserving everywhere.

Because a Möbius transformation is determined by the three ratios $a : b : c : d$ it is easy to see that there is a unique Möbius transformation sending any three distinct points to any three distinct points. In particular, the Möbius transformation

$$z \mapsto \frac{az + b}{cz + d}$$

has this effect:

$$0 \mapsto b/d, \quad 1 \mapsto (a + b)/c + d), \quad \infty \mapsto a/c.$$

Show that the only Möbius transformations that map the set of three points $\{0, 1, \infty\}$ to itself are given by

$$z \mapsto z, \frac{1}{z}, 1 - z, \frac{z - 1}{z}, \frac{z}{z - 1}, \text{ and } \frac{1}{1 - z}.$$

This group of transformations is important in the study of Riemann's P-functions.

The equation of the circle centre (a, b) and radius r is

$$x^2 + y^2 - 2ax - 2by + c = 0,$$

where $r^2 = a^2 + b^2 - c$. Show that the equation can be written as

$$z\bar{z} - \bar{\alpha}z - \alpha\bar{z} + c = 0,$$

where $\alpha = a + ib$ and $c = \alpha\bar{\alpha} - r^2$. The equation of the circle can also be written in the suggestive forms

$$z = \frac{\alpha\bar{z} - c}{\bar{z} - \bar{\alpha}}$$

and

$$z = A\bar{z},$$

where $A = \begin{pmatrix} \alpha & -c \\ 1 & -\bar{\alpha} \end{pmatrix}$. Thus, for example, the unit circle, which has equation $z\bar{z} = 1$, can be written in the form

$$z = A\bar{z},$$

where $A = \begin{pmatrix} 0 & 1 \\ 1 & 0 \end{pmatrix}$.

It is sometimes convenient to speak of the circle (α, c), which has radius r given by $r^2 = \alpha\bar{\alpha} - c$.

F.2 Inversion in a Circle

Inversion in a circle of radius r and centre O is the map that sends a point P to the point Q on the half line OP and such that $OP.OQ = r^2$. In particular, inversion in the unit circle with centre at the origin can be written as

$$z \mapsto 1/\bar{z} = \frac{z}{z\bar{z}}.$$

Show, by using a sequence of scalings and translations, that inversion in the circle centre α and radius r is the map

$$z \mapsto \frac{r^2}{\bar{z} - \bar{\alpha}} + \bar{\alpha} = \frac{r^2}{\bar{z} - \bar{\alpha}} + \alpha = \frac{\alpha\bar{z} - c}{\bar{z} - \bar{\alpha}}.$$

We shall be interested in the effect of inversion on lines and circles and on angles between lines and circles, and without loss of generality, we may assume we are inverting in the unit circle, $z\bar{z} = 1$.

Consider the locus defined by $kz\bar{z} - \bar{\alpha}z - \alpha\bar{z} + c = 0$, where $k = 0$ or 1, which defines a circle when $k = 1$ and a straight line when $k = 0$. Show that under inversion, this transforms to the locus $k\frac{1}{\bar{z}}\frac{1}{z} - \bar{\alpha}\frac{1}{\bar{z}} - \alpha\frac{1}{z} + c = 0$, which simplifies to $k - \bar{\alpha}z - \alpha\bar{z} + cz\bar{z} = 0$. This yields four cases:

(1) $k = 1, c \neq 0$: a circle not through the origin goes to a circle not through the origin;
(2) $k = 1, c = 0$: a circle through the origin goes to a straight line not through the origin;
(3) $k = 0, c \neq 0$: a straight line not through the origin maps to a circle through the origin;
(4) $k = 0, c = 0$: a straight line through the origin maps to itself.

In each case, these statements need to be modified to take note of the fact that the origin has been deleted—we shall henceforth assume that this has been done.

Notice that in case (1), which is the case of most interest, we may write the transformed equation as

$$\frac{1}{c} - \frac{\bar{\alpha}}{c}z - \frac{\alpha}{c}\bar{z} + z\bar{z} = 0.$$

This makes it clear that the image circle has centre $\frac{\alpha}{c}$ and radius $\frac{r}{c}$. Deduce that the image of the centre of the original circle does not go to the centre of the transformed circle.

We also need the concept of inversion in a straight line. A straight line has an equation of the form $ax + by = c$, which can be written as

$$\bar{\alpha}z + \alpha\bar{z} = 2c, \text{ or } z = \frac{-\alpha\bar{z} + 2c}{\bar{\alpha}}.$$

Reflection in this line is the map

$$z \mapsto \frac{-\alpha \bar{z} + 2c}{\bar{\alpha}}.$$

For example, the x-axis, $y = 0$, has the equation $z - \bar{z} = 0$ (in this case α is purely imaginary), and reflection in it is given by $z \mapsto \bar{z}$.

Find the angle between two circles by applying the cosine rule to the triangle formed by their centres and the relevant point of intersection. If the circles are (α, c) and (α', c') with radii r and r', respectively, show that the distance between their centres is $|\alpha - \alpha'|^2$ and that the cosine of the angle between their radii is

$$\frac{\alpha \bar{\alpha}' + \bar{\alpha} \alpha' - (c + c')}{2rr'}.$$

(a) Deduce that the circles are perpendicular if and only if

$$\alpha \bar{\alpha}' + \bar{\alpha} \alpha' = c + c'. \tag{F.1}$$

(b) Deduce that a circle (α', c') is perpendicular to the unit circle if and only if $c' = 1$.

Consider the circles (α, c) and (α', c'), which we assume intersect, with radii r and r', respectively. Show that inversion is angle-preserving (up to sign, so strictly speaking one should say angle reversing) by showing that inversion in the unit circle sends them to the circles $(\alpha/c, 1/c)$ and $(\alpha'/c', 1/c')$, and that their radii are likewise transformed to r/c and r'/c', respectively. Deduce that the angle between the transformed circles is the same as the angle between the original circles, and so inversion is angle preserving.

Because inversion is a Möbius transformation, and Möbius transformations are angle-preserving, inversion in the circle (α, c) maps the circle (α', c') to itself if and only if the circles are at right angles.

We now connect proper Möbius transformations and inversions by showing that every proper Möbius transformation is a product (not in a unique way) of two inversions.

We already know that there is a unique Möbius transformation that maps the points a, b, c in that order to the points $0, 1, \infty$ in that order. So it is enough to find a product of two inversions that has the same effect. Consider the inversion that maps a to 0 and c to ∞ given by

$$A = \begin{pmatrix} \alpha & -c \\ 1 & -\bar{\alpha} \end{pmatrix} = \begin{pmatrix} k/a & -k \\ 1 & -k/\bar{a} \end{pmatrix}.$$

It maps b to b', say. Now we consider the inversion that maps $(0, b', \infty)$ to $(0, 1, \infty)$ given by $z \mapsto z/b'$. The composite of these two has the required effect.

Draw a picture of the map

$$z \mapsto e^{2\pi/3}z.$$

Hint: its fixed points are $z = 0$ and $z = \infty$.

Now draw a picture of the map

$$z \mapsto z', \quad \frac{z' - \alpha}{z' - \beta} = e^{2\pi i/3} \left(\frac{z - \alpha}{z - \beta} \right).$$

Hint: what are the two fixed points of this map?

Can you conjugate the second map into the first by a map that sends α to 0 and β to ∞?

F.2.1 Maps of the Unit Disc to Itself

A map of the unit disc to itself necessarily maps the unit circle to the unit circle, so we restrict our attention to Möbius transformations. As we have seen, a circle has an equation of the form $z = A(\bar{z})$. A Möbius transformation can be written as

$$z' = M(\bar{z}),$$

or $z = \overline{M^{-1}(z')}$. So the equation of the image circle under this Möbius transformation is

$$\overline{M^{-1}(z')} = A M^{-1}(z').$$

So the circle is mapped to itself if

$$M A^{-1} \overline{M^{-1}} = k A$$

for an arbitrary non-zero complex number k.

Applied to the unit circle, this says, in terms of the components of A, that

$$\bar{a} = kd \quad \text{and} \quad \bar{b} = kc.$$

Therefore, the Möbius transformations mapping the unit circle to itself are of the form

$$z \mapsto \frac{az + b}{\bar{b}z + \bar{a}}.$$

The real axis can be regarded as the circle with equation $z = I\bar{z}$, where I is the identity matrix, and the same argument shows that a Möbius transformation maps the upper half-plane to itself if and only if its entries are all real and its determinant is positive.

F.2.1.1   Coaxial Circles

Because a Möbius transformation can map any three distinct points to any three distinct points it can map any circle (or straight line) to any circle (or straight line). So a study of one-parameter family of circles will often apply to circles that are somehow tied to two points, as is the case with what are called coaxial circles.

Coaxial circles come in two main families. The first family consists of all the circles (and straight lines) through two given points. We shall take the points to be $z = \pm 1$, so the circles have their centres on the y-axis and α is purely imaginary ($\alpha + \bar{\alpha} = 0$). The condition that the circles pass through the points ± 1 forces $c = -1$. So circles in the first coaxial family have equations of the form

$$x^2 + y^2 - 2by - 1 = 0, \quad \alpha = ib, \ c = -1. \tag{F.2}$$

The second family seems artificial at first sight. We start from the observation that if

$$S = x^2 + y^2 - 2ax - 2by + c = 0$$

and

$$S' = x^2 + y^2 - 2a'x - 2b'y + c' = 0$$

are the equations of two circles, then the equation

$$\lambda S + \mu S' = (\lambda + \mu)(x^2 + y^2) + 2(\lambda a + \mu a')x + 2(\lambda b + \mu b')y + (\lambda c + \mu c') = 0$$

is also the equation of a circle.

We now regard the points $z = -1$ and $z = 1$ as circles of zero radius, and consider what circles we get by the above routine. The equations of the point circles, as they are called, are

$$S = x^2 + y^2 + 2x + 1 = 0$$

and

$$S = x^2 + y^2 - 2x + 1 = 0.$$

So the family of circles that we obtain has equations of the form

$$\lambda S + \mu S' = (\lambda + \mu)(x^2 + y^2) + 2(\lambda - \mu)x + (\lambda + \mu) = 0,$$

which we write in the form

$$x^2 + y^2 + 2\frac{\lambda - \mu}{\lambda + \mu}x + 1,$$

or, setting $\dfrac{\lambda - \mu}{\lambda + \mu} = a$, as

$$x^2 + y^2 + 2ax + 1 = 0, \quad \alpha' = a, \ c' = 1. \tag{F.3}$$

If we now apply the rule for determining if two circles meet at right angles, (F.1), we find that

$$\alpha\bar{\alpha}' + \bar{\alpha}\alpha' = iba - iba = 0 \text{ and } c + c' = 0,$$

so every member of the first coaxial family is perpendicular to every member of the second coaxial family.

A striking version of this result is obtained by considering, as we may, all the circles and straight lines through the points $z = 0$ and $z = \infty$. The first family of coaxial circles in this case consists of all lines through the origin, and the second of all the circles whose centres are at the origin. Because the first picture can be conjugated into the second, they are equivalent for the purposes of inversion.

There is a third family of coaxial circles, which is obtained by starting with two coincident points, but we shall not need it.

Appendix G
Lipschitz and Picard

In 1877 the German mathematician Rudolf Lipschitz published a paper [190], in which he observed that the question of the existence of solutions to a system of ordinary differential equations had been solved in the complex analytic case by Weierstrass [266] and Briot and Bouquet ([24], 49), and solutions shown to exist at least on some suitable domain. The method first obtained a formal power series that "solves" the equation, and then shows that on some domain around the initial point the series converges. However, there was no proof that a system of ordinary differential equations can be solved in the real case, and this was a gap he proposed to fill.

The system of equations for functions $y^1, y^2, \ldots y^n$ has the form

$$\frac{dy^\alpha}{dx} = f^\alpha(x, y^1, y^2, \ldots y^n) \quad (\alpha = 1, 2 \ldots n).$$

Lipschitz assumed that the functions f^α are defined and continuous on some domain G where they are bounded above by some given quantity. Furthermore, he assumed that

$$|f^\alpha(h, k^1 \ldots k^n) - f^\alpha(h, l^1 \ldots l^n)| < c^{\alpha 1}|k^1 - l^1| + \cdots + c^{\alpha n}|k^n - l^n|, \quad \text{(G.1)}$$

where the quantities $c^{\alpha\beta}$ are positive constants. The initial conditions are that when $x = x_0$ $y^\alpha = y_0^\alpha$, and the point $(x_0, y_0^1, \ldots, y_0^n)$ lies (as we would say) in the interior of G, so that there are positive quantities a_0, b_0^α such that if the point $(x, y^1, \ldots y^n)$ satisfies the inequalities

$$|x - x_0| \leqq a_0, \; |y^\alpha - y_0^\alpha| \leqq b_0^\alpha,$$

then it lies in G.

Lipschitz was then able to show the existence of a domain H lying entirely in G such that there is a system of functions $y^1, y^2, \ldots y^n$ that satisfies initial conditions at x_0 and lies inside H for $|x - x_0| < A_0$ for some $A_0 < a_0$. Lipschitz then divided

© The Author(s), under exclusive license to Springer Nature Switzerland AG 2021
J. Gray, *Change and Variations*, Springer Undergraduate Mathematics Series,
https://doi.org/10.1007/978-3-030-70575-6

the interval $[x_0, x_0 + A_0]$ into p equal pieces (the interval $[x_0 - A_0, x_0]$ is dealt with similarly). In each of these intervals he found quantities η_j^α such that

$$\eta_1^\alpha - y_0^\alpha = f^\alpha(x_0, y_0^1 \dots y_0^n)(x_1 - x_0) \quad \alpha = 1, 2 \dots n$$

and

$$\eta_{a+1}^\alpha - \eta_a^\alpha = f^\alpha(x_a, \eta_a^1 \dots \eta_a^n)(x_{a+1} - x_a) \quad \alpha = 1, 2 \dots n,$$

where $a = 1, 2 \dots p - 1$. The inequality (G.1) shows that all these points lie in H. Lipschitz then proved that as a finer and finer subdivision is produced, the variables tend to limits that establish the existence of solutions to the system of ordinary differential equations.

The novelty of Lipschitz's method is partly that it applies to systems of ordinary differential equations, and partly that inequality (G.1) is weaker than Cauchy's conditions—it is, indeed, the famous Lipschitz condition. That said, it is clear that Lipschitz did not know of Cauchy's earlier paper, which he never mentioned.

G.1   Picard's Method

Let us first take a single first-order equation[18]

$$\frac{dy}{dx} = F(x, y),$$

then, setting $y = y_0$ when $x = x_0$, one can establish the fundamental existence theorem for this equation. To this end, consider the equations

$$\frac{dy_1}{dx} = F(x, y_0),$$

$$\frac{dy_2}{dx} = F(x, y_1),$$

$$\dots$$

$$\frac{dy_n}{dx} = F(x, y_{n-1}),$$

effecting each quadrature[19] in such a way that for $x = x_0$ one has $y = y_0$. The problem is to prove that, as $n \to \infty$ y_n tends to a limit y which represents the desired integral provided that x remains in the neighborhood of x_0. We assume that the function $F(x, y)$ is continuous and defined for values of x and y between $x_0 - a$ and $x_0 + a$ on the one hand and $y_0 - b$ and $y_0 + b$ on the other; moreover, that one can determine a positive constant k such that

$$|F(x, y_2) - F(x, y_1) < k|y_2 - y_1|;$$

we also assume that the function and the variables are real.

---

[18]This is a lightly corrected version of the translation in the Birkhoff *Source Book*, 250–251.
[19]Picard here chooses the constants of integration.

Let M be the maximum modulus of $F(x, y)|$ when x and y remain between the indicated limits. One will have

$$y_1 = \int_{x_0}^{x_1} F(x, y_0)dx + y_0.$$

Let ρ be a quantity at most equal to a: y_1 will stay within the desired limits if $M\rho < b$, and it is evident that the same will be true for $y_2, \ldots, y_n$. Letting δ denote a quantity at most equal to ρ, we will suppose that x remains between $x_0 - \delta$ and $x_0 + \delta$ We then have, on putting $v_n - y_{n-1} = z_n$

$$\frac{dz_1}{dx} = F(x, y_0),$$

$$\frac{dz_2}{dx} = F(x, y_1) - F(x, y_0),$$

$$\cdots$$

$$\frac{dz_n}{dx} = F(x, y_{n-1}) - F(x, y_{n-2}),$$

and all the z vanish at $x = x_0$. One has $|z_1| < M\delta, |z_2| < kM\delta^2, |z_3| < k^2 M\delta^3$ and generally,

$$|z_n| < M\delta(k\delta)^{n-1}.$$

Hence, writing

$$y_n = y_0 + z_1 + z_2 + \cdots + z_n,$$

one sees that y_n tends to a limit if $k\delta < 1$. As a decreasing geometric progression, the series

$$y_n = y_0 + z_1 + z_2 + \cdots + z_n, + \cdots$$

will be convergent. Thus y_n converges to a limit y when x remains between $x_0 - \delta$ and $x_0 + \delta$, δ being the smallest of the quantities $a, b/M, 1/k$. In this interval, y evidently represents a continuous function of x. Thus one also has

$$y_n = \int_{x_0}^{x} F(x, y_{n-1})dx + y_0,$$

and, as y_n and y_{n-1} tend to y, it follows that

$$y = \int_{x_0}^{x} F(x, y)dx + y_0,$$

and hence $dy/dx = F(x, y)$; that, is, the limit y satisfies the differential equation. Thus, the existence of the solution has been established. One can evidently employ the same type of proof if F is an analytic function of the complex variables z and w.

Appendix H
The Assessment

H.1 Introduction

In any historical or reflective essay it's always good to push for more evidence and better arguments. If you want to claim that a book is important because it marks a significant advance, ask yourself: What advance? Why was that important? How does that book do it? When you have answered those questions, ask the next round of questions, such as: What was known before? Who said it was important? If, say, the book displays an improved use of the calculus then what was that improvement? Was it a new technique? An old technique in a new application? And so on.

As for evidence, quotes always help, such as, in the case of the first essay, Euler's or Clairaut's comments on how difficult the *Principia* is to read, or Euler's remark in his *Mechanica* to the effect that even a slight change from one problem to the next can produce great difficulties.

H.2 Assessment 1

Set at the end of week 3, to be handed in at the end of week 4, and returned to the students at the end of week 5.

Question 1 Imagine you are British Professor of mathematics in about the year 1770 who is recommending a good student to spend a year studying mathematics with either Euler or Lagrange. Explain to him or her:

EITHER In what ways is Euler's theory of mechanics an improvement on Newton's *Principia*.

OR What is involved in the study of partial differential equations.

© The Author(s), under exclusive license to Springer Nature Switzerland AG 2021
J. Gray, *Change and Variations*, Springer Undergraduate Mathematics Series,
https://doi.org/10.1007/978-3-030-70575-6

Your answer should describe what has been taken to be important and why is the topic you select.

This question is to be answered in not more than 500 words, which is a single side of A4 in 10-point print, one and a half line spaced.
Please do not use a smaller font size.
Leave room for me to scribble comments.
I will NOT turn over the page—your answer must be on one side of a piece of A4.
Contact me directly or by e-mail if you cannot comply with these requirements.

Advice Think hard about what is the significance of what you report upon. In particular, do not diminish Newton's remarkable achievements in the *Principia*.
When you are offering an opinion or judgement (as in "How important was . . ." or "why was . . . .") give a brief argument in support of your opinion.
Distinguish between contemporary criticisms of someone's work and your own judgements. Don't be afraid of offering your own judgement: you don't have to understand everything you've read or heard, but do not say anything from the perspective of the present day that could not have been said in 1770. Think about what was picked up, what was missing in the actual reception of these ideas.
You do not have very many words, so anything you say about people must be essential—think of the mathematical styles of Euler and Lagrange, not their personalities.
No need to mention anything of a family or personal nature—imagine you've written all that good stuff elsewhere in the letter.

You may find it helpful to write notes for the essay first. Try to use not more than 150 words and observe the following rules:
Each note should be one sentence long and should contain exactly one idea.
The sentences should be organised in groups according to topic.
The sentences, and the topics, should be arranged in a sensible order.
When you have finished, you should be confident that you can write an essay of the required length in which the topics come in this order.

H.3   Assessment 2

EITHER Choose one of the following, and write an essay based on it that demonstrates some understanding of the mathematics, and situates the people and the ideas in a historical context.
Spend roughly three pages describing the most important features of the text, and a page incorporating your analysis into an account of its importance when it was published.

- Cauchy: Note on the integration of first-order partial differential equations in any number of variables (see the translation in Sect. 31.1).
- Thomson and Stokes on the telegraphist's equation.

OR write an article on the hypergeometric equation, covering the work of Gauss, Riemann, Schwarz, and Poincaré.

OR write an article on the works of Schwarz mentioned in the course.

[The course website had pdf copies of all of these essays.]

This question is to be answered in not more than 2200 words, which is four sides of A4 in 11-point print, one and a half line spaced, (or up to four and a half pages in TeX).

Please do not use a smaller font size.

A further page may be used for diagrams, and a *modest* number of extra words (I won't count them).

Bibliographical information should also be given at the end—I won't count the words.

Leave room for me to scribble comments.

Comments on these passages will be found below.

Advice on Choosing an Extract

In order to choose the text you intend to work on, I suggest that you read all the texts quickly over once and find (at least) one you want to proceed with. None of them are altogether easy to read. Let the obscure bits wash over you and wait for something more comprehensible to turn up. You may find that an offending paragraph has an easier second part, or that it is followed by an easier one. Try to form some sense of what the text is about. Terminology can be unclear. List the terms you don't know the meaning of and e-mail me if there's a problem.

Once you have chosen a text or set of texts, a good general strategy would be to look quickly at each of its sections and write down what is claimed in each of them without at this stage worrying about how anything was proved. This will give you a skeleton to work on and enable you to see the general argument.

Grapple with as much of one of these texts as you can. If parts are too hard, be sure that you need to understand and describe them—you may or may not. You may make a modest use of footnotes to alert me that you did not understand something.

The essays by Thomson[20] can be found in the Digital Mathematics Library and the Internet Archive on the web, which send you to Gallica, the site of the Bibliothèque Nationale in Paris. You may find these easier to read than a printed copy, but the graph in Thomson's paper is not at all clear. The copy here is from the original publication in the *Proceedings of the Royal Society* for 1855.[21]

---

[20] Enter http://gallica.bnf.fr/ark:/12148/bpt6k95119q for Thomson.

[21] Enter http://rspl.royalsocietypublishing.org/content/7/382.full.pdf+html.

Advice on writing your essay The main point of this exercise is to put you in the situation of a mathematician (student or professional) who has just studied a mathematical topic (one whose history we are studying in this course) and for you to show that you understand the mathematics and its importance.

For some of the extracts, you may well want to say that the author's reasoning is odd, even wrong; if you do so, prove your claim.

It is better to demonstrate a real understanding of a piece of the mathematics than a superficial understanding of all of the extract.

Your secondary task is to say something historical about the extract and its author that uses your analysis of the extract to establish its importance in the context of its time.

You may stray beyond the Lecture Notes and draw on information accessible in a good library (for example, the *Dictionary of Scientific Biography* and standard histories of mathematics such as Kline's) or on the Web (keep a critical edge), but maximum marks are available for information entirely drawn from the Notes.

A comment on How the Assignment Will Be Marked

I am looking for well-written essays, so an extra mark will be given for essays that are well organised and literate. So, for example, a coherent account that is mathematically correct and insightful but presented in ungrammatical prose will get one mark less than the same account in grammatical prose. If you want this mark, avoid English that is too conversational, flippant, or childish. Address to impress!

If you want to get marks for good writing in your final essay but aren't sure how, let me know. I can also look (but only quickly) at specific requests for help, brief outlines, and the like.

H.3.1 *Cauchy*

Two things are required here.

- I want you to grapple with Cauchy's argument and to compare it with Monge's account and with a modern account, such as the one in the appendix that forms Chap. C.
- I want you to explain the crucial difference between the (quasi)-linear case and the general case.
- The hardest part of the paper is seeing how Cauchy handled the initial conditions. He succeeds, but with unfortunate notation and less clarity than one would like. You may find it easier to work back from the modern account (see Appendix C).
- You may find it helpful to use the following partial differential equation as your worked example if you give one

$$F(x, y, z, p, q) = xp^2 + yq^2 - 2z = 0,$$

with initial conditions that along the initial curve

$$(x(0, s), y(0, s), z(0, s), p(0, s), q(0, s)) = (1 + s, 1 - s, s^2, s, -s), \quad 1/4 < s < 1.$$

Or you may prefer to be clear what is involved in solving this equation, and use your knowledge to assess the ideas of Cauchy. Do not attempt to reach the complete solution in parametric form $(x(s, t) =, \text{etc.})$ but indicate the integrals that would have to be evaluated, and assuming that they have evaluated indicate how the initial conditions are used.

Ideally, you will finish up understanding the subject of first-order partial differential equations much better as a result.

H.3.2 Thomson

These are some of the actual documents that led to the creation of the first successful trans-Atlantic telegraph, so they are genuinely applied mathematics, and I thought it might be interesting for you to work out some of the thinking behind it. So perhaps we want a greater sense of the difficulties and the uncertainties involved in the work as well of the mathematics that was used, and an indication of what was good about it.

H.3.3 The Hypergeometric Equation

I suggest that this essay has an introduction (where you write down the hypergeometric equation and comment on its key properties) and a conclusion, and in between one page on what Gauss did and rather more on the contributions of Riemann, Schwarz, and Poincaré.

For Gauss, state what he discovered about solutions of the hypergeometric equation in the neighbourhood of the points $z = 0, 1, \infty$, and explain the distinction he drew between a (possibly many-valued) function and its power series expansions.

For Riemann, explain what a P-function is locally by definition, and what the key properties of such functions are. Compare this with the usual situation for solutions of a linear second-order ordinary differential equation, and indicate the key steps in Riemann's argument that his function is the general solution of the hypergeometric equation studied by Gauss. (You can read Riemann's paper if you are comfortable with the idea of analytic continuation around a branch point, or willing to become so. I have omitted Section 5 of the paper, which dealt with a technical point not needed for a sufficient appreciation.)

For Poincaré, use Schwarz's work to explain how special cases of the hyper-geometric equation lead to triangles with particularly simple properties, and then indicate how non-Euclidean geometry naturally enters the story (you may take the case where the branch points have orders 2, 3, and 7—very attractive figures for this case are on the web, and see also Fig. 16.3).

H.3.4 Schwarz

His work is strikingly coherent and this essay calls for a detailed analysis of both the texts and their context. Your essay should discuss his alternating method, explain how it is connected to the so-called Schwarz–Christoffel theorem, and why mapping the half-plane to a triangle, to a square, and to a general quadrilateral (each with vertices specified in advance) are problems of increasing difficulty and, if possible, show how to solve them.

[I note here that students can also read Green's "An Essay on the application of mathematical analysis to the theories of electricity and magnetism", which is reprinted in his *Mathematical Papers* (pp. 356–374), consult the Digital Mathematics Library and the Internet Archive on the web. There are two proofs (Sects. 4 and 5) where Green slips between mathematics and physics; it is good to bring them out, to make sense of Sect. 6, and then to compare his paper with the ideas of, for example, Gauss or Dirichlet.]

H.4 Assessment 3

The history of partial differential equations in the nineteenth century belongs to applied mathematics, not pure mathematics. To what extent do you agree, and why?

H.4.1 Advice

Claims like this can have several different kinds of answers. Clearly it depends on what is meant by such terms as *pure* and *applied* mathematics, about which there is legitimate disagreement.

You might reply, for example, that the claim

- is clearly true.
- is clearly false.
- is true in certain respects but not in others, say because the story is better seen as some mix of pure and applied.

- is only true because what happened is some mix of pure and applied mathematics, but it's better seen as simply mathematics underneath.
- is silly, perhaps because you don't accept the terms of the debate, and find the terms pure and applied are superficial (they may mislead) and it's all simply mathematics.

Therefore, it is essential to have an opinion, to state it clearly at the start, to argue for it, to address such counter-arguments as seem to you to have merit, and to reach a substantial conclusion.

To organise your thoughts, you should think about the aims, methods, and results of several mathematicians of the period, and about how they might have answered the question. Think also about the significant theoretical advances made in the study of partial differential equations in the nineteenth century, as well as the significant problems that were solved and applications that were made.

Your essay should respond to the work of most of the leading mathematicians mentioned in the course. The nineteenth century is here defined to cover the period discussed from Lecture 10 to the end of the course.

When you are ready to start writing, structure your essay so that it is easy to appreciate.

- The Introduction should be very brief, but indicate the key points that you will make in your essay. In particular, state the opinion that you are going to argue for.
- You should decide what you mean by such terms as *pure* mathematics, *applied* mathematics, *geometry*, and even *physics*. You may find it helpful to define what you mean by these terms early in your essay.
- The body of your essay should be rich but clear.
- Every paragraph must support your argument, even the ones where you are showing why your evidence doesn't support a different opinion.
- Be prepared to defend a complicated position (the first and second halves of the nineteenth century were different, for example, if you think they were) or an extreme position if you have one.
- Your answer is a measure of the extent of your agreement or disagreement, it is not a string of facts about partial differential equations.
- Facts about partial differential equations are grist to your agreement or disagreement and must be presented clearly in their own right and as supporting that agreement or disagreement.
- If you are arguing for a complicated position, it might be a good idea to give a paragraph to each single aspect of it.
- If you are arguing for a simple position (such as "clearly true", "clearly false", or "silly") be sure you explain why and offer rebuttals of opposing positions.
- The Conclusion should restate the position of the Introduction, but in a way that recalls subtleties and complications acknowledged in the body of the essay.

You have to decide what is right for you, state that position, and defend it like a lawyer. Channel your favourite court room drama, call your expert witnesses. Give clear references to all the sources you use, so that they can be checked; for web-based sources give me the URL.

Full marks are available for essays that draw only on course material, but you are welcome to move beyond.

H.4.2 How the Essays Will Be Marked

To obtain a mark appropriate to a **First**:
Quality of argument: The argument is convincing, supported with relevant facts, and well organised.
The judgements reached are, when necessary, subtle, and balanced.
The coverage is broad—there are no potentially damaging omissions.
There are no unnecessary digressions.
Ideally, and without being cranky, the essay is original (in its emphases, or its conclusions).
Good use of quotations.
Accuracy: The historical facts are indeed correct and to the point.
The mathematics is correct, clear, and relevant.
The written English, as a piece of prose, is well written, and of the right length.
And remember that extra mark for essays that are genuinely well written.

Upper Second:
Falls below the above in one or two significant ways.

Lower Second:
Falls below what is required for an Upper Second in one or two significant ways. An unconvincing or desultory argument.

Third:
Shows a bare knowledge of the topic, but is poorly organised and/or at times inaccurate.

Fail:
Does not demonstrate a knowledge of the topic.

Borderline distinctions It can be hard to tell a First from a good Upper Second. Roughly speaking, a First-class essay is something you (and I!) should be proud of and you (or I) could put with confidence in front of anyone, whereas varieties of Second go to good, and even very good, work. On any borderline, a good original point can push you up, a significant error can push you down. At the other end, a Third says "Yes, you know some things you didn't know before but only just enough" and a Lower Second says that you are either generally, if intangibly, better than that, or at some identifiable point clearly better than that. A Fail mark goes to an essay that doesn't do more than recycle facts but lacks coherence or an argument, or fails to address the question of importance.

Plagiarism The taking of information and arguments from sources you do not acknowledge is theft. It will result in a Fail.

References

1. Archibald, T. 1996. From Attraction Theory to Existence Proofs: The evolution of potential-theoretic methods in the study of boundary-value problems, 1860–1890, *Revue d'Histoire des mathématiques* 2, 67–93.
2. Ampère, A.-M. 1826. *Mémoire sur la théorie mathématique des phénomènes électrodynamiques, uniquement déduite de l'expérience,* extract from *Mémoires de l'Académie Royale des Sciences de l'Institut de France,* Vol. 6, 175–388, 1823, Paris, Firmin Didot.
3. Barrow-Green, J.E. 1997. *Poincaré and the Three Body Problem,* American and London Mathematical Societies, HMath 11.
4. Barrow-Green, J.E., J.J. Gray, and R. Wilson, 2019. *The History of Mathematics: A Source-Based Approach,* Vol. 1, American Mathematical Society, MAA Press, and Vol. 2, 2021.
5. Beltrami, E. 1868. Saggio di interpretazione della geometria non–euclidea, *Giornale di Matematiche* 6, 285–315, in *Opere Matematiche* 1, 374–405. English trans. J. Stillwell, *Sources of Hyperbolic Geometry,* American and London Mathematical Societies, HMath 10.
6. Beltrami, E. 1868. Teoria fondamentale degli spazii di curvatura costante, *Annali di Matematiche* (2) 2, 232–255, in *Opere Matematiche* 1, 406–430.
7. Bernoulli, D. 1738. *Hydrodynamica,* Basel.
8. Bernoulli, Johann. 1691. Solutio problematis funicularii, *Acta Eruditorum,* 274–276, in *Opera Johannis Bernoulli* I, 48–51.
9. Bernoulli, Johann. 1696. Problema novum ad cujus solutionem Mathematici invitantur, *Acta Eruditorum,* 269, in *Opera Johannis Bernoulli* I, 161.
10. Bernoulli, Johann. 1697. Curvatura radii in diaphanis non uniformibus, etc. *Acta Eruditorum* 201–211, in *Opera Johannis Bernoulli* I, 187–193.
11. Bernoulli, Johann. 1702. Solution d'un problème concernant le calcul intégral, Groningen, in *Opera Johannis Bernoulli* I, 393–397.
12. Bernoulli, Johann. 1706. Problema mechanico-geometricum de linea celerrimi descensus, in *Opera Johannis Bernoulli* I, 166–169.
13. Bernoulli, Johann. 1742. *Hydraulica.*
14. Bernstein, S. 1905. Sur la nature analytique des solutions des équations aux derivées partielles du second ordre, *Mathematische Annalen* 59, 20–76.
15. Bernstein, S. 1906. Sur la déformation des surfaces. *Mathematische Annalen* 60, 434–436.
16. Biermann, K.-R. 1973. *Die Mathematik und ihre Dozenten an der Berliner Universität 1810–1920: Stationen auf dem Wege eines mathematischen Zentrums von Weltgeltung,* Akademie-Verlag, Berlin.

© The Author(s), under exclusive license to Springer Nature Switzerland AG 2021
J. Gray, *Change and Variations,* Springer Undergraduate Mathematics Series,
https://doi.org/10.1007/978-3-030-70575-6

17. Birkhoff, G. 1973. *A Source Book in Classical Analysis*, Harvard U.P.
18. Bois-Reymond, P. 1889. Ueber lineare partielle Differentialgleichungen zweiter Ordnung, *Journal für die reine und angewandte Mathematik* 104, 241–301.
19. Bolza, O. 1904. *Lectures on the Calculus of Variations*, University of Chicago Press, Dover repr.
20. Bonnet O. 1860. Mémoire sur l'emploi d'un nouveau systeme de variables dans l'étude des propriétés des surfaces courbes, *Journal de mathématique*, (2) 5, 153–266.
21. Bottazzini, U. 1986. *The Higher Calculus: A History of Real and Complex Analysis from Euler to Weierstrass*, Springer.
22. Bottazzini, U. and J.J. Gray. 2013. *Hidden Harmony – Geometric Fantasies: The Rise of Complex Function Theory*, Springer.
23. Briot, Ch. and J.-C. Bouquet. 1856. Étude des fonctions d'une variable imaginaire. *Journal de l'École Polytechnique* 21, 85–132; 21, 133–198; and 21, 199–254.
24. Briot, Ch. and J.-C. Bouquet. 1859. *Théorie des fonctions doublement périodiques et, en particulier, des fonctions elliptiques*, Paris, Mallet-Bachelier.
25. Buchwald, J.Z. 1985. *From Maxwell to Microphysics*, Chicago U.P.
26. Bukowski, J. 2008. Christiaan Huygens and the Problem of the Hanging Chain, *The College Mathematical Journal* 39, 1–11.
27. Cannell, M. 1993. *George Green, Mathematician and Physicist, 1793–1841. The Background to his Life and Work*, The Athlone Press.
28. Cannon, J.T. and S. Dostrovsky. 1981. *The Evolution of Dynamics: Vibration Theory from 1687 to 1742*, Springer.
29. Capobianco, G., M.R. Enea, and G. Ferraro. 2013. Geometry and analysis in Euler's integral calculus, *Archive for History of Exact Sciences* 71, 1–38.
30. Carathéodory, C. (ed.) 1914. *Mathematische Abhandlungen Hermann Amandus Schwarz, zu seinem funfzigjähren Doktorjubiläum*, Springer.
31. Carathéodory, C. 1965/1967. *Calculus of variations and partial differential equations of the first order*, 2nd. revised English edn., Chelsea.
32. Carlson, J., A. Jaffe, and A. Wiles (eds.) 2006. *The Millennium Prize Problems*, AMS and Clay Mathematics Institute.
33. Cartwright, M.L. 1965. Jacques Hadamard. 1865–1963. *Biographical Memoirs of Fellows of the Royal Society* 11, 75–99.
34. Cauchy, A.-L. 1819. Note sur l'intégrations des équations aux différences partielles du premier ordre à un nombre quelconque de variables, *Bulletin de la Société Philomatique* 10–21, in *Oeuvres Complètes* (2) 2, 238–252.
35. Cauchy, A.-L. 1835. *Mémoire sur l'intégration des équations différentielles*, (lith.) Prague, in *Oeuvres Complètes* (2) 11, 399–465.
36. Cauchy, A.-L. 1842. Mémoire sur un théorème fondamental dans le calcul intégral, *Comptes Rendus* 14, 1020–1026, in *Oeuvres Complètes* (1) 6, 461–466.
37. Cauchy, A.-L. 1842. Mémoire sur l'emploi du calcul des limites dans l'intégration des équations aux dérivées partielles, *Comptes Rendus* 15, 44–59, in *Oeuvres Complètes* (1) 7, 17–33.
38. Cauchy, A.-L. 1851. Rapport sur un mémoire présenté à l'Académie par M. Puiseux et intitulé: 'Recherches sur les fonctions algébriques'. *Comptes Rendus* 32, 276–284 in *Oeuvres Complètes* (1) 11, 325–335.
39. Cauchy, A.-L. 1981. *Équations différentielles ordinaires. Cours inédit*, Ch. Gilain, (ed.). Études vivantes, Paris and Johnson Reprint Co., New York and London.
40. Child, J.M. 1920. *The early mathematical manuscripts of Leibniz*, Open Court, Dover repr. 2005.
41. Christoffel, E.B. 1877. Untersuchungen über die mit dem Fortbestehen linearer partieller Differentialgleichungen verträglichen Unstetigkeiten, *Annali di Matematiche pura et applicata* (2) 8, 81–113.
42. Clairaut, A.C. 1736. Solution de plusieurs Problemes où il s'agit de trouver des Courbes dont la propriété consiste dans une certaine relation entre leurs branches, exprimée par une Équation donnée, *Histoire de l'Académie Royale des Sciences*, 196–215.

43. Clairaut, A.C. 1739. Recherches générales sur le calcul intégral, *Mémoires de l'Académie Royale des Sciences* (1741) 425–436.
44. Clairaut, A.C. 1743. *Théorie de la figure de la terre, tirée des principes de l'hydrostatique*, David fils, Paris.
45. Clairaut, A.C. 1749. Du système du monde dans les principes de la gravitation universelle, *Mémoires de l' Académie des Sciences*, Paris, 329–364, 528.
46. Cohen, I.B. 1980. *The Newtonian Revolution*, Cambridge U.P.
47. Cooke, R. 1984. *The Mathematics of Sonya Kovalevskaya*, Springer.
48. Corry, L. 2004. *David Hilbert and the Axiomatization of Physics (1898–1918)*, Archimedes, Springer.
49. Courant, R. and D. Hilbert, 1937. *Methods of Mathematical Physics*, 2 vols., Springer, 2nd. ed., Wiley, 1989.
50. D'Alembert, J. le Rond, 1743. *Traité de dynamique*, David, Paris.
51. D'Alembert, J. le Rond, 1747. *Réflexions sur la cause générale des vents*, Paris.
52. D'Alembert, J. le Rond, 1749. Recherches sur la courbe que forme une corde tenduë mise en vibration, *Histoire de l'Académie de Berlin* 3, 214–219, and Suite des recherches, 220–249.
53. D'Alembert, J. le Rond, 1752. Addition au mémoire sur la courbe que forme une corde tenduë mise en vibration, *Histoire de l'Académie de Berlin* 6, 355–360.
54. D'Alembert, J. le Rond, 1752. *Essai d'une nouvelle théorie de la résistance des fluides*, David, Paris.
55. D'Alembert, J. le Rond, 1768. Recherches de calcul intégral, *Opuscules* 4, 225–253, Paris.
56. D'Alembert, J. le Rond, 1768. Supplement au Mémoire précédent, *Opuscules* 4, 254–281, Paris.
57. Darboux, G. 1875. Mémoire sur l'existence de l'intégrale dans les équations aux dérivées partielles contenant un nombre quelconque de fonctions et de variables, *Comptes Rendus* 80, 101–104, 317–319.
58. Darboux, G. 1888. *Leçons sur la Théorie Générale des Surfaces*, Gauthier-Villars, 2nd edn. 1914.
59. Darboux, G. 1912. Notice historique sur le général Meusnier, *Éloges Académiques et Discours*, Paris, 218–262.
60. Darboux, G. 1916. Poincaré, in Poincaré, *Oeuvres* 2, vii–lxxi.
61. Demidov, S. 1982. The Study of Partial Differential Equations of the First Order in the 18th and 19th Centuries, *Archive for History of Exact Sciences* 26, 325–350
62. Descartes, R. 1637. *Discours de la méthode pour bien conduire la raison et chercher la verité dans les sciences. Plus la dioptrique, les météores et la géométrie, qui sont des essays de cette méthode*, Leyden; *Geometry* transl. D.E. Smith and M.L. Latham, Open Court Publishing Co., 1925, Dover reprint, 1954.
63. Dirichlet, P.G.L. 1829. Sur la convergence des séries trigonométriques, *Journal für die reine und angewandte Mathematik* 4, 157–169, in *Werke*, I, 117–132.
64. Dirichlet, P.G.L. 1876. *Vorlesungen über die im ungekehrten Verhältniss des Quadrats der Entfernung wirkenden Kräfte*. F. Grube, (ed.). Teubner, Leipzig.
65. Dugac, P. 1973. Éléments d'analyse de Karl Weierstrass. *Archive for History of Exact Sciences* 10, 41–176.
66. Eneström, G. 1905. Der Briefwechsel zwischen Leonhard Euler und Johann I Bernoulli, *Bibliotheca Mathematica* 6.
67. Engelsman, S.B. 1980. Lagrange's early contributions to the theory of first-order partial differential equations, *Historia Mathematica* 7, 7–23.
68. Engelsman, S.B. 1984. *Families of curves and the origins of partial differentiation*, Elsevier.
69. Euler, L. 1732. De linea brevissima in superficie quacunque duo quaelibet puncta jungente, *Commentarii Academiae Scientiarum Petropolitanae* 3, 110–124, *Opera Omnia* (1) 25, 1–12 (E9).
70. Euler, L. 1736. *Mechanica sive motus scientia analytice exposita*, 2 vols. St Petersburg *Opera Omnia* (2) 2, (E15, E16).

408   References

71. Euler, L. 1740. De minimis oscillationibus corporum tam rigidorum quam flexibilium. Methodus nova et facilis, *Commentarii Academiae Scientiarum Petropolitanae* 7, 99–122, *Opera Omnia* 2 (10) 17–34 (E40).

72. Euler, L. 1740. De infinitis curvis eiusdem generis seu methodus inveniendi aequationes pro infinitis curvis eiusdem generis, *Commentarii Academiae Scientiarum Petropolitanae* 7, 174–189, *Opera Omnia* 1 (22) 36–56 (E44).

73. Euler, L. 1743. De integratione aequationum differentialium altiorum graduum, *Miscellanea Berolinensia* 7, 193–242, *Opera Omnia* (1) 22, 108–149 (E62).

74. Euler, L. 1744. *Methodus Inveniendi Lineas Curvas Maximi Minimivi Proprietate Gaudentes sive Solutio Problematis Isoperimetrici Latissimo Sensu Accepti*, Lausanne and Geneva, *Opera Omnia* (1) 24 (E65).

75. Euler, L. 1750. Sur la vibration des cordes, *Mémoires de l'Académie des Sciences de Berlin* 4, 69–85, *Opera Omnia* (2) 10, 63–77 (E140).

76. Euler, L. 1750. Methodus aequationes differentialis altiorum graduum integrandi ulterius promota, *Novi Commentarii Academiae Scientiarum Petropolitanae* 3 (1753) 3–35, *Opera Omnia* (1) 22, 181–213 (E188).

77. Euler, L. 1750. Animadversiones in rectificationem ellipsis, *Opuscula varii argumenti* 2, 121–166, *Opera Omnia* (1), 20, 21–55 (E154).

78. Euler, L. 1751. De la controverse entre Mrs. Leibniz et Bernoulli sur les logarithmes des nombres negatifs et imaginaires, *Mémoires de l'Académie des Sciences de Berlin* 5, 139–179, in *Opera Omnia* (1), 17, 195–232 (E168).

79. Euler, L. 1752. Découverte d'un nouveau principe de Mécanique, *Mémoires de l'Académie des Sciences de Berlin* 6, 185–217, *Opera Omnia* (2) 5, 81–108 (E177).

80. Euler, L. 1755. Remarques sur les mémoires précédens de M. Bernoulli, *Mémoires de l'Académie des Sciences de Berlin* 9, 196–222, *Opera Omnia* (2) 10, 233–254 (E213).

81. Euler, L. 1757. Principes généraux de l'état d'equilibre des fluides, *Mémoires de l'académie des sciences de Berlin* 11, 217–273, *Opera Omnia*, (2) 12, 2–53 (E225).

82. Euler, L. 1757. Principes généraux du mouvement des fluides, *Mémoires de l'Académie des Sciences de Berlin* 11, 274–315, *Opera Omnia*, (2) 12, 54–91 (E226).

83. Euler, L. 1757. Continuation des recherches sur la théorie du mouvement des fluides, *Mémoires de l'Académie des Sciences de Berlin* 11, 316–361, *Opera Omnia*, (2) 12, 92–132 (E227).

84. Euler, L. 1758. Exposition de quelques paradoxes dans le calcul integral, *Mémoires de l'Académie des Sciences de Berlin* 12, 300–321, *Opera Omnia* (1) 22, 214–236 (E236).

85. Euler, L. 1761. Observationes de comparatione arcuum curvarum irrectificibilium, *Novi Commentarii Academiae Scientiarum Petropolitanae* 6, 58–84, *Opera Omnia* (1), 20, 80–107 (E252).

86. Euler, L. 1761. De integratione aequationis differentialis $\frac{m\,dx}{\sqrt{1-x^4}} = \frac{n\,dy}{\sqrt{1-y^4}}$, *Novi Commentarii Academiae Scientiarum Petropolitanae* 6, 37–57, *Opera Omnia* (1) 20, 58–79 (E251).

87. Euler, L. 1761. Principia motus fluidorum, *Novi Commentarii Academiae Scientiarum Petropolitanae* 6, 271–311, *Opera Omnia* (2) 12, 133–168 (E258).

88. Euler, L. 1763. De integratione aequationum differentialium *Novi Commentarii Academiae Scientiarum Petropolitanae* 8, 3–63, *Opera Omnia* (1) 22, 334–394 (E269).

89. Euler, L. 1764. Investigatio functionum ex data differentialium conditione, *Novi Commentarii Academiae Scientiarum Petropolitanae* 9, 170–212, *Opera Omnia* (1) 23, 133–168 (E285).

90. Euler, L. 1765. Du mouvement de rotation des corps solides autour d'un axe variable, *Mémoires de l'Académie des Sciences de Berlin* 14, 154–193, *Opera Omnia* (2) 8, 200–235 (E292).

91. Euler, L. 1765. *Theoria Motus Corporum Solidorum seu Rigidorum*, *Opera Omnia* (2) 3 (E289).

92. Euler, L. 1766. De la propagation du son, *Mémoires de l'Académie des Sciences de Berlin* 15, 185–209, *Opera Omnia* (3) 1, 428–451 (E305).

93. Euler, L. 1766. Supplement aux recherches sur la propagation du son, *Mémoires de l'Académie des Sciences de Berlin* 15, 210–240, *Opera Omnia* (3) 1, 452–483 (E306).

94. Euler, L. 1768. *Institutionum Calculi integralis* Vol. 1, *Opera Omnia* (1) 11 (E342).
95. Euler, L. 1769. *Institutionum Calculi integralis* Vol. 2, *Opera Omnia* (1) 12 (E366).
96. Euler, L. 1770. *Institutionum Calculi integralis* Vol. 3, *Opera Omnia* (1) 13 (E385).
97. Euler, L. 1772. Methodus nova et facilis calculum variationum tractandi, *Novi Commentarii Academiae Scientiarum Petropolitanae* 16, 35–70, *Opera Omnia* (1) 25, 208–235 (E420).
98. Euler, L. 1794. Specimen transformationis singularis serierum, *Nova acta academiae scientiarum Petropolitanae* 12, 58–70, *Opera Omnia* (2) 16, 41–55 (E710).
99. Euler, L. 1998. *Leonhard Euler: Briefwechsel mit Johann (I) Bernoulli und Niklaus (I) Bernoulli,* Emil A. Fellmann and Gleb K. Mikhajlov, (eds.) *Opera omnia* Series quarta A: Commercium epistolicum Vol. II, Birkhäuser, Basel.
100. Evans, L.C. 1998. *Partial Differential Equations,* American Mathematical Society.
101. Fagnano, G. 1750. *Produzioni Matematiche,* 2 vols. Stamperia Gavelliana, Pesaro in *Opere Matematiche.* V. Volterra, G. Loria, and D. Gambioli (eds). 3 vols. Dante Alighieri, Milano, 1911.
102. Fauvel, J. and J.J. Gray. 1987. *The History of Mathematics: A Reader,* Macmillan, in association with the Open University.
103. Fellmann, E.A. 2007. *Leonhard Euler,* Birkhäuser.
104. Fourier, J. 1822. *Théorie Analytique de la Chaleur,* transl. A. Freeman, *The Analytical Theory of Heat,* Dover repr. 1955.
105. Fraser, C. 1983. J.L. Lagrange's early contributions to the principles and methods of mechanics, *Archive for History of Exact Sciences* 28, 197–241.
106. Fraser, C. 1985. J.L. Lagrange's changing approach to the foundations of the calculus of variations, *Archive for History of Exact Sciences* 32, 151–191.
107. Fraser, C. 2003. The Calculus of Variations: A Historical Survey, in *A History of Analysis,* (ed.) H. N. Jahnke, American Mathematical Society, 355–384.
108. Fraser, C. and M. Nakane. 2002. The early history of Hamilton–Jacobi Dynamics, 1834–1837, *Centaurus* 44, 161–227.
109. Fregulia, P. and M. Giaquinta. 2016. *The Early Period of the Calculus of Variations,* Birkhäuser.
110. Fuchs, L.I. 1865. Zur Theorie der linearen Differentialgleichungen mit veränderlichen Coefficienten, *Jahresberichte Gewerbeschule Berlin* in *Ges. Math. Werke* 1, 111–158.
111. Fuchs, L.I. 1866. Zur Theorie der linearen Differentialgleichungen mit veränderlichen Coefficienten, *Journal für die reine und angewandte Mathematik* 66, 121–160 in *Ges. Math. Werke* 1, 159–204.
112. Fuchs, L.I. 1880. Über eine Klasse von Functionen mehrerer Variabeln, welche durch Umkehrung der Integrale von Lösungen der linearen Differentialgleichungen mit rationalen Coefficienten entstehen, *Journal für die reine und angewandte Mathematik* 89, 151–169 in *Ges. Math. Werke* 1, 191–212.
113. Galileo, Galilei, 1638. *Discorsi e Dimonstrazioni Matematiche Intorno a Due Nuove scienze,* Leyden, *Two New Sciences,* transl. H. Crew and A. de Salvio, MacMillan, 1914, Dover repr.
114. Gauss, C.F. 1812. Disquisitiones generales circa seriem infinitam $1 + \frac{\alpha.\beta}{1.\gamma}x + \frac{\alpha(\alpha+1)\beta(\beta+1)}{1.2.\gamma(\gamma+1)}xx + etc.$ Pars prior, *Comm. Soc. Göttingen* 2, 46 pp. in *Werke* 3, 123–162.
115. Gauss, C.F. 1812. Determinatio series nostrae per Aequationem Differentialem Secundi Ordinis, in *Werke* 3, 207–230.
116. Gauss, C.F. 1828. *Disquisitiones generales circa superficies curvas,* in *Werke* IV, 217–258, ed. P. Dombrowski, in *astérisque* 62, 1979, Latin original, with a reprint of the English translation by A. Hiltebeitel and J. Morehead 1902 and as *General investigations of curved surfaces,* P. Pesic (ed.) Dover Books, New York, 2005.
117. Gauss, C.F. 1840. *Allgemeine Lehrsätze in Beziehung auf die im verkehrten Verhältnisse des Quadrats der Entfernung wirkenden Anziehungs- und Abstossungs-Kräfte,* Leipzig.
118. Genocchi, A. 1875. Observations relatives à une communication de M. Darboux sur l'existence de l'intégrale etc. Lettre à M. Bertrand, *Comptes Rendus* 80, 315–317.
119. Germain, S. 1831. Mémoire sur la courbure des surfaces, *Journal für die reine und angewandte Mathematik* 7, 1–29.

120. Gillispie, C.G. 1997. *Pierre Simon Laplace 1749–1827. A Life in Exact Science*, Princeton U.P.
121. Goldstine, H. 1980. *A History of the Calculus of Variations from the 17th through the 19th Century*. Studies in the History of Mathematics and Physical Sciences, 5, Springer.
122. Grattan-Guinness, I. and S. Engelsman. 1982. The manuscripts of Paul Charpit, *Historia Mathematica* 9, 65–75.
123. Grattan-Guinness, I. and G. Ravetz. 1972. *Joseph Fourier 1768–1830*, MIT Press.
124. Gray, J.J. 2000. *Linear Differential Equations and Group Theory from Riemann to Poincaré*, 2nd ed. Birkhäuser.
125. Gray, J.J. 2000. *The Hilbert Challenge*, Oxford U.P.
126. Gray, J.J. 2013. *Henri Poincaré: A Scientific Biography*, Princeton U.P.
127. Gray, J.J. 2015. *The Real and the Complex: A History of Analysis in the 19th Century*, Springer.
128. Green, G. 1828. *An Essay on the application of mathematical analysis to the theories of electricity and magnetism*, also in *Journal für die reine und angewandte Mathematik*, 1850 and 1854, in *Mathematical Papers*, 1–82, N.M. Ferrers (ed.), Macmillan 1871, Chelsea repr. 1970.
129. Greenberg, J.L. 1995. *The Problem of the Earth's Shape from Newton to Clairaut: The Rise of Mathematical Science in Eighteenth–Century Paris and the Fall of 'Normal' Science*, Cambridge U.P.
130. Guicciardini, N. 1999. *Reading the Principia. The Debate on Newton's mathematical Methods for Natural Philosophy from 1687 to 1736*, Cambridge U.P.
131. Guilbaud, A. and G. Jouve. 2009. La Résolution des Équations aux Dérivées Partielles dans les *Opuscules Mathématiques* de d'Alembert (1761–1783), *Revue d'histoire de mathématiques* 15, 59–122
132. Hadamard, J. 1902. Sur les problèmes aux dérivées partielles et leur signification physique, *Princeton University Bulletin*, 49–52, in *Oeuvres* 3, 1099–1105 and Hadamard *Selecta*, Gauthier-Villars, 1935, 214–220.
133. Hadamard, J. 1923. *Lectures on Cauchy's problem in linear partial differential equations*, Yale U.P. repr. Dover 1952.
134. Hadamard, J. 1968. *Oeuvres de Jacques Hadamard*, 4 vols. CNRS, Paris.
135. Hamilton, W.R. 1834. On a General Method in Dynamics, Part 2, *Philosophical Transactions* Part 2 for 1834, 4–308, in *Mathematical Papers* 2, 103–161.
136. Hamilton, W.R. 1835. Second Essay on a General Method in Dynamics, Part 1, *Philosophical Transactions* 95–144, in *Mathematical Papers* 2, 162–211.
137. Hamilton, W.R. *The Mathematical Papers of Sir William Rowan Hamilton, 2, "Dynamics,"* A.W. Conway and A.J. McConnell (eds.), Cunningham Memoir no. 14, Cambridge U.P.
138. Harnack, A. 1887. *Grundlagen der Theorie des logarithmischen Potentiales [etc.]*, Teubner, Leipzig.
139. Hawkins, T. 1991. Jacobi and the Birth of Lie's Theory of Groups, *Archive for History of Exact Sciences* 42, 187–278.
140. Heaviside, O. 1876. On duplex telegraphy, *Philosophical Magazine* (5) 1, 32–43, in *Electrical Papers* 1, 53–65.
141. Hertz, H. 1894. *Die Prinzipien der Mechanik in neuen Zusammenhange dargestellt*, Leipzig, in *Gesammelte Werke*, 3.
142. Hilbert, D. 1900. Ueber das Dirichlet'sche Princip, *Jahresbericht der Deutschen Mathematiker-Vereinigung* 8, 184–188. Repr. in *Journal für die reine und angewandte Mathematik* 129, 63–67, in *Gesammelte Abhandlungen* 3, 10–14.
143. Hilbert, D. 1901. Mathematische Probleme. *Archiv für Mathematik und Physik* 1, 44–63 and 213–237, in *Gesammelte Abhandlungen* 3, 290–329. English transl. in *Mathematical Developments arising from Hilbert Problems*, Proceedings of the Symposium in Pure Math., 28, Amer. Math. Soc., Providence, R. I., 1976, repr,. in (Gray 2000).
144. Hilbert, D. 1906. Zur Variationsrechnung, *Mathematische Annalen* 62, 351–370, in *Gesammelte Abhandlungen* 3, 38-55.
145. Heilbert, D. and S. Cohn-Vossen. 1952. *Geometry and the Imagination*, Chelsea.

146. Holmgren, E. 1903. Über eine Klasse von partiellen Differentialgleichungen zweiter Ordnung, *Mathematische Annalen* 57, 409–420.
147. Hugoniot, H. 1887. Mémoire sur la propagation du mouvement dans un fluide indéfini, 1, *Journal de Mathématiques* (4) 3, 477–494.
148. Hugoniot, H. 1888. Mémoire sur la propagation du mouvement dans un fluide indéfini, 2, *Journal de Mathématiques* (4) 4, 153–167.
149. Huygens, C. 1646. Two letters to Mersenne, in Huygens, *Oeuvres Complètes* 1, Nijhoff, 1888.
150. Jacobi, G.C.J. 1837. Zur Theorie der Variations-Rechnung und der Differential-Gleichungen, *Journal für Mathematik* 17, 68–82, in *Werke* 4, 39–55.
151. Jacobi, G.C.J. 1837. Uber die Reduction der Integration der partiellen Differentialgleichungen erster Ordnung zwischen irgend einer Zahl Variabeln auf die Integration eines einzigen Systems gewöhnlicher Differentialgleichungen, *Journal für Mathematik* 17, 97–162, in *Werke* 4, 57–127.
152. Jacobi, G.C.J. 1866. *Vorlesungen über analytischen Mechanik*, A. Clebsch (ed.) 2nd. edn., H. Pulte (ed.), Vieweg, 1999.
153. John, F. 1971. *Partial Differential Equations*, Springer.
154. Jordan, C. 1870. *Traité des Substitutions et des Équations Algébriques*, Gauthier-Villars.
155. Kirchhoff, G. 1857. Ueber die Bewegung der Elektricität in Drähten, *Annalen der Physik und Chemie* (4) 100, 193–217, in *Gesammelte Abhandlungen* 131–155.
156. Klainerman, S. 2000: PDE as a unified subject, *Geometric and Functional Analysis*, special volume, 1–37.
157. Klein, C.F. 1894. *Vorlesungen über die hypergeometrische Functionen*, Göttingen.
158. Klein, C.F. 1967. *Vorlesungen über die Entwicklung der Mathematik im 19. Jahrhundert*, 2 vols in 1, Chelsea.
159. Kline, M. 1972. *Mathematical Thought from Ancient to Modern Times*, Oxford U.P.
160. Kneser, A. 1900. *Lehrbuch der Variationsrechnung*, Braunschweig, F. Vieweg und Sohn.
161. Kneser, A. 1903. Beiträge zur Theorie und Anwendung der Variationsrechnung, *Math. Ann.* 56, 169–232.
162. Koenigsberger, L. 1904 *Hermann von Helmholtz*, transl. F.A. Welby, repr. Dover, 1965.
163. Kopelevich, Y.K. 1966. The Petersburg Academy contest in 1751, *Soviet Astronomy* 9, 653–661, and American Institute of Physics, NASA Astrophysics Data System.
164. Körner, T.W. 2002. *Fourier Analysis*, Cambridge U.P.
165. Kovalevskaya, S. 1875. Zur Theorie der partiellen Differentialgleichungen, *Journal für die reine und angewandte Mathematik* 80, 1–32.
166. Kronecker, L. 1894. *Vorlesungen über die Theorie der einfachen und vielfachen Integrale*, (ed.) E. Netto, Teubner, Leipzig.
167. Kummer, E.E. 1836. Über die hypergeometrische Reihe [etc.]. *Journal für die reine und angewandte Mathematik* 15, 39–83; 127–172 in *Collected Papers* 2, 75–166. A. Weil, (ed.) Springer, New York.
168. Lacroix, S.F. 1798, *Traité du calcul différentiel et du calcul intégral*, Paris.
169. Lagrange, J.-L. 1760. Nouvelle recherches sur la nature et propagation du son, *Miscellanea Taurensis*, 2, in *Oeuvres* 1, 151–317.
170. Lagrange, J.-L. 1760. Essai d'une nouvelle méthode pour determiner les maxima et les minima des formules intégrales indéfinies, *Miscellanea Taurensis* 2, in *Oeuvres* 1, 335–364.
171. Lagrange, J.-L. 1762. Solution de différentes problèmes de calcul intégral, *Miscellanea Taurensis* 3, in *Oeuvres* 1, 471–670.
172. Lagrange, J.-L. 1765. Recherches sur la libration de la Lune, Prix de l'Académie Royale des Sciences, Paris, 1764, *Oeuvres* 6, 5–66.
173. Lagrange, J.-L. 1774. Sur l'intégration des équations à différences partielles du premier ordre, *Nouveaux Mémoires de l'Académie Royale des Sciences et Belles-Lettres de Berlin* 1772, in *Oeuvres* 3, 549–575.
174. Lagrange, J.-L. 1776, Sur les intégrales particulières des équations différentielles, *Nouveaux Mémoires de l'Académie Royale des Sciences et Belles-Lettres de Berlin*, année 1774, in *Oeuvres* 4, 5–111.

175. Lagrange, J.-L. 1779. Sur différentes questions d'analyse relatives à la théorie des intégrales particulières, *Nouveaux Mémoires de l'Académie Royale des Sciences et Belles-Lettres de Berlin*, année 1779, in *Oeuvres* 4, 585–636.
176. Lagrange, J.-L. 1788. *Méchanique Analitique*, Paris. 2nd. edn., *Méchanique analytique*, 1811.
177. Lagrange, J.-L. 1806 *Leçons sur le Calcul des Fonctions*, in *Oeuvres* 10.
178. Lagrange, J.-L. 1808. *Mécanique Analytique* vol. 1 (1808), in *Oeuvres* 11 and vol. 2 (1815) in *Oeuvres* 11, 12.
179. Laplace, P.S. 1777. Recherches sur le calcul intégral aux différences partielles, *Mémoires de l'Acdémie Royale des Sciences de Paris 1773*, in *Oeuvres* 9, Gauthier-Villars, Paris, 1878, 5–68.
180. Laplace, P.S. 1784/1787. Mémoire sur les inégalités séculaires des planètes et des satellites, *Mémoires de l'Académie Royale des Sciences de Paris* 1–50, in *Œuvres* 11, 49–92.
181. Laplace, P.S. 1796. *Exposition du Système du Monde*.
182. Laplace, P.S. 1799. *Mécanique Celeste*, 5 vols., *Mécanique Céleste by the Marquis de Laplace, translated with commentary*, N. Bowditch (ed. and transl.) 4 vols., Boston (1829–39).
183. Legendre, A.-M. 1786. Mémoire sur la manière de distinguer les maxima des minima dans le calcul des variations, *Mémoires de l'Académie des Sciences*.
184. Legendre, A.-M. 1811–1817. *Exercises de Calcul intégral*, 3 vols. Courcier, Paris.
185. Legendre, A.-M. 1825–1832. *Traité des Fonctions Elliptiques et des Intégrales Euleriennes*. 3 vols. [Vol. 3 consists of 3 supplements dated 1828, 1829, 1832], Huzard–Courcier, Paris.
186. L'Hôpital, G. de. 1696. *Analyse des Infiniment Petits, pour l'Intelligence des Lignes Courbes*, Paris.
187. Liouville, J. 1856. Expression remarquable de la quantité qui, dans le mouvement d'un système de points matériels à liaisons quelconques, est un minimum en vertu du principe de la moindre action, *Comptes Rendus* 42, 1146–1154.
188. Lipschitz, R. 1872. Untersuchung eines Problems der Variationsrechnung in welchem das Problem der Mechanik enthalten ist, *Journal für die reine und angewandte Mathematik* 74, 116–149.
189. Lipschitz, R. 1873. Extrait de six mémoires publiés dans le Journal de Mathématiques de Borchardt, *Bulletin des Sciences Mathématiques et Astronomique* 4, 97–110, 142–157, 212–224, 297–320.
190. Lipschitz, R. 1877. Sur la possibilité d'intégrer complétement un système donné d'équations différentielles, *Bulletin des Sciences Mathématiques* 10, 149–159.
191. Lützen, J. 1982. *The Prehistory of the Theory of Distributions*, Springer.
192. Lützen, J. 1990. *Joseph Liouville, 1809–1882. Master of Pure and Applied Mathematics*, Springer.
193. Lützen, J. 1995. Interactions between Mechanics and Differential Geometry in the 19th Century, *Archive for History of Exact Sciences* 49, 1–72.
194. Lützen, J. 2005. *Mechanistic Images in Geometric Form: Heinrich Hertz's Principles of Mechanics*, Oxford U.P.
195. Maupertuis, P.L. 1737. *La Figure de la Terre*, Paris, transl. *The Figure of the Earth*, 1738.
196. Maupertuis, P.L. 1746. Les Loix du mouvement et du repos déduites d'un principe métaphysique, *Histoire de l'Acaémie Royale des Sciences et des Belles Lettres*, 267–294.
197. Maz'ya, V. and T. Shaposhnikova. 1998. *Jacques Hadamard, a Universal Mathematician*, American and London Mathematical Societies, HMath, 14.
198. Mersenne, M. 1946. *Correspondance* VIII, P. Tannery, D. de Waard (eds.) Paris: Presses Universitaires de France.
199. Meusnier, J.B. 1785. Mémoire sur la courbure des surfaces, *Mémoires Mathématiques et Physiques des Savants Étrangers* 10, 477–510.
200. Moigno, Abbé F. 1840–1844. *Leçons de Calcul Différentiel et de Calcul Intégral*, 2 vols., Bachelier, Paris.
201. Monge, G. 1787. Mémoire sur le Calcul Intégral des Équations aux Différentialles Partielles, *Histoire de l'Académie royale des sciences avec les mémoires de mathématique et de physique tirés des registres de cette Académie, 1784*, 118–192.

202. Monge, G. 1809. *Applications de l'Analyse à la Géométrie*, also 5th ed. 1850, Bachelier, Paris

203. Nash, O. 2014. On Klein's icosahedral solution of the quintic, *Expositiones Mathematicae* 32, 99–120.

204. Nehari, Z. 1975. *Conformal mapping*, repr. 1952 edn. Dover Publications.

205. Newton, Isaac. 1671. *De Methodis Serierum et Fluxionum* (The Method of Series and Fluxions) in D.T. Whiteside *The Mathematical Papers of Isaac Newton* III, 32–353.

206. Newton, Isaac. 1687. *PhilosophiæNaturalis Principia Mathematica* London, 2nd ed. 1713, 3rd ed. 1726. English translations of the 3rd edition: *Sir Isaac Newton's Mathematical Principles of Natural Philosophy and his System of the World*, Andrew Motte 1729, revised F. Cajori, University of California Press, Berkeley, 1934, and *Isaac Newton's Philosophiae Naturalis Principia Mathematica. The third edition, 1726, with variant readings*, ed. I.B. Cohen and A. Whitman, Cambridge U.P. 1999.

207. Nitsche, J.C.C. 1989. *Lectures on Minimal Surfaces*, Cambridge U.P.

208. Osgood, W.F. 1901. Sufficient Conditions in the Calculus of Variations, *Annals of Mathematics*, (2) 2, 105–129.

209. Pfaff, J.F. 1797. *Disquisitiones Analyticae Maxime ad Calculum Integralem et Doctrinam Serierum Pertinentes* vol. 1, Helmstädt.

210. Pfaff, J.F. 1814–1815. Methodus generalis, aequationes differentiarum particularum, necnon aequationes differentiales vulgares, utrasque primi ordinis inter quotcumque variabiles, complete integrandi, *Abhandlungen der Königlichen Akademie der Wissenschaften zu Berlin*.

211. Picard, É. 1890. Mémoire sur la théorie des équations aux derivées partielles et la méthode des approximations successives, *Journal de Mathématiques*, 145–210.

212. Poincaré, H. 1879. *Sur les propriétés des fonctions définies par les équations aux différences partielles*, Thèse. Gauthier–Villars, in *Oeuvres* 1, xlix–cxxix.

213. Poincaré, H. 1881. Mémoire sur les courbes définies par une équation différentielle, *Journal de Mathématiques* 7, 375–422, in *Oeuvres* 1, 3–84.

214. Poincaré, H. 1885. Sur les courbes définies par les équations différentielles, *Journal de Mathématiques* 4, 167–244, in *Oeuvres* 1, 90–163.

215. Poincaré, H. 1890. Sur le problème des trois corps et les équations de la dynamique, *Acta Mathematica* 13, 1–270, in *Oeuvres* 7, 262–479.

216. Poincaré, H. 1890. Sur les équations aux dérivées partielles de la physique mathématique, *American Journal of Mathematics* 12, 211–294, in *Oeuvres* 9, 28–113.

217. Poincaré, H. 1893. Sur la propagation de l'électricité, *Comptes Rendus de l'Académie des Sciences* 117, 1027–1032, in *Oeuvres* 9, 278–283.

218. Poincaré, H. 1894. *Cours sur les oscillations électriques*, Paris.

219. Poincaré, H. 1897. Sur les rapports de l'analyse pure et de la physique mathématique, *Verhandlungen des internationaler Mathematiker-Congresses Zürich* 1, 81–90.

220. Poincaré, H. 1905. *La Valeur de la Science*. Flammarion, Paris. Many subs. reprints and translations. Engl. transl. as *The Value of Science*. The Science Press, New York 1907. Rep. Dover, 1958, 2003.

221. Poincaré, H. 1908. *Science et Méthode*. Flammarion, Paris. Many subs. reprints and translations. Engl. transl. as *Science and Method*. T. Nelson & Sons, London and New York, 1914. Repr. Dover, New York 1952, 2003.

222. Poincaré, H. 1912–1955. *Oeuvres*, 11 vols., Gauthier–Villars, Paris.

223. Poincaré, H. 1923. Extrait d'un mémoire inédit de Henri Poincaré sur les fonctions fuchsiennes, *Acta Mathematica* 39, 1923, 58–93, in *Oeuvres*, I, 578–613.

224. Poincaré, H. 1985. *Papers on Fuchsian Functions*, Stillwell, J. (ed.), Springer, New York.

225. Poincaré, H. 1997. *Three Supplementary Essays on the Discovery of Fuchsian Functions*, J.J. Gray and S.A. Walter (eds.), Akademie Verlag, Berlin and Blanchard, Paris.

226. Poisson, S.-D. 1823. Mémoire sur la distribution de la chaleur dans les corps solides, *Journal de l'École Polytechnique* 12, 1–162, 249–403.

227. Poisson, S.-D. 1835. *Théorie Mathématique de la Chaleur*, Paris.

228. Prym, F.E. 1871. Zur Integration der Differentialgleichung $\frac{\partial^2 u}{\partial x^2} + \frac{\partial^2 u}{\partial y^2} = 0$, *Journal für die reine und angewandte Mathematik* 73, 340–364.

229. Puiseux, V. 1850. Recherches sur les functions algébriques, *Journal de mathématiques* 15, 365–480.
230. Pulte, H. 1998. Jacobi's criticism of Lagrange: the changing role of mathematics in the foundations of classical mechanics, *Historia Mathematica*, 25.2, 154–184.
231. Pulte, H. 2005. Joseph Louis Lagrange, *Méchanique analitique*, in *Landmark Writings in Western Mathematics 1640–1940*, I. Grattan-Guinness (ed.) Elsevier, 208–224.
232. Rankine, W.J.M. 1870. On the thermodynamic theory of waves of finite longitudinal disturbance, *Phil. Trans. Royal Society* 160, 277–288.
233. Rayleigh, J.W. 1877. On progressive waves, *Proceedings of the London Mathematical Society* 9, 21–26.
234. Riemann, G.F.B. 1851. Grundlagen für eine allgemeine Theorie der Functionen einer veränderlichen complexen Grösse (Inaugural dissertation), Göttingen, in *Werke* 3–45
235. Riemann, G.F.B. 1857. Beiträge zur Theorie der durch Gauss'sche Reihe $F(\alpha, \beta, \gamma, x)$ darstellbaren Functionen, *Göttingen Abh.* 7, 3–22, in *Werke*, 99–115, Engl. transl. in Riemann *Collected Papers* 2004, 57–76.
236. Riemann, G.F.B. 1860. Ueber die Fortpflanzung ebener Luftwellen von endlicher Schwingungsweite, *Göttingen Abh.* 8, 43–65, in *Werke*, 157–175. Engl. transl. in Riemann *Collected Papers* 2004, 145–165.
237. Riemann, G.F.B. 1870. *Partielle Differentialgleichungen und deren Anwendung auf physikalische Fragen*, K. Hattendorff (ed.), Vieweg. 3rd edn. 1882.
238. Riemann, G.F.B. 1875. *Schwere, Elektricität und Magnetismus*, K. Hattendorff (ed.), Rümpler, Hannover.
239. Riemann, G.F.B. 1876. *Gesammelte Mathematische Werke*, 1st ed. R. Dedekind and H. Weber (eds.), 3rd edn. R. Narasimhan (ed.), Springer, 1990. Engl. transl. R. Baker, C. Christenson, H. Orde, *Collected Papers*, Kendrick Press, 2004.
240. Schering, E. 1870. Die Schwerkraft im Gaussischen Raume, *Nachrichten Königl. Ges. Wiss. Göttingen*, 311–321.
241. Schwarz, H.A. 1865. Ueber die Minimalfläche, deren Begrenzung als ein von vier Kanten eines regulären Tetraeders gebildetes räumliches Viersiet gegeben ist. *Monatsberichte der Königlichen Akademie der Wissenchaften zu Berlin*, 149–153, in *Ges. Math. Abh.* 1, 1–5.
242. Schwarz, H.A. 1869. Ueber einige Abbildungsaufgaben, *Journal für die reine und angewandte Mathematik* 70, 105–120, in *Ges. Math. Abh.* 2, 65–83.
243. Schwarz, H.A. 1870. *Zur Theorie der Abbildung*, Programme der Eidgenössischen Polytechnischen Schule in Zürich, in *Ges. Math. Abh.* 2, 108–132.
244. Schwarz, H.A. 1870. Ueber einen Grenzübergang durch alternirendes Verfahren, *Natur. Gesell. Zürich* 15, 272–286, in *Ges. Math. Abh.* 2, 133–143.
245. Schwarz, H.A. 1870. Ueber die integration der partiellen Differentialgleichung $\frac{\partial^2 u}{\partial x^2} + \frac{\partial^2 u}{\partial y^2} = 0$ unter vorgeschriebenen Grenz– und Unstetigkeitsbedingungen. *Monatsberiche Berlin*, 767–795, in *Ges. Math. Abh.* 2, 144–171.
246. Schwarz, H.A. 1872. Zur Integration der partiellen Differentialgleichung $\frac{\partial^2 u}{\partial x^2} + \frac{\partial^2 u}{\partial y^2} = 0$, *Journal für die reine und angewandte Mathematik* 74, 218–253, in *Ges. Math. Abh.* 2, 175–210.
247. Sommerfeld, A. 1900. Randwertaufgaben in der Theorie der partiellen Differentialgleichungen, *Encylopädie der Mathematischen Wissenschaften*, II.1.1, 504–570.
248. Sommerfeld, A. 1943. *Vorlesungen über Theoretische Physik; Partielle Differentialgleichungen in der Physik*, Klemm Verlag. English transl. *Partial Differential Equations in Physics: Lectures in Theoretical Physics*, Vol. 6, Academic Press, 1964.
249. Struik, D.J. 1969. *A Source Book in Mathematics, 1200–1800*, Harvard U.P.
250. Sturm, C. 1836. Mémoire sur une classe d'équations à differences partielles, *Journal de Mathématiques*, 1, 373–444.
251. Taylor, B. 1714. *De motu nervi tensi, Philosophical Transactions of theRoyal Society*, 28, 26–32.
252. Taylor, B. 1715. *Methodus Incrementorum Directa et Inversa*, London.

253. Terrall, M. 2002. *The Man Who Flattened the Earth*, Chicago U.P.
254. Thompson, S.P. 1910. *The life of William Thomson, baron Kelvin of Largs*, MacMillan.
255. Thomson, W. 1847. Note sur une équation aux différences partielles qui se présente dans plusieurs questions de physique mathématique, *Journal de Mathématiques pures et appliquées* 12, 493–496.
256. Thomson, W. 1855. On the theory of the electric telegraph, *Proceedings of the Royal Society* 7, 382–399, in *Mathematical and Physical Papers*, 2, 61–76, Cambridge U.P., 1884.
257. Thomson, Sir W. and P. G. Tait. 1879. *Treatise on Natural Philosophy*, 2 vols. Cambridge U.P.
258. Truesdell, C.A. 1955. The rational mechanics of flexible or elastic bodies 1638–1788, in Euler *Opera Omnia* (2) 11 part 2.
259. Truesdell, C.A. 1984. *An Idiot's Fugitive Essays in Science*, Springer.
260. Verhulst, F. 2012. *Henri Poincaré: Impatient Genius*, Springer.
261. Villani, C. 2016. *Birth of a Theorem*, transl. M. De Bevoise, Bodley Head, London.
262. Voltaire, 1738. *Elements of Newton's Philosophy*, Stephen Austen, London.
263. Voltaire, 1753. *Diatribe du docteur Akakia, médicin du Pape*, many editions.
264. Weber, E. von, 1915. Partielle Differentialgleichungen, *Encylopädie der Mathematischen Wissenschaften*, II.I.5, 296–401.
265. Webster, A.G. 1955. *Partial Differential Equations of Mathematical Physics*, 2nd. edn. Dover.
266. Weierstrass K.T.W. 1842. Definition analytischer Functionen einer Veränderlichen vermittelst algebraischer Differentialgleichungen, Ms. in *Math. Werke* 1, 75–84.
267. Weierstrass, K.T.W. 1866. Untersuchungen über die Flächen, deren mittlere Krümmung überall gleich Null ist, Umarbeitung einer am 25. Juni 1866 in der Akademie der Wissenschaften zu Berlin gelesen, in der *Monatsberichte der Akademie*, 1866, 612–625, auszugeweise abgedruckten Abhandlung, in (Weierstrass 1903, 39–52) (pub. 1867)
268. Weierstrass, K.T.W. 1866. Fortsetzung der Untersuchung über die Minimalflächen, *Monatsberichte der Akademie der Wissenschaften zu Berlin*, 1866, 855–856, in (Weierstrass 1903, 219–220)
269. Weierstrass, K.T.W. 1867. Analytische Bestimmung einfach zusammenhängender Minimalflächenstücke, deren Begrenzung aus geradlinigen, ganz im Endlichen liegenden Strecke besteht, in (Weierstrass 1903, 221–238, Anmerkung 239)
270. Weierstrass, K.T.W. 1903. *Mathematische Werke* III, J. Knoblauch (ed.) Berlin, Mayer and Müller
271. Weierstrass, K.T.W. 1870. Über das sogenannte Dirichlet'sche Princip, Ms. in *Math. Werke* 2, 49–54.
272. Weierstrass, K.T.W. 1885. Ueber die analytische Darstellbarkeit sogenannter willkürlicher Functionen einer reellen Veränderlichen, *Berlin Berichte*, 633–640, 789–806, in *Math. Werke* 3, 1–37.
273. Whiteside, D.T. (ed.) 1967–1981. *The Mathematical Papers of Isaac Newton*, 8 vols., Cambridge U.P.
274. Wilson, C. 2002. Newton and celestial mechanics. *The Cambridge Companion to Newton*, 202–226, I.B. Cohen and G. Smith (eds.), Cambridge U.P.
275. Yandell, H. 2002. *The Honors Class: Hilbert's Problems and their Solvers*, A.K. Peters.
276. Yavetz, I. 1995. *From Obscurity to Enigma: The Work of Oliver Heaviside, 1872–1889*, Birkhäuser.
277. Zermelo, E., and H. Hahn, 1904. Weiterentwicklung der Variationsrechnung in den letzten Jahren, *Encylopädie der Mathematischen Wissenschaften* II.1.1, 626–641.
278. Zinsser, J.P. 2001. Translating Newton's *Principia*: the Marquise de Châtelet's revisions and additions for a French audience, *Notes and Records of the Royal Society* 55, 227–245.

Index

© The Author(s), under exclusive license to Springer Nature Switzerland AG 2021   417
J. Gray, *Change and Variations*, Springer Undergraduate Mathematics Series,
https://doi.org/10.1007/978-3-030-70575-6

Printed in the United States
by Baker & Taylor Publisher Services